Emerging Paradigms in Insulin Resistance

Emerging Paradigms in Insulin Resistance

Editor

Susan J. Burke

MDPI • Basel • Beijing • Wuhan • Barcelona • Belgrade • Manchester • Tokyo • Cluj • Tianjin

Editor
Susan J. Burke
Pennington Biomedical
Research Center
USA

Editorial Office
MDPI
St. Alban-Anlage 66
4052 Basel, Switzerland

This is a reprint of articles from the Special Issue published online in the open access journal *Biomedicines* (ISSN 2227-9059) (available at: https://www.mdpi.com/journal/biomedicines/special_issues/Paradigms_Insulin_Resistance).

For citation purposes, cite each article independently as indicated on the article page online and as indicated below:

LastName, A.A.; LastName, B.B.; LastName, C.C. Article Title. *Journal Name* **Year**, *Volume Number*, Page Range.

ISBN 978-3-0365-5365-8 (Hbk)
ISBN 978-3-0365-5366-5 (PDF)

Contents

About the Editor

Susan J. Burke

Susan J. Burke (Assistant Professor) studies the role of lipid metabolism in islet beta-cell function using rodent models of obesity and diabetes. Her research is conducted at the Pennington Biomedical Research Center in Baton Rouge, Louisiana.

biomedicines

MDPI

Editorial

Special Issue: Emerging Paradigms in Insulin Resistance

J. Jason Collier [1,*] and Susan J. Burke [2,*]

1 Laboratory of Islet Biology and Inflammation, Pennington Biomedical Research Center, Baton Rouge, LA 70808, USA
2 Laboratory of Immunogenetics, Pennington Biomedical Research Center, Baton Rouge, LA 70808, USA
* Correspondence: jason.collier@pbrc.edu (J.J.C.); susan.burke@pbrc.edu (S.J.B.); Tel.: +1-225-763-2884 (J.J.C.); +1-225-763-2532 (S.J.B.)

This *Biomedicines* Special Issue was designed to attract articles that focused on different facets of biology relating to insulin resistance, defined as reduced cellular and organismal response to the insulin hormone, and its underlying mechanisms. Studies that centered around the relationship of insulin resistance to other conditions relevant to human disease were also welcomed. Collectively, this Special Issue was intended to contain both review articles, as well as studies with novel data, that offered insights into mechanisms, treatments, and perspectives on this prevalent condition. Indeed, insulin resistance impacts most, if not all, human disease conditions. These include, but are not limited to, aging, cancer, diabetes, dementia, and infectious disease [1–4].

Elevated fasting insulin is one of the best predictors of eventual progression to Type 2 Diabetes (T2D) [5,6]. Cooper and colleagues present the idea that the regulation of basal insulin secretion involves osteocalcin working in conjunction with GLP-1 [7]. This is proposed to be distinct from the paradigm of glucose-stimulated insulin secretion, which arises through increased glucose metabolism, production of metabolites that regulate membrane ion channels, and promote exocytosis. They also suggest that this possible pathway helps to regulate glucagon output from pancreatic alpha cells. This review article puts forth the idea that regulation of both basal and stimulated insulin secretion is important to understanding hyperinsulinemia, metabolic dysregulation, and progression to T2D.

There are several drugs used to treat insulin resistance in humans, with the intent of preventing the metabolic sequalae that ultimately leads to T2D. Some examples include metformin and pioglitazone [8,9], which are both indicated for use to decrease the risk of progression from insulin resistance and prediabetes to overt T2D. These pharmaceuticals act via distinct mechanisms, but each improves overall health when administered to people with insulin resistance and other symptoms of metabolic syndrome. A study in this issue of *Biomedicines* reveals interesting similarities and key differences between adipose tissue depots from mice and humans given pioglitazone, a thiazolidinedione (TZD) drug.

Two different adipose tissue depots, femoral (representing subcantaeous adipose tissue) and abdominal (representing visceral adipose tissue) taken from human clinical trial participants were compared and also examined alongside similar depots isolated from *db/db* mice [10]. In both mice and humans, the subcutaneous white adipose tissue depots responded to pioglitzone by increasing the expression of *Ucp1*, often used as a marker of brown adipose tissue or 'browning' of white adipose tissue [11]. In addition, the eWAT in mice and abdominal depot from humans each showed downregulation of the gene encoding interleukin-1 beta, a pro-inflammatory cytokine produced by macrophages, in response to pioglitazone. In mice, the *Cd68* gene, expressed in macrophages, was also decreased after pioglitazone intervention [10]. We interpret these data to indicate that at least part of the mechanism underlying the insulin sensitizing effects of pioglitazone involves remodeling of adipose tissue at the gene expression level, leading to reduced inflammation and increased reesterification of lipid into triglyceride storage. These observations are consistent with

Citation: Collier, J.J.; Burke, S.J. Special Issue: Emerging Paradigms in Insulin Resistance. *Biomedicines* **2022**, *10*, 1471. https://doi.org/10.3390/biomedicines10071471

Received: 13 June 2022
Accepted: 17 June 2022
Published: 22 June 2022

improved insulin sensitivity despite weight gain, which is a phenotype observed in both mouse models of obesity (e.g., *db/db*) and in humans exposed to TZDs [10,12–15].

Another study using both mice and human subjects to understand insulin resistance was conducted by Naryzhnaya and colleagues [16]. In this study, the authors report that epicardial adipose tissue hypertrophy is associated with elevated fasting insulin. In addition, these data correlate with severity of coronary artery disease. Moreover, lower serum adiponectin was linked with increased adipocyte size. Thus, alterations in fat tissue surrounding the heart, which appears to be connected with alterations in the circulating hormones insulin and adiponectin, may be a critical factor in determining risk for, or severity of, coronary artery disease.

Following up on factors that regulate insulin abundance, Gonzalez-Casimiro et al., review the historical and contemporary literature surrounding the insulin-degrading enzyme (IDE) [17]. They present the conserved nature of IDE, which is present in microorganisms (e.g., viruses) to mammals (including humans). One of the major functions of IDE is to enzymatically cleave insulin into smaller fragments, based on the tertiary structure of the insulin protein as opposed to sequence-specific recognition of precise amino acids. In addition to degradation of insulin, IDE has several other targets, which include pancreatic proteins glucagon, somatostatin, and amylin. The authors go on to describe the controversies that exist in the literature regarding gene deletion of IDE in mice and the in vivo relevance of IDE to circulating insulin in both standard chow feeding as well as high-fat diet conditions.

Both diet and genetics contribute to the development of obesity and insulin resistance [18]. One gene reported to contribute to obesity and have a strong link with body mass index (BMI) is neuronal growth regulator 1 (NEGR1). In addition to high expression in the brain, NEGR1 is also expressed in adipose tissue [19]. Kaare et al. report that NEGR1 deficient male, but not female, mice on HFD display impaired glucose tolerance [20]. The NEGR1 deficient mice also trend towards increased weight gain, while eating significantly less food than wild-type mice. Interestingly, female, but not male, NEGR1 deficient mice have elevated basal levels of glucose after 6 weeks of high-fat feeding. NEGR1 was shown to play a role in regulating circulating lipids, hepatic lipid content (increased in KOs), as well as a reduction in the cross-sectional area of muscle fibers in NEGR1 deficient male, but not female, mice. In a separate study, NEGR1 deficient mice were shown to also have increased lipid storage in liver, which was consistent with their hyperinsulinemia when compared with wild-type mice [21].

Insulin resistance and hyperinsulinemia are often found together, although there is not universal agreement over which arises first [22]. However, it is clear that prolonged, elevated levels of circulating insulin are associated with metabolic disease and cancer [23]. In addition, hyperinsulinemia is associated with increased risk for, and is predictive of, T2D [5]. In a timely review, Cooper et al., outline the links between bone fragility, increased risk of fractures, and hyperinsulinemia [24]. The authors present an interesting argument that alterations in bone mineralization can be driven by changes in circulating insulin, with part of the mechanism being adipocyte sequestration of Vitamin D.

In addition to changes in bone fragility, osteoarthritis (OA), insulin resistance, and T2D are more common in postmenopausal women than in premenopausal women. To address whether a current drug used to treat OA could also be effective against insulin resistance, Chen et al., used raloxifene in glucosamine exposed, ovariectomized rats [25]. Raloxifine is an estrogen receptor modulator, and by virtue of its ability to activate transcriptional targets of the estrogen receptor, may be effective against the inflammation associated with OA and T2D. In this rat model, markers of improved insulin sensitivity were observed after raloxifene intervention, including restoration of skeletal muscle GLUT-4, reduced liver expression of PEPCK, and normalization of serum glucose and insulin concentrations. Thus, estrogen receptor modulators may be highly effective in multiple conditions associated with diseases of aging.

In a separate study using both rodent models and data from human study participants, Solares et al., noted hyperinsulinemia in patients with acute intermittent porphyria

(AIP) [26]. AIP is a metabolic disease that results from reduced activity of key hepatic enzyme (porphobilinogen deaminase) in the heme synthesis pathway. This disease has been associated with hyperinsulinemia for decades [27] and the present study noted a significant prevalence of hyperinsulinemia in Spanish patients with a mutation in the gene encoding porphobilinogen deaminase. A proof of principle study indicated that a fusion protein of insulin/ApoAI reduced the expression of *Alas1*, which encodes an ezyme that synthesizes a metabolic precursor associated with the phenotype of AIP. Thus, the mouse model used in this study suggests that liver-specific targeting of insulin may be one strategy to reduce symptoms associated with AIP.

In summary, hyperinsulinemia and insulin resistance are associated with, and often predictive of, metabolic disease in rodent models and humans. In addition, these two highly prevalent conditions worsen the outcomes of many human diseases. This issue provides a variety of perspectives, as well as novel data, that center around insulin resistance, hyperinsulinemia, and distinct conditions impacted by these ever-increasing disorders. Using mice as experimental models with an eye on translational components offers new insights into both mechanisms and possible new therapies on the horizon.

Author Contributions: J.J.C. and S.J.B. each contributed to the writing and editing of this manuscript. All authors have read and agreed to the published version of the manuscript.

Funding: Work in the authors' laboratories is supported by NIH grants R01 DK123183 (J.J.C.), R03 AI151920 (J.J.C.) and P20 GM135002 (S.J.B.).

Institutional Review Board Statement: Not applicable.

Informed Consent Statement: Not applicable.

Conflicts of Interest: The authors declare no conflict of interest.

References

1. Finucane, F.M.; Davenport, C. Coronavirus and Obesity: Could Insulin Resistance Mediate the Severity of COVID-19 Infection? *Front. Public Health* **2020**, *8*, 184. [CrossRef] [PubMed]
2. Facchini, F.S.; Hua, N.; Abbasi, F.; Reaven, G.M. Insulin resistance as a predictor of age-related diseases. *J. Clin. Endocrinol. Metab.* **2001**, *86*, 3574–3578. [CrossRef] [PubMed]
3. Arcidiacono, B.; Iiritano, S.; Nocera, A.; Possidente, K.; Nevolo, M.T.; Ventura, V.; Foti, D.; Chiefari, E.; Brunetti, A. Insulin resistance and cancer risk: An overview of the pathogenetic mechanisms. *Exp. Diabetes Res.* **2012**, *2012*, 789174. [CrossRef] [PubMed]
4. Ferreira, L.S.; Fernandes, C.S.; Vieira, M.N.; De Felice, F.G. Insulin Resistance in Alzheimer's Disease. *Front. Neurosci.* **2018**, *12*, 830. [CrossRef]
5. Dankner, R.; Chetrit, A.; Shanik, M.H.; Raz, I.; Roth, J. Basal-state hyperinsulinemia in healthy normoglycemic adults is predictive of type 2 diabetes over a 24-year follow-up: A preliminary report. *Diabetes Care* **2009**, *32*, 1464–1466. [CrossRef] [PubMed]
6. Haffner, S.M.; Stern, M.P.; Mitchell, B.D.; Hazuda, H.P.; Patterson, J.K. Incidence of type II diabetes in Mexican Americans predicted by fasting insulin and glucose levels, obesity, and body-fat distribution. *Diabetes* **1990**, *39*, 283–288. [CrossRef]
7. Cooper, I.D.; Brookler, K.H.; Kyriakidou, Y.; Elliott, B.T.; Crofts, C.A.P. Metabolic Phenotypes and Step by Step Evolution of Type 2 Diabetes: A New Paradigm. *Biomedicines* **2021**, *9*, 800. [CrossRef]
8. Pavo, I.; Jermendy, G.; Varkonyi, T.T.; Kerenyi, Z.; Gyimesi, A.; Shoustov, S.; Shestakova, M.; Herz, M.; Johns, D.; Schluchter, B.J.; et al. Effect of pioglitazone compared with metformin on glycemic control and indicators of insulin sensitivity in recently diagnosed patients with type 2 diabetes. *J. Clin. Endocrinol. Metab.* **2003**, *88*, 1637–1645. [CrossRef]
9. DeFronzo, R.A.; Inzucchi, S.; Abdul-Ghani, M.; Nissen, S.E. Pioglitazone: The forgotten, cost-effective cardioprotective drug for type 2 diabetes. *Diab. Vasc. Dis. Res.* **2019**, *16*, 133–143. [CrossRef]
10. Collier, J.J.; Batdorf, H.M.; Merrifield, K.L.; Martin, T.M.; White, U.; Ravussin, E.; Burk, D.H.; Cooley, C.R.; Karlstad, M.D.; Burke, S.J. Pioglitazone Reverses Markers of Islet Beta-Cell De-Differentiation in db/db Mice While Modulating Expression of Genes Controlling Inflammation and Browning in White Adipose Tissue from Insulin-Resistant Mice and Humans. *Biomedicines* **2021**, *9*, 1189. [CrossRef]
11. Nedergaard, J.; Cannon, B. The browning of white adipose tissue: Some burning issues. *Cell Metab.* **2014**, *20*, 396–407. [CrossRef]
12. Defronzo, R.A.; Tripathy, D.; Schwenke, D.C.; Banerji, M.; Bray, G.A.; Buchanan, T.A.; Clement, S.C.; Gastaldelli, A.; Henry, R.R.; Kitabchi, A.E.; et al. Prevention of diabetes with pioglitazone in ACT NOW: Physiologic correlates. *Diabetes* **2013**, *62*, 3920–3926. [CrossRef] [PubMed]

13. DeFronzo, R.A.; Tripathy, D.; Schwenke, D.C.; Banerji, M.; Bray, G.A.; Buchanan, T.A.; Clement, S.C.; Henry, R.R.; Hodis, H.N.; Kitabchi, A.E.; et al. Pioglitazone for diabetes prevention in impaired glucose tolerance. *N. Engl. J. Med.* **2011**, *364*, 1104–1115. [CrossRef] [PubMed]
14. Bajaj, M.; Suraamornkul, S.; Pratipanawatr, T.; Hardies, L.J.; Pratipanawatr, W.; Glass, L.; Cersosimo, E.; Miyazaki, Y.; DeFronzo, R.A. Pioglitazone reduces hepatic fat content and augments splanchnic glucose uptake in patients with type 2 diabetes. *Diabetes* **2003**, *52*, 1364–1370. [CrossRef]
15. Guan, H.P.; Li, Y.; Jensen, M.V.; Newgard, C.B.; Steppan, C.M.; Lazar, M.A. A futile metabolic cycle activated in adipocytes by antidiabetic agents. *Nat. Med.* **2002**, *8*, 1122–1128. [CrossRef] [PubMed]
16. Naryzhnaya, N.V.; Koshelskaya, O.A.; Kologrivova, I.V.; Kharitonova, O.A.; Evtushenko, V.V.; Boshchenko, A.A. Hypertrophy and Insulin Resistance of Epicardial Adipose Tissue Adipocytes: Association with the Coronary Artery Disease Severity. *Biomedicines* **2021**, *9*, 64. [CrossRef] [PubMed]
17. Gonzalez-Casimiro, C.M.; Merino, B.; Casanueva-Alvarez, E.; Postigo-Casado, T.; Camara-Torres, P.; Fernandez-Diaz, C.M.; Leissring, M.A.; Cozar-Castellano, I.; Perdomo, G. Modulation of Insulin Sensitivity by Insulin-Degrading Enzyme. *Biomedicines* **2021**, *9*, 86. [CrossRef]
18. Bray, G.A.; Bouchard, C. The biology of human overfeeding: A systematic review. *Obes Rev.* **2020**, *21*, e13040. [CrossRef]
19. Walley, A.J.; Jacobson, P.; Falchi, M.; Bottolo, L.; Andersson, J.C.; Petretto, E.; Bonnefond, A.; Vaillant, E.; Lecoeur, C.; Vatin, V.; et al. Differential coexpression analysis of obesity-associated networks in human subcutaneous adipose tissue. *Int. J. Obes.* **2012**, *36*, 137–147. [CrossRef]
20. Kaare, M.; Mikheim, K.; Lillevali, K.; Kilk, K.; Jagomae, T.; Leidmaa, E.; Piirsalu, M.; Porosk, R.; Singh, K.; Reimets, R.; et al. High-Fat Diet Induces Pre-Diabetes and Distinct Sex-Specific Metabolic Alterations in Negr1-Deficient Mice. *Biomedicines* **2021**, *9*, 1148. [CrossRef]
21. Joo, Y.; Kim, H.; Lee, S.; Lee, S. Neuronal growth regulator 1-deficient mice show increased adiposity and decreased muscle mass. *Int. J. Obes.* **2019**, *43*, 1769–1782. [CrossRef] [PubMed]
22. Shanik, M.H.; Xu, Y.; Skrha, J.; Dankner, R.; Zick, Y.; Roth, J. Insulin resistance and hyperinsulinemia: Is hyperinsulinemia the cart or the horse? *Diabetes Care* **2008**, *31* (Suppl. S2), S262–S268. [CrossRef] [PubMed]
23. Zhang, A.M.Y.; Wellberg, E.A.; Kopp, J.L.; Johnson, J.D. Hyperinsulinemia in Obesity, Inflammation, and Cancer. *Diabetes Metab. J.* **2021**, *45*, 622. [CrossRef] [PubMed]
24. Cooper, I.D.; Brookler, K.H.; Crofts, C.A.P. Rethinking Fragility Fractures in Type 2 Diabetes: The Link between Hyperinsulinaemia and Osteofragilitas. *Biomedicines* **2021**, *9*, 1165. [CrossRef] [PubMed]
25. Chen, C.-H.; Cheng, T.-L.; Chang, C.-F.; Huang, H.-T.; Lin, S.-Y.; Wu, M.-H.; Kang, L. Raloxifene Ameliorates Glucosamine-Induced Insulin Resistance in Ovariectomized Rats. *Biomedicines* **2021**, *9*, 1114. [CrossRef] [PubMed]
26. Solares, I.; Izquierdo-Sanchez, L.; Morales-Conejo, M.; Jerico, D.; Castelbon, F.J.; Cordoba, K.M.; Sampedro, A.; Lumbreras, C.; Moreno-Aliaga, M.J.; Enriquez de Salamanca, R.; et al. High Prevalence of Insulin Resistance in Asymptomatic Patients with Acute Intermittent Porphyria and Liver-Targeted Insulin as a Novel Therapeutic Approach. *Biomedicines* **2021**, *9*, 255. [CrossRef]
27. Sixel-Dietrich, F.; Verspohl, F.; Doss, M. Hyperinsulinemia in acute intermittent porphyria. *Horm. Metab. Res.* **1985**, *17*, 375–376. [CrossRef]

biomedicines

Article

Pioglitazone Reverses Markers of Islet Beta-Cell De-Differentiation in *db/db* Mice While Modulating Expression of Genes Controlling Inflammation and Browning in White Adipose Tissue from Insulin-Resistant Mice and Humans

J. Jason Collier [1,*], Heidi M. Batdorf [1], Kaelan L. Merrifield [1], Thomas M. Martin [1], Ursula White [2], Eric Ravussin [3], David H. Burk [4], Chris R. Cooley [5], Michael D. Karlstad [5] and Susan J. Burke [6,*]

1 Laboratory of Islet Biology and Inflammation, Pennington Biomedical Research Center, Baton Rouge, LA 70808, USA; heidi.batdorf@pbrc.edu (H.M.B.); Kmerri5@uic.edu (K.L.M.); thomas.martin@pbrc.edu (T.M.M.)
2 Physiology of Human Adipose Tissue, Pennington Biomedical Research Center, Baton Rouge, LA 70808, USA; ursula.white@pbrc.edu
3 Human Physiology, Pennington Biomedical Research Center, Baton Rouge, LA 70808, USA; eric.ravussin@pbrc.edu
4 Cell Biology and Bioimaging Core, Pennington Biomedical Research Center, Baton Rouge, LA 70808, USA; david.burk@pbrc.edu
5 Department of Surgery, Graduate School of Medicine, University of Tennessee Health Science Center, Knoxville, TN 37920, USA; cooley.35@wright.edu (C.R.C.); mkarlsta@utmck.edu (M.D.K.)
6 Laboratory of Immunogenetics, Pennington Biomedical Research Center, Baton Rouge, LA 70808, USA
* Correspondence: Jason.collier@pbrc.edu (J.J.C.); susan.burke@pbrc.edu (S.J.B.); Tel.: +1-225-763-2884 (J.J.C.); +1-225-763-2532 (S.J.B.)

Citation: Collier, J.J.; Batdorf, H.M.; Merrifield, K.L.; Martin, T.M.; White, U.; Ravussin, E.; Burk, D.H.; Cooley, C.R.; Karlstad, M.D.; Burke, S.J. Pioglitazone Reverses Markers of Islet Beta-Cell De-Differentiation in *db/db* Mice While Modulating Expression of Genes Controlling Inflammation and Browning in White Adipose Tissue from Insulin-Resistant Mice and Humans. *Biomedicines* **2021**, *9*, 1189. https://doi.org/10.3390/biomedicines9091189

Academic Editors: Antonio Andrés and Christian Dani

Received: 25 June 2021
Accepted: 2 September 2021
Published: 10 September 2021

Abstract: Obesity, insulin resistance, and type 2 diabetes contribute to increased morbidity and mortality in humans. The *db/db* mouse is an important mouse model that displays many key features of the human disease. Herein, we used the drug pioglitazone, a thiazolidinedione with insulin-sensitizing properties, to investigate blood glucose levels, indicators of islet β-cell health and maturity, and gene expression in adipose tissue. Oral administration of pioglitazone lowered blood glucose levels in *db/db* mice with a corresponding increase in respiratory quotient, which indicates improved whole-body carbohydrate utilization. In addition, white adipose tissue from *db/db* mice and from humans treated with pioglitazone showed increased expression of glycerol kinase. Both *db/db* mice and humans given pioglitazone displayed increased expression of *UCP-1*, a marker typically associated with brown adipose tissue. Moreover, pancreatic β-cells from *db/db* mice treated with pioglitazone had greater expression of insulin and Nkx6.1 as well as reduced abundance of the de-differentiation marker Aldh1a3. Collectively, these findings indicate that four weeks of pioglitazone therapy improved overall metabolic health in *db/db* mice. Our data are consistent with published reports of human subjects administered pioglitazone and with analysis of human adipose tissue taken from subjects treated with pioglitazone. In conclusion, the current study provides evidence that pioglitazone restores key markers of metabolic health and also showcases the utility of the *db/db* mouse to understand mechanisms associated with human metabolic disease and interventions that provide therapeutic benefit.

Keywords: diabetes; inflammation; obesity; thiazolidinedione

1. Introduction

Obesity and insulin resistance are predictors of the development of type 2 diabetes (T2D) [1–3]. Importantly, the progression to T2D requires the loss of islet β-cell mass, function, or both [4–6]. Strategies to protect total islet β-cell mass, insulin production, and insulin secretion are therefore sought to prevent onset of such metabolic diseases. Both pharmacological and lifestyle interventions can be successful at preventing or restoring metabolic tissue function to combat onset of hyperglycemia, a critical defining feature of T2D [7,8]. Lifestyle interventions typically target weight reduction, leading to decreases in tissue lipid content that restore organ function [7,9]. Alternatively, weight reduction is not typically required for the therapeutic effects of many pharmaceutical approaches, such as administration of metformin or thiazolidinediones (TZDs).

FDA-approved TZDs, such as rosiglitazone and pioglitazone, often promote weight gain despite strong insulin-sensitizing properties. However, this weight gain appears to be preferentially in subcutaneous regions, which likely contributes to the improved metabolic health despite increased total fat mass [10]. Indeed, the power of TZDs to prevent progression to T2D was revealed in several clinical trials, even outperforming lifestyle interventions [11,12]. The TZD class of drugs act as agonists of the transcription factor peroxisome-proliferator-activated receptor gamma (PPARγ) [13]. PPARγ is important for adipogenesis [14] and also displays anti-inflammatory activity [15]. Thus, the therapeutic actions of TZDs are likely to result from a multitude of regulatory actions at the gene expression level via PPARγ activation.

In the present study, we investigated whether the TZD pioglitazone could reverse existing hyperglycemia in *db/db* mice, a genetic model of obesity and T2D [16]. We found that pioglitazone rapidly restored glycemia to levels observed in non-diabetic lean littermate control mice. This complete restoration in blood glucose concentration was associated with shifts in respiratory quotient to reflect greater whole-body carbohydrate oxidation, an observation consistent with increased glucose utilization and improved insulin sensitivity. Circulating insulin also returned to the amounts observed in lean mice while adiponectin, an insulin-sensitizing hormone, was markedly increased. Markers of browning were present in white adipose tissue of *db/db* mice receiving pioglitazone. We also found that this expression pattern was recapitulated in the femoral depot of human white adipose tissue from subjects given pioglitazone. Strikingly, pioglitazone-enhanced insulin gene expression in isolated pancreatic islets, with reductions in the de-differentiation marker Aldh1a3. In pancreatic sections, Aldh1a3 protein was decreased in mice receiving pioglitazone concomitant with increased abundance of the transcription factor Nkx6.1, a marker of mature β-cells. Therefore, we conclude that oral administration of pioglitazone in a mouse model of obesity and T2D restores several key markers of metabolic health.

2. Materials and Methods

2.1. Experimental Animals

Male C57BL/6J (Jax number 000664), *db/+* and *db/db* mice (B6.BKS(D)-*Lepr*^{db}/J; Jax number 000697) were purchased from the Jackson Laboratory (Bar Harbor, ME, USA) at seven weeks of age. All animals were allowed to acclimate to the Pennington Biomedical Research Center or University of Tennessee Medical Center facilities for at least seven days to allow for normalization of physiological parameters following transport [17]. During the acclimation period, animals were given ad libitum access to Teklad 8640 Rodent Diet (Envigo, Indianapolis, IN, USA) and water. Prior to beginning each study, *db/db* mice were randomized into two dietary groups and fed Teklad 8640 Rodent Diet (supplemented or not with pioglitazone) after stratification based on body weight and blood glucose to avoid any significant differences between groups at baseline. Pioglitazone hydrochloride was purchased from Sigma Aldrich (St. Louis, MO, USA; Cat # E6910) and blended into Teklad 8640 Rodent Diet at a concentration of 105 mg/kg. The dose of PIO in the food based on the 105 mg/kg diet provides approximately 15 mg/kg per day per mouse; this is less than or very near to what has been reported for other studies [18]. Animals were placed on the control (CON) or pioglitazone-supplemented (PIO) diets for 11 to 28 days.

Three cohorts of mice were required to complete the studies described herein. For cohort 1, non-fasting blood glucose and body mass were measured on study days 0, 4, 7, 11, 14, 18, 21, 25, and 28. On day 28, following a 4 h fast, animals were sacrificed by CO_2 asphyxiation and cervical dislocation. Blood was collected by cardiac puncture and the serum fraction was subsequently extracted. Fat depots were snap frozen in liquid nitrogen. Pancreata were perfused and islets were isolated using our previously published protocol [19]. For cohort 2, measurements of energy expenditure, respiratory quotient, activity, and caloric intake were conducted using Promethion metabolic cages (Sable Systems, North Las Vegas, NV, USA). One week before administering diets, *db/db* mice were moved from their home cages to single-housed metabolic training cages to allow for acclimation. On study day 0, animals were moved to the Promethion cages and the dietary protocol was initiated at the start of the metabolic cage measurements. On day 7, animals were removed from the metabolic cages and returned to their home cages. Non-fasting blood glucose, body mass, and body composition were thus assessed in this cohort on study days 0, 7, 14, and 28. Measurements of body composition (fat, lean, and fluid mass) were made by NMR using a Bruker Minispec LF110 Time-Domain NMR system. Cohort 3 used lean mice given either control or pioglitazone-supplemented diets. Insulin tolerance was measured using i.p. injection of Humulin R at 1 U/kg body weight after a 4 h fast. Upon completion of cohorts 2 and 3, animals were sacrificed by CO_2 asphyxiation and decapitation following a 4 h fast. Trunk blood was collected for serum extraction. Fat depots were snap frozen in liquid nitrogen. Pancreata were fixed in 10% neutral-buffered formalin (NBF). The number of animals used is stated in the figure legend. All animal procedures described herein were approved by the Institutional Care and Use Committees of Pennington Biomedical Research Center (IACUC protocol # 1021; approved 05/02/2018) and University of Tennessee Health Science Center (IACUC protocol # 2171; approved 02/26/2016).

2.2. *Pancreas Immunohistochemistry*

Following fixation in 10% NBF for 24–48 h, pancreatic tissue was embedded in paraffin, sectioned, stained, and analyzed as previously described [20,21]. Primary antibodies used were as follows: guinea pig anti-insulin (Invitrogen, Grand Island, NY, USA; #18-0067; 1:800), glucagon (Cell Signaling Technology, Danvers, MA, USA; #2760; 1:300), Nkx6.1 (Developmental Studies Hybridoma Bank, Iowa City, IA, USA; #F55A12; 1:100); and Aldh1a3 (Novus Biologicals, Centennial, CO, USA; #NBP2-15339; 1:100).

2.3. *Pioglitazone-Treated Human Study Participants and RNA Isolation from Human Adipose Tissue*

2.3.1. Study Participant Characteristics

The Apple & Pear trial ("Cellular Dynamics of Subcutaneous Fat Distribution in Obese Women"; ClinicalTrials.gov ID- NCT01748994) was a randomized, double-blind, placebo-controlled, parallel-arm trial conducted at Pennington Biomedical Research Center (PBRC). Details of the study design have been reported [22]. Briefly, healthy women, with overweight or obesity, who were 18–40 years of age and had a body mass index (BMI) of 27–38 kg/m^2 were recruited for this study. Participants were absent of diabetes or any major organ disease, weight stable for $\geq$3 months ($\pm$3.2 kg), had no significant changes in diet or physical activity in the previous month, and had no chronic use of medications to cause weight gain, weight loss, or other potential metabolic effects (e.g., glucocorticoids, adrenergic agents, and thiazolidinediones).

After screening for eligibility, women completed baseline metabolic assessments, including adipose tissue biopsy collections, and were then randomized (1:1 allocation ratio) to consume 30 mg/day of pioglitazone (PIO group) or to a placebo group for 16 weeks. PIO (30 mg), purchased from an outside pharmacy, was repackaged into capsules by the pharmacist at PBRC, and the placebo capsules were packaged in similar capsules. The PIO and placebo were administered with visits every 4 weeks at PBRC. To monitor compliance, participants were required to return unused pills for counting. After 16 weeks, the metabolic assessments were repeated. Pennington Biomedical Research Center's IRB approved (Protocol # 10039 -PBRC) all procedures from the originally published study [22] and all study participants gave written, informed consent. Adipose tissue biopsy samples used in this study were de-identified prior to RNA isolation outlined below.

2.3.2. Adipose Tissue Biopsies and RNA Isolation

Adipose tissue biopsies were collected with the Bergstrom and the Mercedes lipoaspirate techniques under sterile conditions and local anesthesia at baseline and post-intervention. Samples were taken from the subcutaneous abdominal region, between one- and two-thirds of the distance from the iliac spine to the umbilicus, and from the subcutaneous femoral region, on the anterior aspect of the thigh, one- to two-thirds of the distance from the superior iliac spine to the patella. The tissue was immediately frozen in liquid nitrogen and stored at −80 °C. Total RNA was extracted using the miRNeasy kit (Qiagen, Germantown, MD, USA), and the yield determined by spectrophotometry (NanoDrop Technologies, Wilmington, DE, USA). From each RNA sample, 500 ng was reverse transcribed to cDNA by using the iScript cDNA Reverse Transcription kit (Bio-Rad, Hercules, CA, USA). Real-time PCR was performed using the CFX real-time PCR system (Bio-Rad).

2.4. Gene Expression Analysis

Mouse epididymal (eWAT) and inguinal white adipose tissue (iWAT) depots were powdered and 50 mg aliquots were homogenized in TRIzol. Total RNA was extracted from adipose tissue and isolated islets using the RNeasy Mini RNA kit (Qiagen, Germantown, MD, USA). RNA quality and quantity was assessed using a Nanodrop spectrophotometer (NanoDrop Technologies, Wilmington, DE, USA). cDNA was generated from total RNA using the iScript cDNA synthesis kit (Bio-Rad). Relative mRNA abundance was measured by real-time RT-PCR using the iTaq Universal SYBR Green Supermix (Bio-Rad) on a CFX96 instrument (Bio-Rad). Transcript levels were normalized to the housekeeping gene Rs9 [23]. Primer pairs were designed using the Primer3Plus software and sequences are available upon request.

2.5. Serum Analyses

Mouse ELISA Kits from Mercodia (Uppsala, Sweden) were used to measure serum insulin and glucagon. Corticosterone was measured using an ELISA kit (Cat number ADI-900-097) from Enzo Life Sciences (Farmingdale, NY, USA). Triacylglycerol was determined using the Triglyceride Determination Kit (Sigma Aldrich; cat. no. TR0100-1KT). Mouse HMW and Total Adiponectin ELISA kit (Cat number 47-ADPMS-E01) was from Alpco (Salem, NH, USA). Manufacturers' recommended protocols were used for all serum measurements.

2.6. Statistical Analysis

Statistical analyses were performed using GraphPad Prism 6.07 (GraphPad Software, La Jolla, CA, USA). Data were analyzed by two-tailed Student's t-test, one-way analysis of variance (ANOVA) using a Tukey's test for post hoc analysis, or repeated-measures ANOVA (for longitudinal measures of blood glucose and body mass). Datasets were tested for outliers using the Rout method (Q = 1%). Data are presented as the means ± SEM.

3. Results

3.1. Pioglitazone (PIO) Lowers Blood Glucose in Obese Diabetic Mice

db/db mice are genetically obese and exhibit key features of human T2D, including insulin resistance, hyperglycemia, and alterations in islet β-cell markers [2,16,24–26]. We observed that oral delivery of the TZD pioglitazone (PIO), an insulin-sensitizer prescribed to patients with prediabetes or existing T2D [11,27,28], counteracted these pathological metabolic outcomes. Blood glucose in obese *db/db* mice (mean = 357 mg/dL at baseline) was restored to concentrations observed in lean littermate (*db*/+) controls four days after PIO administration (Figure 1a–c). These data are consistent with previous observations [29]. Pioglitazone has no effect on blood glucose, circulating insulin, or insulin tolerance in lean normoglycemic mice (Supplementary Figure S1). We did note a modest but significant increase in insulin positive area, islet fraction, and major and minor axis length in the islets of lean mice receiving pioglitazone. However, circulating insulin was not changed (Supplementary Figure S1). Because we were interested in the impact of pioglitazone during the obese, hyperglycemic state, a condition for which it is prescribed to humans, we did not study lean normoglycemic mice on pioglitazone any further.

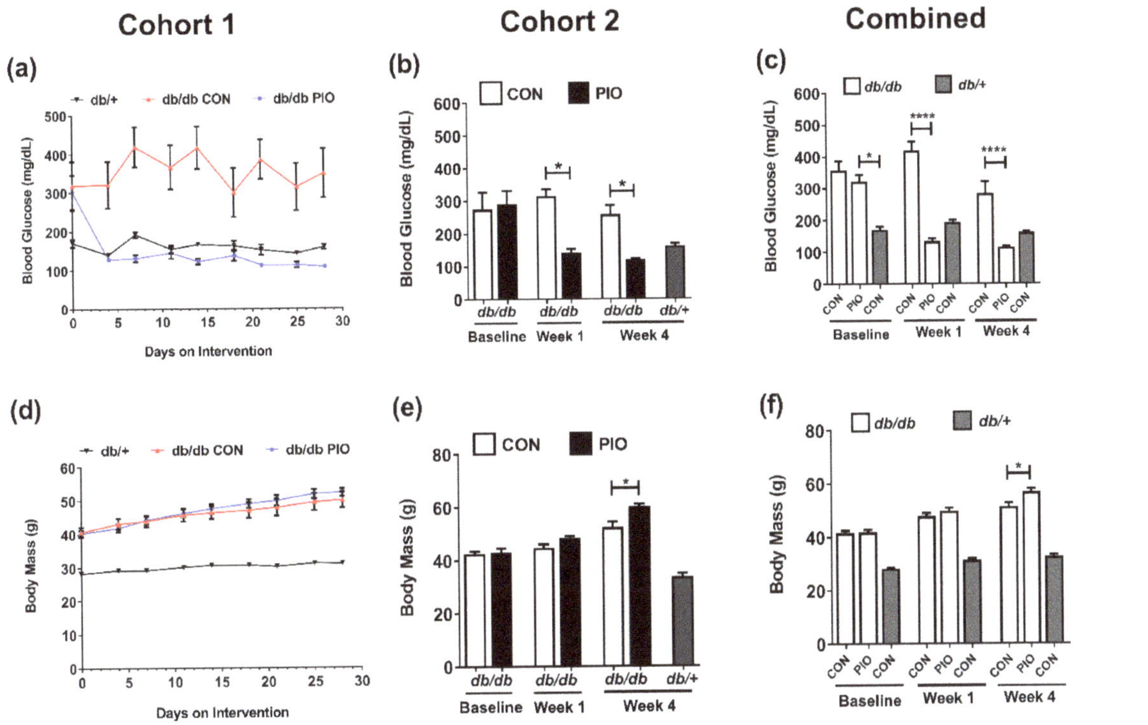

Figure 1. Dietary supplementation of pioglitazone (PIO) alleviates hyperglycemia in *db/db* mice. (**a–c**) Blood glucose in *db/+* mice on a control diet, and *db/db* mice on either a control (CON) or pioglitazone (PIO)-supplemented diet for 28 d starting at 8 weeks of age. (**d–f**) Body mass in *db/+* mice on a control diet, and *db/db* mice on either a CON or PIO-supplemented diet for 28 d. Data from cohort 1 are represented in Figure 1a,d (n = 6 per group), cohort 2 is shown in Figure 1b,e (n = 8 per group), and data compiled from both cohorts are shown in Figure 1c,f (n = 14 per group). Data are represented as the means ± SEM. * $p < 0.05$, **** $p < 0.0001$.

Blood glucose remained in the normal range in PIO-treated *db/db* mice for the duration of the 28 day dietary study with no evidence of hypoglycemia (Figure 1a–c). Compared to lean control *db*/+ mice, *db/db* mice displayed a body mass of 41.7 g (obese) versus 28.3 g (lean *db*/+; Figure 1d). PIO-treated *db/db* mice did not differ in body mass after one week on the drug (Figure 1e). However, there was a significant increase in body mass at the end of the 4 week study between *db/db* mice on PIO versus *db/db* mice consuming the control diet (56.7 g vs. 51.3 g body mass; Figure 1f; week 4).

3.2. Four Weeks of PIO Therapy Increases Fat and Fluid Mass in db/db Mice

At baseline, body composition was not different between the two groups of *db/db* mice (Baseline; compare white bar to light grey bar; Figure 2a–c). No significant difference in body composition was observed after one week of pioglitazone administration (Figure 2a–c); however, the mice receiving PIO display an 28% increase in fat mass after four weeks (Figure 2a) with no significant difference in lean mass (Figure 2b). Consistent with increased fat mass, there was also a 24% increase in fluid mass compared to animals receiving the control diet (Figure 2c). The lean control mouse (*db/+*) is shown for comparison (darker grey bars; Figure 2a–c).

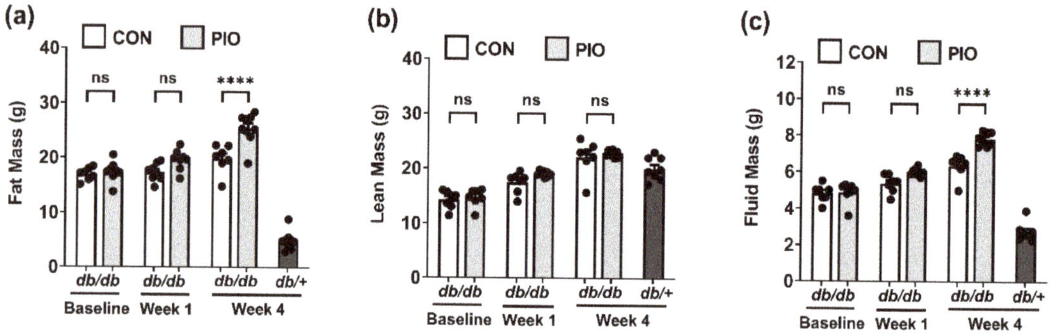

Figure 2. Four week dietary supplementation of PIO increases fat and fluid mass in *db/db* mice. (**a**) Fat mass, (**b**) lean mass, and (**c**) fluid mass in *db/+* mice on a control diet, and *db/db* mice on either a CON or PIO-supplemented diet for 28 d. n = 8 per group. Data are represented as the means ± SEM. **** $p < 0.0001$. ns, not significant.

3.3. *Pioglitazone Increases Respiratory Quotient (RQ) and Energy Expenditure (EE), but Does Not Alter Locomotor Activity or Food Intake*

Because blood glucose concentrations were rapidly restored to normal values in obese mice receiving pioglitazone (Figure 1), we conducted a separate study where *db/db* mice were placed into metabolic cages and given a PIO-supplemented diet or a control diet at the start of the metabolic cage measurements. We found a rapid increase in respiratory quotient (RQ), reflecting enhanced whole-body carbohydrate utilization, in mice receiving PIO when compared to *db/db* CON mice (Figure 3a–c). Over a period of 7 days, mean RQ was significantly higher across both light (day) and dark (night) cycles in PIO-fed *db/db* mice (Figure 3a,b), with an overall increase in RQ from 0.84 to 0.91 in the PIO group relative to CON-fed animals (Figure 3c). These data, representing increased whole-body glucose utilization, are consistent with the decrease in blood glucose shown in Figure 1.

Daily energy expenditure was similar between groups (Figure 3d) with clear differences between light and dark cycle (Figure 3e; white bar to grey bar). We noted a cumulative 4.74% increase in mean energy expenditure after seven days in PIO-supplemented *db/db* mice relative to CON animals (Figure 3f). No significant alterations in physical activity (Figure 3g), food consumption (Figure 3h), or liquid intake (Figure 3i) were observed between CON or PIO groups. Furthermore, mean sleep time was similar between dietary groups (data not shown).

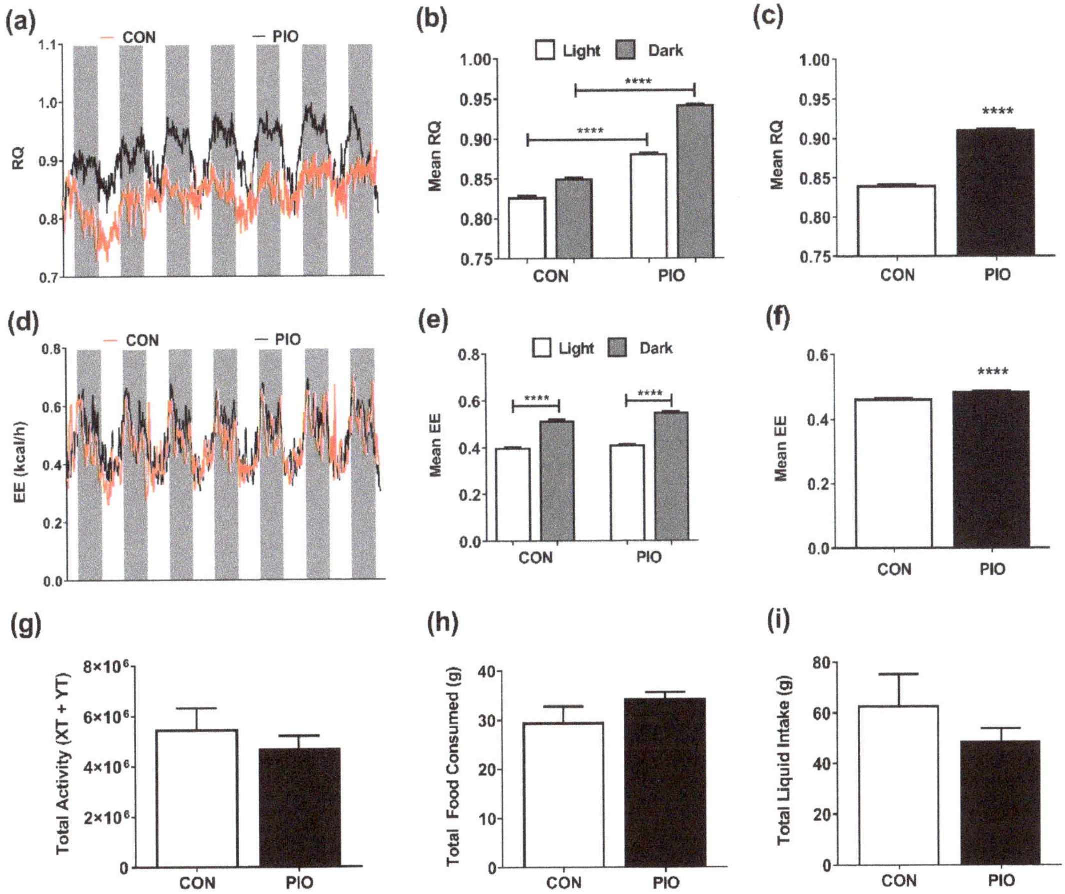

Figure 3. Pioglitazone increases respiratory quotient (RQ) and energy expenditure (EE), but does not impact activity or caloric intake. (**a**) RQ at daily intervals showing light (daytime; white bars) and dark (night time; grey bars) cycles, (**b**) mean RQ across light cycles, (**c**) mean RQ across all time points, (**d**) EE at daily intervals showing light (day) and dark (night) cycles, (**e**) mean EE across light cycles, (**f**) mean EE across all time points, (**g**) total activity, (**h**) total food consumed, (**i**) total liquid consumed in *db/db* mice on either a CON or PIO-supplemented diet for 7 d. n = 8 per group. Data are represented as the means ± SEM. **** $p < 0.0001$.

3.4. PIO Therapy Restores the Majority of Circulating Hormones in Obese Mice to Values Observed in Lean Controls

As shown in Figures 1–3, *db/db* mice receiving PIO display normal blood glucose levels with an increase in RQ despite elevations in total body mass and fat mass. Thus, we next examined circulating hormones to investigate whether they help to explain the metabolic changes. Corticosterone promotes insulin resistance and increases blood glucose levels when elevated chronically [21,30,31]. Corticosterone quantities in serum were reduced by intervention with PIO, although not back to control levels (Figure 4a). Circulating insulin concentrations are greater in untreated *db/db* mice when compared with either *db/db* mice receiving PIO or lean control mice (Figure 4b). In addition, circulating glucagon in *db/db* mice administered PIO also returned to values observed in lean control mice (Figure 4c). The changes in circulating hormones were consistent with the reduction in blood lipid (measured as triacylglycerols; Figure 4d). Further, PIO therapy increased both total serum adiponectin (Figure 4e) and its high-molecular-weight (HMW) form (Figure 4f).

Taken together, these data are consistent with improved blood glucose levels in obese mice receiving oral PIO therapy.

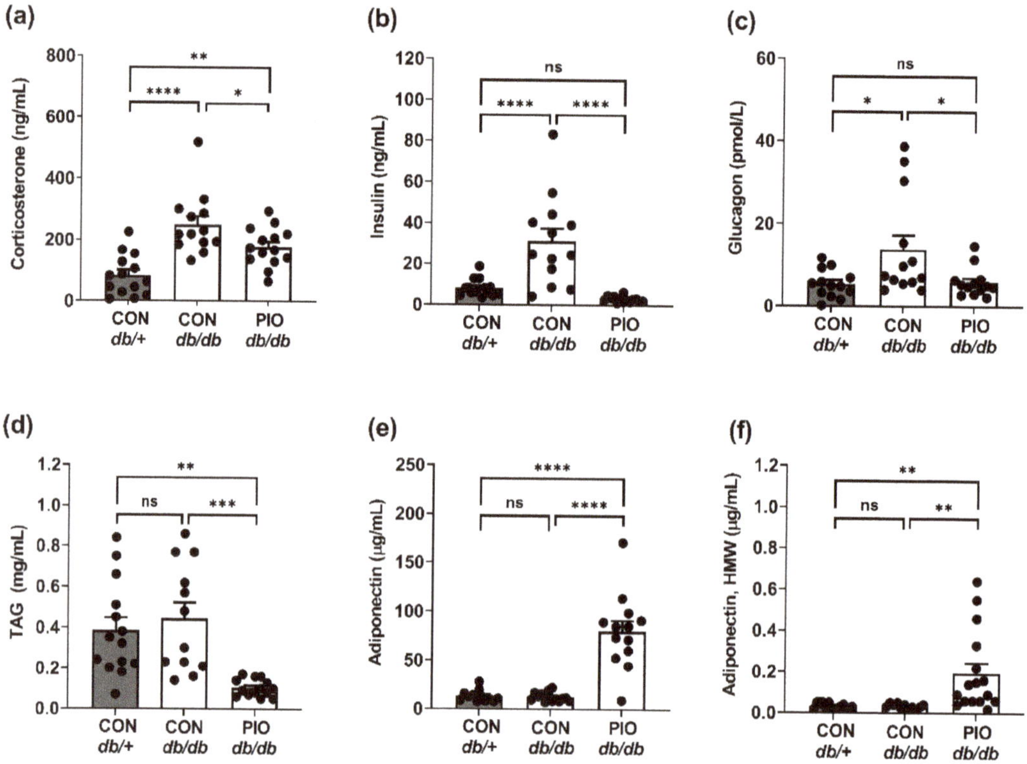

Figure 4. Dietary supplementation of PIO alters circulating levels of endocrine hormones, triglyceride, and adipokines in *db/db* mice. Serum levels of (**a**) corticosterone, (**b**) insulin, (**c**) glucagon, (**d**) TAG, (**e**) adiponectin, and (**f**) high-molecular-weight (HMW) adiponectin, in *db/+* mice on a CON diet (gray bars) and *db/db* mice on either a CON or PIO-supplemented diet for 28 d (white bars). n = 8 per group. Data are represented as the means ± SEM. * $p < 0.05$, ** $p < 0.01$, ***, $p < 0.001$; **** $p < 0.0001$. ns, not significant.

3.5. PIO Supplementation Alters Gene Expression Patterns in White Adipose Tissue from db/db Mice

Exposure to TZDs typically promotes patterns of gene expression consistent with 'browning' of white adipose tissue [32]. Indeed, we found that *Ucp1* (Figure 5a) and *Cidea* (Figure 5b) expression were enhanced in iWAT in response to pioglitazone treatment in db/db mice. We observed no significant difference in the expression of genes typically associated with brown fat development (e.g., PRDM16) in inguinal (iWAT) or epididymal white adipose tissue (eWAT) in *db/db* mice compared to lean *db/+* controls (data not shown). However, the genes *Ppara*, *Elovl3*, and *Cpt1b* were markedly elevated in iWAT from mice receiving pioglitazone (Figure 5c–e). These genes encode the transcription factor PPARα and two enzymes involved in lipid metabolism, respectively. In addition, expression of the gene encoding glycerol kinase (*Gk*), an enzyme important for reesterification of fatty acids, was increased in response to pioglitazone in iWAT (Figure 5f). Expression of the *Cd68* gene, a marker of macrophages, was increased in iWAT from obese mice relative to lean, but was not regulated by pioglitazone (Figure 5g). The mRNA levels of *Il1b*, encoding a pro-inflammatory cytokine, was reduced in *db/db* mice receiving pioglitazone when compared with lean (*db/+*) but not when compared with untreated *db/db* mice (Figure 5h).

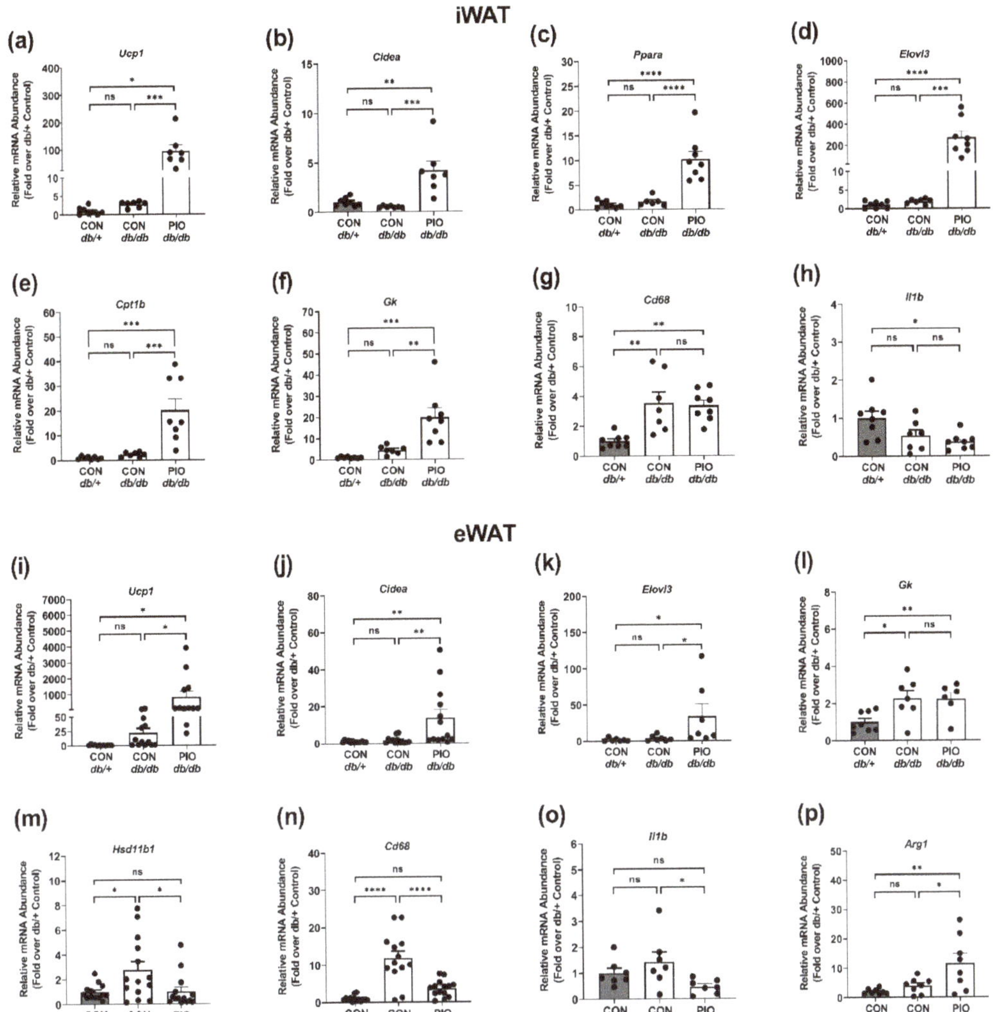

Figure 5. PIO alters expression of genes involved in browning, fatty acid oxidation, and inflammation in white adipose tissue of *db/db* mice. mRNA abundance of (**a**) *Ucp1*, (**b**) *Cidea*, (**c**) *Ppara*, (**d**) *Elovl3*, (**e**) *Cpt1b*, (**f**) *Gk*, (**g**) *Cd68*, and (**h**) *Il1b* in iWAT from *db/+* mice on a CON diet (gray bars) and *db/db* mice on either a CON- or PIO-supplemented diet (white bars) for 4 w. Gene expression analysis of (**i**) *Ucp1*, (**j**) *Cidea*, (**k**) *Elovl3*, (**l**) *Gk*, (**m**) *Hsd11b1*, (**n**) *Cd68*, (**o**) *Il1b*, and (**p**) *Arg1* in eWAT from *db/+* mice on a CON diet (gray bars) and *db/db* mice on either a CON- or PIO-supplemented diet (white bars) for 4 w. n = 8–14 per group. Values are represented as the means ± SEM. * $p < 0.05$, ** $p < 0.01$, *** $p < 0.001$, **** $p < 0.0001$. ns, not significant.

In comparison with iWAT, we note that the gene *Ucp1* (Figure 5i) was upregulated in eWAT of PIO-fed *db/db* mice when compared with untreated mice (52 fold in eWAT). These data are similar to what we observed in iWAT between PIO-treated *db/db* versus untreated *db/db* mice (Figure 5a; 56 fold). A similar outcome was seen with expression of the *Cidea* gene (Figure 5j; 8.4 fold increase in eWAT in PIO-exposed *db/db* versus untreated *db/db*). Additionally, similar to what was observed in iWAT, the gene encoding *Elovl3* was enhanced in eWAT from *db/db* mice receiving receiving PIO therapy (Figure 5k). By contrast, expression of *Gk* was elevated in both untreated *db/db* and *db/db* mice receiving PIO relative to lean control (*db/+*) mice (Figure 5l).

Increased availability of cortisol in adipose tissue impairs glucose and fat metabolism in individuals with metabolic syndrome and promotes insulin resistance in mice [33,34]. *Hsd11b1*, the gene that encodes the enzyme that converts inactive cortisone in humans (and corticosterone in rodents) to active cortisol, was restored to the levels seen in lean control mice (Figure 5m). Expression of *Cd68*, encoding a marker of activated macrophages [35], was markedly suppressed in eWAT by PIO exposure (Figure 5n). This was not observed in iWAT (Figure 5g). Similarly, expression of *Il1b*, a cytokine associated with pro-inflammatory macrophages, was also reduced 66% in *db/db* mice receiving PIO therapy (Figure 5o). The expression of *Arg1*, a gene associated with tissue repair and resolution of inflammation type macrophages [36], was enhanced in *db/db* mice receiving PIO (Figure 5p). Taken together, there are similarities as well as clear depot specific differences in the pioglitazone ability to regulate expression of certain targets genes in iWAT compared with eWAT.

3.6. Oral Pioglitazone Administration to Human Study Participants Alters White Adipose Tissue Gene Expression

Subcutaneous abdominal and femoral adipose tissues from seven women from the Apple & Pear study who had baseline (CON) and post-intervention (PIO) assessments were analyzed (26 ± 5 years; BMI 32.2 ± 3.2 kg/m^2). The main study outcomes were previously reported [22]. These depots were chosen for their known relationships to metabolic health [37].

In femoral depots of human white adipose tissue, we found that *UCP1* expression in the group receiving pioglitazone was increased 5.1-fold over control (Figure 6a), while the expression of the *DLK1* gene was not changed under these conditions (Figure 6b). *DLK1* encodes a transmembrane protein that can be cleaved and regulates adipogenesis [38]. The glycerol kinase (*GK*) gene was upregulated 3.3-fold over control (Figure 6c) while the expression of *HSD11B1* and *IL1B* were not significantly altered by PIO (Figure 6d,e).

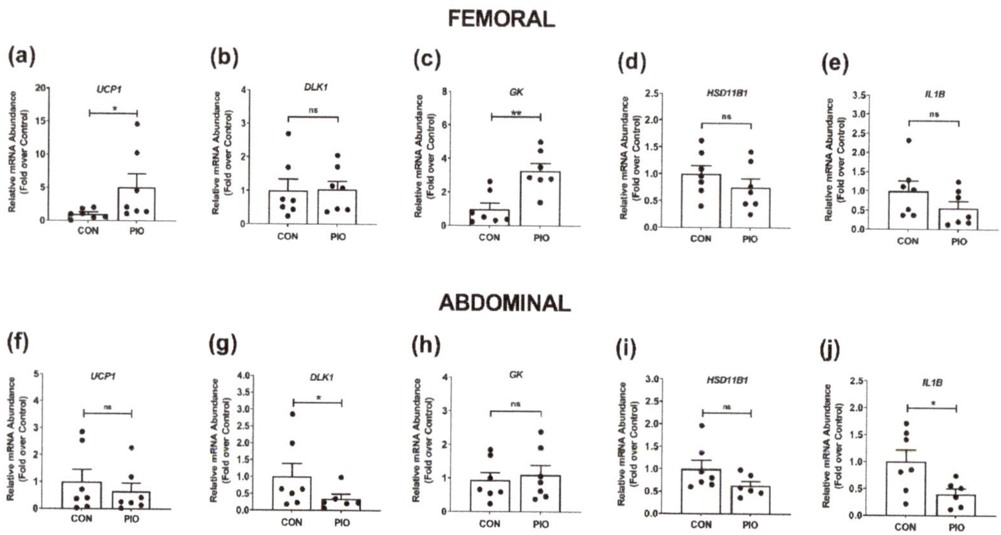

Figure 6. Pioglitazone reduces inflammation while also promoting markers of browning and fatty acid reesterification in white adipose tissue of human subjects. Relative mRNA abundance of (**a**) UCP1, (**b**) DLK1, (**c**) GK, (**d**) HSD11B1, and (**e**) IL1B in femoral adipose tissue from human subjects. Gene expression analysis of (**f**) UCP1, (**g**) DLK1, (**h**) GK, (**i**) HSD11B1, and (**j**) IL1B in abdominal adipose tissue from humans with the indicated conditions. n = 7 per group. Values are represented as the means ± SEM. * $p < 0.05$, ** $p < 0.01$. ns, not significant.

In contrast to the femoral depot, *UCP1* expression was not altered by PIO (Figure 6f), while *DLK1* expression was significantly reduced (Figure 6g). *GK* expression was not induced in the abdominal depot (Figure 6h). We note that *CD68* expression decreased by 51% in response to pioglitazone in the femoral depot (data not shown). The gene encoding 11β-HSD1 (*HSD11B1*), a key enzyme regulating glucocorticoid action [33], trended towards a decrease in abdominal adipose tissue with pioglitazone (p value = 0.11; Figure 6i). The gene encoding interleukin-1beta (*IL1B*) was reduced by 60% (Figure 6j). Note that *IL1B* was not significantly changed in the femoral depot (Figure 6e). We note that the pattern of glycerol kinase expression is similar in mouse iWAT (Figure 5f), analogous to human subcutaneous adipose tissue [39], when compared with the femoral adipose tissue in humans (Figure 6c). The abdominal depot from humans displayed patterns most comparable with mouse eWAT (compare Figure 5l with Figures 6h and 5o with Figure 6j).

3.7. db/db Mice on a PIO-Enhanced Diet Display Increased Expression of the Insulin Genes and Decreased Expression of the Aldh1a3 Gene

Islets from humans with T2D show clear evidence of de-differentiation as measured by loss of key β-cell transcription factors (e.g., *MafA* and *Nkx6.1*) and gain of *Aldh1a3* [25,26]. The *db/db* mouse recapitulates many features of human T2D, including obesity, insulin resistance, hyperglycemia, and the aforementioned changes in markers of mature β-cells (e.g., Aldh1a3, insulin, and Nkx6.1) [16,24]. After four weeks of PIO administration, islets isolated from *db/db* mice had greater expression of both *Ins1* and *Ins2* genes (Figure 7a,b). In addition, *MafA* expression was also increased (Figure 7c). Moreover, expression of *Aldh1a3* was reduced alongside increased expression of *Ffar1* (Figure 7e), *Gpr119* (Figure 7f), and *Ffar4* (Figure 7g). We did note a mild increase in the expression of *Ddit3*, a gene linked with ER stress, in islets isolated from PIO-exposed mice (Figure 7h).

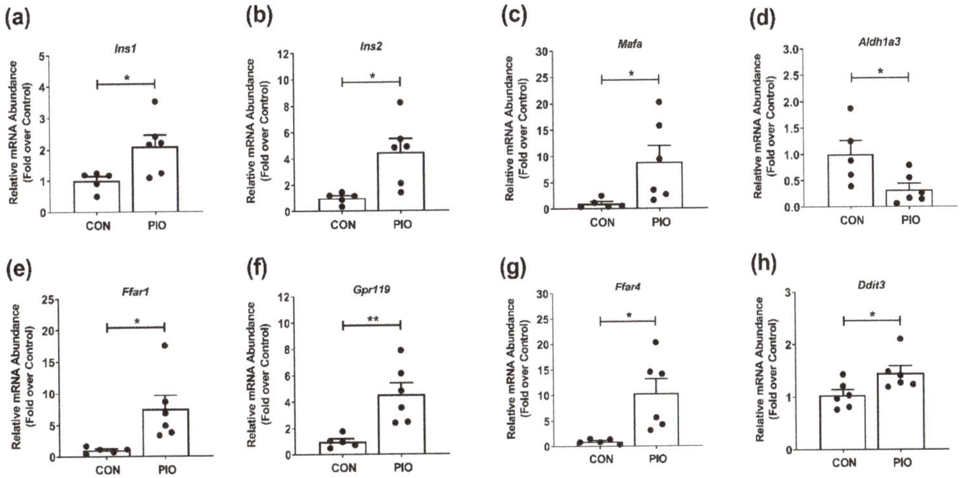

Figure 7. Islet gene expression reflects improvements in markers of the differentiated state in *db/db* mice receiving pioglitazone. Gene expression of (**a**) *Ins1*, (**b**) *Ins2*, (**c**) *Mafa*, (**d**) *Aldh1a3*, (**e**) *Ffar1*, (**f**) *Gpr119*, (**g**) *Ffar4*, and (**h**) *Ddit3* in islets isolated from *db/db* mice fed either a CON or PIO-supplemented diet for 28 d. n = 6 per group. Data are represented as the means ± SEM. * $p < 0.05$, ** $p < 0.01$.

3.8. PIO-Supplemented Diet Restores Pancreatic Nkx6.1 Abundance and Decreases Abundance of the De-Differentiation Marker Aldh1a3 in db/db Mice

We next examined islet histology of *db/db* mice fed either control or PIO-supplemented diets as well as lean *db/+* mice fed the control diet. Congruent to the gene expression observations in Figure 7a,b, islets from *db/db* mice given the PIO-supplemented diet displayed more intense staining of insulin (Figure 8; top row). In addition, the immunodetection of Aldh1a3 protein was reduced (Figure 8; middle row; compare middle panel with right hand panel). Finally, we found that immunoreactive Nkx6.1 was markedly enhanced in *db/db* receiving PIO when compared with *db/db* mice receiving the control diet (Figure 8; bottom row; compared middle panel with right hand panel).

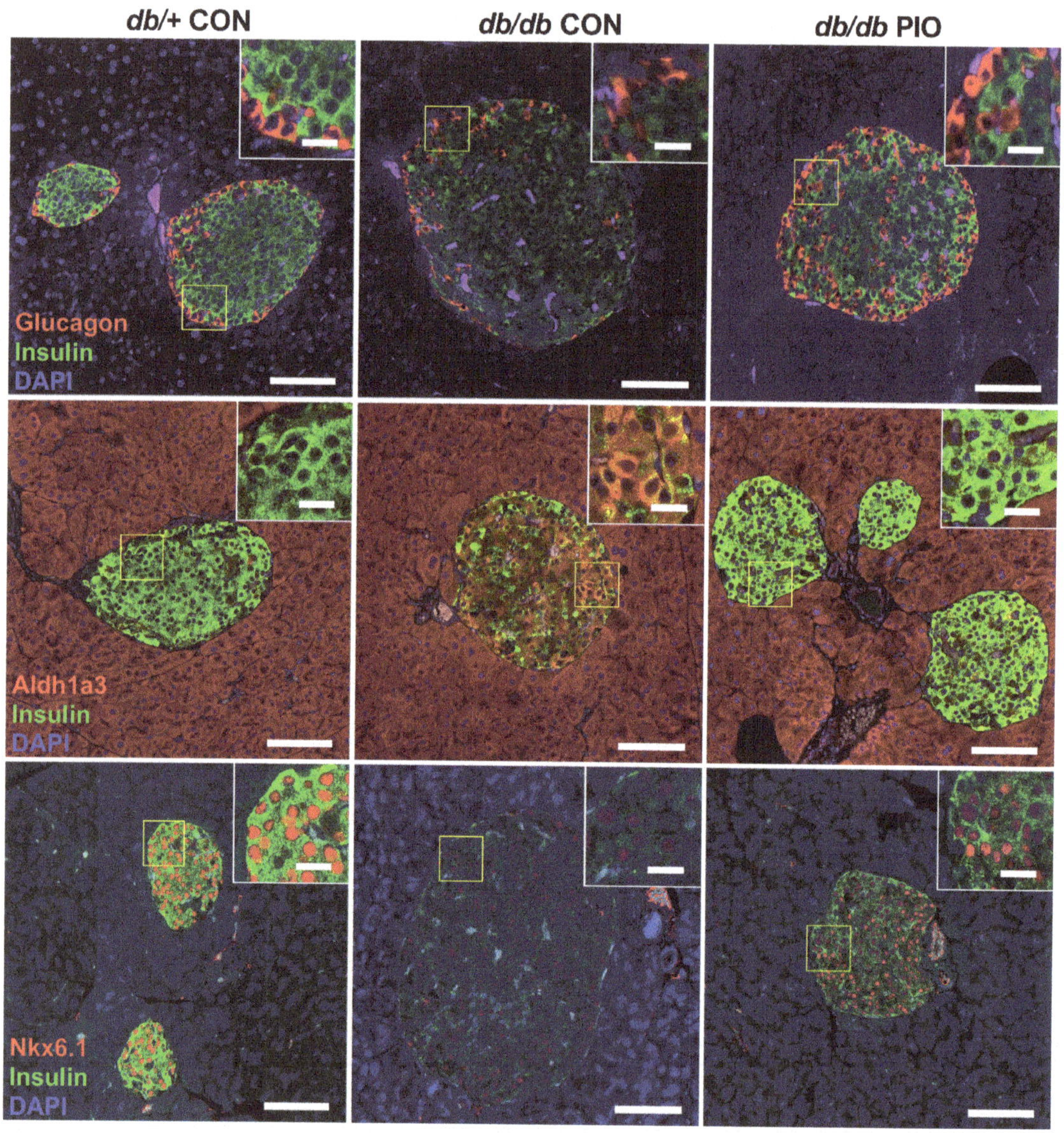

Figure 8. PIO-supplemented diet restores pancreatic Nkx6.1 and decreases abundance of the de-differentiation marker Aldh1a3 in *db/db* mice. Triple-fluorescence staining of fixed pancreatic tissue from *db/+* mice on a control diet (CON), or *db/db* mice on either a CON or PIO-supplemented diet for 28 d. Insulin staining shown in green and DAPI in blue. The red stain indicates glucagon (top row), Aldh1a3 (middle row), and Nkx6.1 (bottom row). Sections were stained from four animals per group and representative images were chosen from each group. Scale bars = 100 μm for large image, 20 μm for inset.

4. Discussion

Pioglitazone is an FDA-approved PPARγ agonist used to treat metabolic diseases, such as T2D [40–43]. The benefits of pioglitazone are through insulin sensitization, improved lipid metabolism, and regulation of inflammation [40,41]. In the present study, the reduction in circulating glucocorticoids (Figure 4a) and lipids (Figure 4d), as well as the rise in adiponectin (Figure 4e,f), are consistent with changes likely to reflect improvements in whole-body insulin sensitivity. Along these lines, improved insulin sensitivity was indirectly reflected by reduced serum insulin (Figure 4b) and glucagon concentration (Figure 4c) as well as by improved blood glucose levels in *db/db* mice receiving pioglitazone (Figure 1a). Importantly, we note that blood glucose levels in *db/db* mice receiving pioglitazone return to the level of lean control (*db*/+) mice without any evidence of hypoglycemia (Figure 1). In addition, pioglitazone has little to no effect on blood glucose, circulating insulin, or insulin tolerance in the lean, normoglycemic mouse (Supplementary Figure S1). We did note a slight but significant increase in insulin positive area in the islets of lean mice receiving pioglitazone, suggesting a possible direct effect of this drug to promote increased β-cell mass under these conditions. However, circulating insulin was not changed (Supplementary Figure S1).

The TZD class of drugs promotes increases in adipose tissue mass in vivo (see Figure 2a and [44]), providing a reservoir to lower lipid levels in circulation. An additional explanation for the lowered circulating lipid levels is the increased expression of adipose tissue glycerol kinase (Figure 5f), an enzyme that promotes retention and re-esterification of fatty acids in cultured adipocytes [45]. Distinct TZDs, such as rosiglitazone and ciglitazone, promote glycerol kinase expression in cultured mouse and human adipocytes as well as in adipose tissue from *ob/ob* mice [45]. Here, we extend those findings to both male *db/db* mice and female human subjects given pioglitazone (Figures 5f and 6c), suggesting an important lipid lowering mechanism for the TZD class of drugs during insulin-resistant states that is relevant to both rodents and humans. We do note that while the phenotype of PIO intervention appears similar between males and females [46], it is possible that mechanisms associated with these beneficial phenotypes could be different due to differences in sex hormones. Nonetheless, our findings are congruent with improved glucose utilization as measured by the decrease in blood glucose concentration (Figure 1a) and the increase in whole-body respiratory quotient (RQ) (Figure 3a–c). The findings herein using *db/db* mice are also consistent with improved insulin sensitivity in humans [22,47].

The enhanced whole-body glucose utilization observed in metabolic cage studies shown in Figure 3 is also accompanied by reduced markers of islet β-cell de-differentiation and restored presence of proteins necessary to maintain mature β-cell identity (Figures 7 and 8). Whether the reduction in blood lipid or blood glucose is the key variable explaining improved β-cell markers of health and maturity is unclear at present. Our best explanation is that collectively lowering serum glucose and lipids removes stress from islet β-cells, allowing them to recover. This is a postulate supported by other studies [29,48]. It is also conceivable that the effects of pioglitazone occur directly on the β-cell as well as on islet resident macrophages; these combined possibilities, along with reductions in blood glucose and blood lipid, promote increased production and storage of insulin in the islet. Interestingly, we found that pioglitazone reduces *Ald1a3* gene expression (Figure 7d) and protein abundance (Figure 8; middle row; *db/db* CON vs. *db/db* PIO) in pancreatic islets. This is important because *Aldh1a3* is a marker of islet β-cell de-differentiation in multiple different mouse models [16,20,24] and in humans [25]. These observations were also consistent with increased presence of *Ins1* and *Ins2* mRNA (Figure 7a,b) and augmented immunoreactive insulin and Nkx6.1 proteins (Figure 8).

Pioglitazone has partial PPARα agonist activity [49], which may be one reason why this TZD is effective in the present study while rosiglitazone was unable to suppress Aldh1a3 expression in mouse islets in a previous study [50]. This new finding may add to the possibility of pioglitazone having underappreciated properties for treating diseases associated with obesity and insulin resistance. Our findings also provide additional pre-clinical metabolic information to aid in understanding the therapeutic potential of pioglitazone when compared with other drugs in the TZD category [42,51]. The comparison with tissues from humans in the present study support the conclusions drawn in the pre-clinical model.

Indeed, we observed that gene expression markers typically associated with brown adipose tissue (e.g., *UCP-1*) were upregulated in both mouse (Figure 5a,i) and the femoral (Figure 6a), but not the abdominal human adipose tissue (Figure 6f). In addition, there is a reduction in abdominal (Figure 6g), but not femoral *DLK1/Pref-1* (Figure 6b) in adipose tissue from humans given pioglitazone. An important observation was the enhanced expression of glycerol kinase (*Gk*) in both mouse iWAT (Figure 5f) and human femoral adipose tissue (Figure 6c). These findings are consistent with redistribution of lipid to subcutaneous adipose tissue and overall increases in BMI in response to TZD therapy [46]. Thus, pioglitazone promotes expansion of, and lipid storage within, specific adipose tissue depots as needed to decrease lipid accumulation in lean tissues and reduce circulating fatty acids. These outcomes likely arise, at least in part, through enhancing the re-esterification of fatty acids within specific adipose tissue depots in mice and humans with glycerol kinase as a key component of the mechanism (present data and [45]). Finally, we observed a reduction in IL-1β gene expression in both mouse eWAT (Figure 5o) and in human abdominal adipose tissue (Figure 6j) in response to pioglitazone. While it would be reasonable to speculate that expression of each of these genes correlates (either positively or negatively) with significant improvements in metabolic health, further in-depth studies are required to provide conclusive statistical evidence. In summary, the present data, and new evidence that pioglitazone does not have the cardiovascular risks that are observed with other TZDs [42], make it clear that pioglitazone has likely been undervalued as a practical therapeutic option for conditions associated with obesity, insulin resistance, and hyperglycemia.

Supplementary Materials: The following are available online at https://www.mdpi.com/article/10.3390/biomedicines9091189/s1, Figure S1: Pioglitazone increases insulin positive area but does not alter circulating glucose or insulin concentrations in lean normoglycemic mice.

Author Contributions: Conceptualization: J.J.C., M.D.K. and S.J.B.; investigation: J.J.C., H.M.B., K.L.M., T.M.M., U.W., D.H.B., C.R.C., M.D.K. and S.J.B.; data curation: J.J.C., M.D.K. and S.J.B.; writing—original draft preparation: J.J.C. and S.J.B.; writing—review and editing: J.J.C., H.M.B., K.L.M., T.M.M., U.W., E.R., D.H.B., M.D.K. and S.J.B.; project administration: J.J.C. and S.J.B.; funding acquisition: J.J.C., U.W. and S.J.B. All authors have read and agreed to the published version of the manuscript.

Funding: Research in the Collier laboratory is supported by NIH grants R03 AI151920, R21 AI138136, and R01 DK123183. Research in the Burke laboratory is supported by NIH grant P20 GM135002. Research in the White laboratory is supported by NIH grant R01 DK121944. This study used PBRC core facilities (Genomics, Comparative Biology, Animal Metabolism and Behavior, Cell Biology and Bioimaging) that are supported in part by COBRE (P20 RR021945 and P30 GM118430) and NORC (P30 DK072476) center grants from the National Institutes of Health, as well as equipment purchased with funds from a shared instrumentation grant (NIH S10 OD023703).

Institutional Review Board Statement: All animal procedures described herein were approved by the Institutional Care and Use Committees of Pennington Biomedical Research Center (IACUC protocol # 1021; approved 05/02/2018) and University of Tennessee Health Science Center (IACUC protocol # 2171; approved 02/26/2016).

Informed Consent Statement: Not applicable.

Data Availability Statement: The data presented in this study are available upon reasonable request by contacting the corresponding authors.

Conflicts of Interest: The authors declare no conflict of interest.

References

1. Gonzalez-Muniesa, P.; Martinez-Gonzalez, M.A.; Hu, F.B.; Despres, J.P.; Matsuzawa, Y.; Loos, R.J.F.; Moreno, L.A.; Bray, G.A.; Martinez, J.A. Obesity. *Nat. Rev. Dis. Primers* **2017**, *3*, 17034. [CrossRef] [PubMed]
2. DeFronzo, R.A.; Ferrannini, E.; Groop, L.; Henry, R.R.; Herman, W.H.; Holst, J.J.; Hu, F.B.; Kahn, C.R.; Raz, I.; Shulman, G.I.; et al. Type 2 diabetes mellitus. *Nat. Rev. Dis. Primers* **2015**, *1*, 15019. [CrossRef] [PubMed]
3. Bray, G.A. Obesity increases risk for diabetes. *Int. J. Obes. Relat. Metab. Disord.* **1992**, *16* (Suppl. 4), S13–S17.
4. Kahn, S.E.; Zraika, S.; Utzschneider, K.M.; Hull, R.L. The beta cell lesion in type 2 diabetes: There has to be a primary functional abnormality. *Diabetologia* **2009**, *52*, 1003–1012. [CrossRef]
5. Doria, A.; Patti, M.E.; Kahn, C.R. The emerging genetic architecture of type 2 diabetes. *Cell Metab.* **2008**, *8*, 186–200. [CrossRef]
6. Burke, S.J.; Karlstad, M.D.; Collier, J.J. Pancreatic Islet Responses to Metabolic Trauma. *Shock* **2016**, *46*, 230–238. [CrossRef]
7. Taylor, R. Type 2 Diabetes and Remission: Practical Management Guided by Pathophysiology. *J. Intern. Med.* **2020**, *289*, 754–770. [CrossRef]
8. Upadhyay, J.; Polyzos, S.A.; Perakakis, N.; Thakkar, B.; Paschou, S.A.; Katsiki, N.; Underwood, P.; Park, K.H.; Seufert, J.; Kang, E.S.; et al. Pharmacotherapy of type 2 diabetes: An update. *Metabolism* **2018**, *78*, 13–42. [CrossRef]
9. Unger, R.H. Lipotoxic diseases. *Annu. Rev. Med.* **2002**, *53*, 319–336. [CrossRef]
10. Eldor, R.; DeFronzo, R.A.; Abdul-Ghani, M. In vivo actions of peroxisome proliferator-activated receptors: Glycemic control, insulin sensitivity, and insulin secretion. *Diabetes Care* **2013**, *36* (Suppl. 2), S162–S174. [CrossRef]
11. Defronzo, R.A.; Tripathy, D.; Schwenke, D.C.; Banerji, M.; Bray, G.A.; Buchanan, T.A.; Clement, S.C.; Gastaldelli, A.; Henry, R.R.; Kitabchi, A.E.; et al. Prevention of diabetes with pioglitazone in ACT NOW: Physiologic correlates. *Diabetes* **2013**, *62*, 3920–3926. [CrossRef]
12. DeFronzo, R.A.; Abdul-Ghani, M.A. Preservation of beta-cell function: The key to diabetes prevention. *J. Clin. Endocrinol. Metab.* **2011**, *96*, 2354–2366. [CrossRef]
13. Lehmann, J.M.; Moore, L.B.; Smith-Oliver, T.A.; Wilkison, W.O.; Willson, T.M.; Kliewer, S.A. An antidiabetic thiazolidinedione is a high affinity ligand for peroxisome proliferator-activated receptor gamma (PPAR gamma). *J. Biol. Chem.* **1995**, *270*, 12953–12956. [CrossRef]
14. Cristancho, A.G.; Lazar, M.A. Forming functional fat: A growing understanding of adipocyte differentiation. *Nat. Rev. Mol. Cell Biol.* **2011**, *12*, 722–734. [CrossRef]
15. Croasdell, A.; Duffney, P.F.; Kim, N.; Lacy, S.H.; Sime, P.J.; Phipps, R.P. PPARgamma and the Innate Immune System Mediate the Resolution of Inflammation. *PPAR Res.* **2015**, *2015*, 549691. [CrossRef]
16. Burke, S.J.; Batdorf, H.M.; Burk, D.H.; Noland, R.C.; Eder, A.E.; Boulos, M.S.; Karlstad, M.D.; Collier, J.J. db/db Mice Exhibit Features of Human Type 2 Diabetes That Are Not Present in Weight-Matched C57BL/6J Mice Fed a Western Diet. *J. Diabetes Res.* **2017**, *2017*, 8503754. [CrossRef]
17. Obernier, J.A.; Baldwin, R.L. Establishing an appropriate period of acclimatization following transportation of laboratory animals. *ILAR J.* **2006**, *47*, 364–369. [CrossRef] [PubMed]
18. Sims, E.K.; Hatanaka, M.; Morris, D.L.; Tersey, S.A.; Kono, T.; Chaudry, Z.Z.; Day, K.H.; Moss, D.R.; Stull, N.D.; Mirmira, R.G.; et al. Divergent compensatory responses to high-fat diet between C57BL6/J and C57BLKS/J inbred mouse strains. *Am. J. Physiol. Endocrinol. Metab.* **2013**, *305*, E1495–E1511. [CrossRef] [PubMed]
19. Burke, S.J.; Karlstad, M.D.; Regal, K.M.; Sparer, T.E.; Lu, D.; Elks, C.M.; Grant, R.W.; Stephens, J.M.; Burk, D.H.; Collier, J.J. CCL20 is elevated during obesity and differentially regulated by NF-kappaB subunits in pancreatic beta-cells. *Biochim. Biophys. Acta* **2015**, *1849*, 637–652. [CrossRef] [PubMed]
20. Burke, S.J.; Batdorf, H.M.; Burk, D.H.; Martin, T.M.; Mendoza, T.; Stadler, K.; Alami, W.; Karlstad, M.D.; Robson, M.J.; Blakely, R.D.; et al. Pancreatic deletion of the interleukin-1 receptor disrupts whole body glucose homeostasis and promotes islet beta-cell de-differentiation. *Mol. Metab.* **2018**, *14*, 95–107. [CrossRef]
21. Burke, S.J.; Batdorf, H.M.; Eder, A.E.; Karlstad, M.D.; Burk, D.H.; Noland, R.C.; Floyd, Z.E.; Collier, J.J. Oral Corticosterone Administration Reduces Insulitis but Promotes Insulin Resistance and Hyperglycemia in Male Nonobese Diabetic Mice. *Am. J. Pathol.* **2017**, *187*, 614–626. [CrossRef]
22. White, U.; Fitch, M.D.; Beyl, R.A.; Hellerstein, M.K.; Ravussin, E. Adipose depot-specific effects of 16 weeks of pioglitazone on in vivo adipogenesis in women with obesity: A randomised controlled trial. *Diabetologia* **2021**, *64*, 159–167. [CrossRef] [PubMed]
23. Scott, D.K.; Collier, J.J.; Doan, T.T.; Bunnell, A.S.; Daniels, M.C.; Eckert, D.T.; O'Doherty, R.M. A modest glucokinase overexpression in the liver promotes fed expression levels of glycolytic and lipogenic enzyme genes in the fasted state without altering SREBP-1c expression. *Mol. Cell. Biochem.* **2003**, *254*, 327–337. [CrossRef]
24. Kim-Muller, J.Y.; Fan, J.; Kim, Y.J.; Lee, S.A.; Ishida, E.; Blaner, W.S.; Accili, D. Aldehyde dehydrogenase 1a3 defines a subset of failing pancreatic beta cells in diabetic mice. *Nat. Commun.* **2016**, *7*, 12631. [CrossRef] [PubMed]
25. Cinti, F.; Bouchi, R.; Kim-Muller, J.Y.; Ohmura, Y.; Sandoval, P.R.; Masini, M.; Marselli, L.; Suleiman, M.; Ratner, L.E.; Marchetti, P.; et al. Evidence of beta-Cell Dedifferentiation in Human Type 2 Diabetes. *J. Clin. Endocrinol. Metab.* **2016**, *101*, 1044–1054. [CrossRef]
26. Guo, S.; Dai, C.; Guo, M.; Taylor, B.; Harmon, J.S.; Sander, M.; Robertson, R.P.; Powers, A.C.; Stein, R. Inactivation of specific beta cell transcription factors in type 2 diabetes. *J. Clin. Investig.* **2013**, *123*, 3305–3316. [CrossRef] [PubMed]

27. Espinoza, S.E.; Wang, C.P.; Tripathy, D.; Clement, S.C.; Schwenke, D.C.; Banerji, M.A.; Bray, G.A.; Buchanan, T.A.; Henry, R.R.; Kitabchi, A.E.; et al. Pioglitazone is equally effective for diabetes prevention in older versus younger adults with impaired glucose tolerance. *Age* **2016**, *38*, 485–493. [CrossRef] [PubMed]
28. Tripathy, D.; Daniele, G.; Fiorentino, T.V.; Perez-Cadena, Z.; Chavez-Velasquez, A.; Kamath, S.; Fanti, P.; Jenkinson, C.; Andreozzi, F.; Federici, M.; et al. Pioglitazone improves glucose metabolism and modulates skeletal muscle TIMP-3-TACE dyad in type 2 diabetes mellitus: A randomised, double-blind, placebo-controlled, mechanistic study. *Diabetologia* **2013**, *56*, 2153–2163. [CrossRef]
29. Ishida, H.; Takizawa, M.; Ozawa, S.; Nakamichi, Y.; Yamaguchi, S.; Katsuta, H.; Tanaka, T.; Maruyama, M.; Katahira, H.; Yoshimoto, K.; et al. Pioglitazone improves insulin secretory capacity and prevents the loss of beta-cell mass in obese diabetic db/db mice: Possible protection of beta cells from oxidative stress. *Metabolism* **2004**, *53*, 488–494. [CrossRef]
30. Burke, S.J.; Batdorf, H.M.; Huang, T.Y.; Jackson, J.W.; Jones, K.A.; Martin, T.M.; Rohli, K.E.; Karlstad, M.D.; Sparer, T.E.; Burk, D.H.; et al. One week of continuous corticosterone exposure impairs hepatic metabolic flexibility, promotes islet beta-cell proliferation, and reduces physical activity in male C57BL/6J mice. *J. Steroid Biochem. Mol. Biol.* **2019**, *195*, 105468. [CrossRef]
31. Fransson, L.; Franzen, S.; Rosengren, V.; Wolbert, P.; Sjoholm, A.; Ortsater, H. beta-Cell adaptation in a mouse model of glucocorticoid-induced metabolic syndrome. *J. Endocrinol.* **2013**, *219*, 231–241. [CrossRef]
32. Vernochet, C.; Peres, S.B.; Davis, K.E.; McDonald, M.E.; Qiang, L.; Wang, H.; Scherer, P.E.; Farmer, S.R. C/EBPalpha and the corepressors CtBP1 and CtBP2 regulate repression of select visceral white adipose genes during induction of the brown phenotype in white adipocytes by peroxisome proliferator-activated receptor gamma agonists. *Mol. Cell. Biol.* **2009**, *29*, 4714–4728. [CrossRef]
33. Masuzaki, H.; Paterson, J.; Shinyama, H.; Morton, N.M.; Mullins, J.J.; Seckl, J.R.; Flier, J.S. A transgenic model of visceral obesity and the metabolic syndrome. *Science* **2001**, *294*, 2166–2170. [CrossRef] [PubMed]
34. Bujalska, I.J.; Kumar, S.; Stewart, P.M. Does central obesity reflect "Cushing's disease of the omentum"? *Lancet* **1997**, *349*, 1210–1213. [CrossRef]
35. Kurushima, H.; Ramprasad, M.; Kondratenko, N.; Foster, D.M.; Quehenberger, O.; Steinberg, D. Surface expression and rapid internalization of macrosialin (mouse CD68) on elicited mouse peritoneal macrophages. *J. Leukoc. Biol.* **2000**, *67*, 104–108. [CrossRef] [PubMed]
36. Martinez, F.O.; Helming, L.; Gordon, S. Alternative activation of macrophages: An immunologic functional perspective. *Annu. Rev. Immunol.* **2009**, *27*, 451–483. [CrossRef] [PubMed]
37. Tchkonia, T.; Thomou, T.; Zhu, Y.; Karagiannides, I.; Pothoulakis, C.; Jensen, M.D.; Kirkland, J.L. Mechanisms and metabolic implications of regional differences among fat depots. *Cell Metab.* **2013**, *17*, 644–656. [CrossRef]
38. Wang, Y.; Kim, K.A.; Kim, J.H.; Sul, H.S. Pref-1, a preadipocyte secreted factor that inhibits adipogenesis. *J. Nutr.* **2006**, *136*, 2953–2956. [CrossRef]
39. Chusyd, D.E.; Wang, D.; Huffman, D.M.; Nagy, T.R. Relationships between Rodent White Adipose Fat Pads and Human White Adipose Fat Depots. *Front. Nutr.* **2016**, *3*, 10. [CrossRef]
40. Ahmadian, M.; Suh, J.M.; Hah, N.; Liddle, C.; Atkins, A.R.; Downes, M.; Evans, R.M. PPARgamma signaling and metabolism: The good, the bad and the future. *Nat. Med.* **2013**, *19*, 557–566. [CrossRef] [PubMed]
41. Soccio, R.E.; Chen, E.R.; Lazar, M.A. Thiazolidinediones and the promise of insulin sensitization in type 2 diabetes. *Cell Metab.* **2014**, *20*, 573–591. [CrossRef]
42. DeFronzo, R.A.; Inzucchi, S.; Abdul-Ghani, M.; Nissen, S.E. Pioglitazone: The forgotten, cost-effective cardioprotective drug for type 2 diabetes. *Diabetes Vasc. Dis. Res.* **2019**, *16*, 133–143. [CrossRef] [PubMed]
43. DeFronzo, R.A.; Tripathy, D.; Schwenke, D.C.; Banerji, M.; Bray, G.A.; Buchanan, T.A.; Clement, S.C.; Henry, R.R.; Hodis, H.N.; Kitabchi, A.E.; et al. Pioglitazone for diabetes prevention in impaired glucose tolerance. *N. Engl. J. Med.* **2011**, *364*, 1104–1115. [CrossRef] [PubMed]
44. Hallakou, S.; Doare, L.; Foufelle, F.; Kergoat, M.; Guerre-Millo, M.; Berthault, M.F.; Dugail, I.; Morin, J.; Auwerx, J.; Ferre, P. Pioglitazone induces in vivo adipocyte differentiation in the obese Zucker fa/fa rat. *Diabetes* **1997**, *46*, 1393–1399. [CrossRef]
45. Guan, H.P.; Li, Y.; Jensen, M.V.; Newgard, C.B.; Steppan, C.M.; Lazar, M.A. A futile metabolic cycle activated in adipocytes by antidiabetic agents. *Nat. Med.* **2002**, *8*, 1122–1128. [CrossRef]
46. Bajaj, M.; Suraamornkul, S.; Pratipanawatr, T.; Hardies, L.J.; Pratipanawatr, W.; Glass, L.; Cersosimo, E.; Miyazaki, Y.; DeFronzo, R.A. Pioglitazone reduces hepatic fat content and augments splanchnic glucose uptake in patients with type 2 diabetes. *Diabetes* **2003**, *52*, 1364–1370. [CrossRef]
47. Miyazaki, Y.; Matsuda, M.; DeFronzo, R.A. Dose-response effect of pioglitazone on insulin sensitivity and insulin secretion in type 2 diabetes. *Diabetes Care* **2002**, *25*, 517–523. [CrossRef] [PubMed]
48. Evans-Molina, C.; Robbins, R.D.; Kono, T.; Tersey, S.A.; Vestermark, G.L.; Nunemaker, C.S.; Garmey, J.C.; Deering, T.G.; Keller, S.R.; Maier, B.; et al. Peroxisome proliferator-activated receptor gamma activation restores islet function in diabetic mice through reduction of endoplasmic reticulum stress and maintenance of euchromatin structure. *Mol. Cell. Biol.* **2009**, *29*, 2053–2067. [CrossRef] [PubMed]
49. Orasanu, G.; Ziouzenkova, O.; Devchand, P.R.; Nehra, V.; Hamdy, O.; Horton, E.S.; Plutzky, J. The peroxisome proliferator-activated receptor-gamma agonist pioglitazone represses inflammation in a peroxisome proliferator-activated receptor-alpha-dependent manner in vitro and in vivo in mice. *J. Am. Coll. Cardiol.* **2008**, *52*, 869–881. [CrossRef] [PubMed]

50. Ishida, E.; Kim-Muller, J.Y.; Accili, D. Pair Feeding, but Not Insulin, Phloridzin, or Rosiglitazone Treatment, Curtails Markers of beta-Cell Dedifferentiation in db/db Mice. *Diabetes* **2017**, *66*, 2092–2101. [CrossRef]
51. Aronoff, S.; Rosenblatt, S.; Braithwaite, S.; Egan, J.W.; Mathisen, A.L.; Schneider, R.L. Pioglitazone hydrochloride monotherapy improves glycemic control in the treatment of patients with type 2 diabetes: A 6-month randomized placebo-controlled dose-response study. The Pioglitazone 001 Study Group. *Diabetes Care* **2000**, *23*, 1605–1611. [CrossRef] [PubMed]

biomedicines

MDPI

Article

High-Fat Diet Induces Pre-Diabetes and Distinct Sex-Specific Metabolic Alterations in Negr1-Deficient Mice

Maria Kaare [1,2,*], Kaie Mikheim [1,2], Kersti Lilleväli [1,2], Kalle Kilk [2,3], Toomas Jagomäe [1,2,4], Este Leidmaa [5], Maria Piirsalu [1,2], Rando Porosk [2,3], Katyayani Singh [1,2], Riin Reimets [4], Egon Taalberg [2,3], Michael K. E. Schäfer [6], Mario Plaas [4], Eero Vasar [1,2] and Mari-Anne Philips [1,2]

1 Institute of Biomedicine and Translational Medicine, Department of Physiology, University of Tartu, 19 Ravila Street, 50411 Tartu, Estonia; kaie.mikheim@ut.ee (K.M.); kersti.lillevali@ut.ee (K.L.); toomas.jagomae@ut.ee (T.J.); maria.piirsalu@ut.ee (M.P.); singhkat@ut.ee (K.S.); eero.vasar@ut.ee (E.V.); marian@ut.ee (M.-A.P.)

2 Center of Excellence in Genomics and Translational Medicine, University of Tartu, 50411 Tartu, Estonia; kmck@ut.ee (K.K.); rando.porosk@ut.ee (R.P.); egon.taalberg@ut.ee (E.T.)

3 Institute of Biomedicine and Translational Medicine, Department of Biochemistry, University of Tartu, 19 Ravila Street, 50411 Tartu, Estonia

4 Institute of Biomedicine and Translational Medicine, Laboratory Animal Center, University of Tartu, 14B Ravila Street, 50411 Tartu, Estonia; riin.reimets@ut.ee (R.R.); mario.plaas@ut.ee (M.P.)

5 Institute of Molecular Psychiatry, Medical Faculty, University of Bonn, 53129 Bonn, Germany; este.leidmaa@uni-bonn.de

6 Department of Anesthesiology, Focus Program Translational Neurosciences, Research Center for Immunotherapy, University Medical Center of the Johannes Gutenberg-University Mainz, 55131 Mainz, Germany; Michael.Schaefer@unimedizin-mainz.de

* Correspondence: maria303@ut.ee; Tel.: +372-53-488-352

Citation: Kaare, M.; Mikheim, K.; Lilleväli, K.; Kilk, K.; Jagomäe, T.; Leidmaa, E.; Piirsalu, M.; Porosk, R.; Singh, K.; Reimets, R.; et al. High-Fat Diet Induces Pre-Diabetes and Distinct Sex-Specific Metabolic Alterations in Negr1-Deficient Mice. *Biomedicines* **2021**, *9*, 1148. https://doi.org/10.3390/biomedicines9091148

Academic Editor: Susan J. Burke

Received: 29 July 2021
Accepted: 28 August 2021
Published: 3 September 2021

Abstract: In the large GWAS studies, *NEGR1* gene has been one of the most significant gene loci for body mass phenotype. The purpose of the current study was to clarify the role of NEGR1 in the maintenance of systemic metabolism, including glucose homeostasis, by using both male and female *Negr1*$^{-/-}$ mice receiving a standard or high fat diet (HFD). We found that 6 weeks of HFD leads to higher levels of blood glucose in *Negr1*$^{-/-}$ mice. In the glucose tolerance test, HFD induced phenotype difference only in male mice; *Negr1*$^{-/-}$ male mice displayed altered glucose tolerance, accompanied with upregulation of circulatory branched-chain amino acids (BCAA). The general metabolomic profile indicates that *Negr1*$^{-/-}$ mice are biased towards glyconeogenesis, fatty acid synthesis, and higher protein catabolism, all of which are amplified by HFD. *Negr1* deficiency appears to induce alterations in the efficiency of energy storage; reduced food intake could be an attempt to compensate for the metabolic challenge present in the *Negr1*$^{-/-}$ males, particularly during the HFD exposure. Our results suggest that the presence of functional *Negr1* allows male mice to consume more HFD and prevents the development of glucose intolerance, liver steatosis, and excessive weight gain.

Keywords: *Negr1*; obesity; metabolic disease; metabolomics; glucose intolerance; genetic models

1. Introduction

Neuronal growth regulator 1 (*NEGR1*) is a candidate gene regulating human obesity, which encodes a neural cell adhesion and growth protein. *NEGR1* was identified as a member of the IgLON superfamily of neural cell adhesion molecules (CAMs), which also include LSAMP, NTM, OPCML, and IGLON5. GPI-anchored IgLONs have been shown to promote and guide neurite growth [1–4] and act as structural elements in the synapse that stabilize pre- and postsynaptic sides [5].

In GWAS studies, *NEGR1* gene locus has been repeatedly shown to have strong associations with human body mass index (BMI), indicating a role in body weight regulation and obesity [6–11]. Besides SNP markers, it has been found that two deletions (43 kb and 8 kb) upstream of *NEGR1* are strongly associated with early onset of extreme obesity [12].

Strong associations also appeared when genetic polymorphisms in the *NEGR1* gene were linked with dietary intake [13,14] and with psychological features generally associated with eating disorders [15].

In accordance with the initially described role of NEGR1 in the brain, polymorphisms in *NEGR1* have emerged among the most robust associations across different psychiatric disorders, including autism spectrum disorder, major depression, and schizophrenia [16]. The strongest GWAS associations, however, have linked *NEGR1* with depression [17,18]. The most recent data from depression patients suggest that the functional impact of NEGR1 might involve systemic regulation of homeostasis, as significant upregulation of NEGR1 has been shown in the hypothalamus [19] and peripheral blood of depression patients [20].

Due to an earlier described role in neuritogenesis, it has been speculated that *Negr1* also regulates neurite outgrowth in hypothalamic neurons [21,22]. Current evidence implies that Negr1 in the hypothalamic area decreases food intake. Higher levels of NEGR1 in hypothalamic nuclei were linked with lower food intake; administration of NEGR1 ectodomains into the paraventricular nucleus of the hypothalamus induced ~20% decrease in food intake in rats [23]. *Negr1* overexpression in the periventricular hypothalamic region, however, did not affect body weight or food intake, whereas knockdown of *Negr1* expression in the same nucleus increased body weight [24]. Exposure to a restricted feeding schedule has been shown to increase *Negr1* (22%) expression in the arcuate nucleus/ventromedial hypothalamus of rats [25]. Similarly, NEGR1 protein expression is increased in the lateral hypothalamus of fasted chicks [26], further suggesting the role of *Negr1* also as a regulator of a negative energy balance.

Accumulating evidence suggests non-central nervous system function of NEGR1 in intracellular lipid storage. Kim et al. [22] have demonstrated that NEGR1 interacts with cholesterol (CHOL) transporter 2 (NPC2), a key player in intracellular CHOL trafficking, and increases the stability of NPC2 in the late endosomes. Furthermore, Sandholt et al. [27] have shown that NEGR1 tag SNP (rs2568958) has significant associations with LDL cholesterol levels. In comparison to its most abundant expression in the brain, however, *Negr1* is only modestly expressed in the liver in humans and mice [28–30].

Among peripheral tissues, the expression of *Negr1* is high in the adipose tissue, remaining approximately four to five times lower than *Negr1* expression in the brain [29]. Bernhard et al. [31] detected lower NEGR1 expression in the subcutaneous adipose tissue (SAT) compared with visceral adipose tissue, whereas NEGR1 expression was lower in the SAT of obese humans compared to lean subjects. Walley et al. [29] demonstrated that, in the human SAT, NEGR1 appears to be central to the set of functionally related genes most differentially expressed between lean and obese subjects. An et al. [32] have shown that the level of lipid droplets (LD) was reduced in Negr1-overexpressing cells, whereas the intracellular LD level was higher in primary adipocytes obtained from *Negr1*$^{-/-}$ mice than those from wild-type (WT) mice [32]. In addition, the expression level of LD-associated protein perilipin-2/ADRP increased in white adipose tissue of *Negr1*-deficient mice. Evidence from Bernhard et al. [29] suggests that NEGR1, which is upregulated during adipogenesis, is important in the adipocytes from early development. They also found that knockdown of NEGR1 significantly inhibited adipocyte maturation.

Data from mouse models with nonfunctional *Negr1* gene have revealed somewhat conflicting data. In the study of Lee et al. [21], both *Negr1* deficiency and loss-of-function mutation of *Negr1* resulted in a slight but steady decrease in body mass, whereas no change in body weight in *Negr1*$^{-/-}$ mice was found by Joo et al. [33]. However, these works agree that lacking or nonfunctional NEGR1 protein causes alterations in the body composition, namely a decrease in muscle/lean mass. Joo et al. [33] showed a significant increase in fat mass with hypertrophic adipose cells containing enlarged cytosolic lipid droplets in *Negr1*$^{-/-}$ mice compared with the WT mice. Moreover, these mice showed significant hepatic lipid accumulation, and a decrease in muscle mass and capacity [33].

In conclusion, accumulating evidence suggests that the systemic effect of *Negr1* in the regulation of body weight phenotype might be mediated both by the ability of *Negr1* to promote the cell–cell adhesion and neuritogenesis in the hypothalamic area [21,23] and by the independent role of *Negr1* in the regulation of fat trafficking/accumulation in peripheral tissues. Walley et al. [29] have shown that there is a high correlation in the expression levels of *NEGR1* between human subcutaneous adipose and hypothalamic tissues, suggesting a linked function for *NEGR1* across tissues.

The purpose of the present study was to shed light on how *Negr1* is implicated in food intake and systemic metabolism. We studied *Negr1*$^{-/-}$ mice from both sexes receiving standard and high-fat diets. We also aimed to provide initial comparative information about liver steatosis [33] and reduced muscle mass [21,33] that have been described earlier in independently created *Negr1*-deficient mouse strains to further validate the phenotype. We confirmed that *Negr1*-deficient mice are prone to liver steatosis and male mice have signs of muscle atrophy, even when receiving a normal diet. At the same time, we showed that sex-specific metabolic alterations, including glucose intolerance phenotype in the *Negr1*$^{-/-}$ mice, appeared only in mice receiving a high-fat diet, emphasizing the importance of sex and the interaction of genes and environment in maintaining homeostasis.

2. Materials and Methods

2.1. Animals

Male and female wild-type (WT) mice and their homozygous *Negr1*-deficient littermates (*Negr1*$^{-/-}$), described previously [21] in F2 background ((129S5/SvEvBrd × C57BL/6N) × (129S5/SvEvBrd × C57BL/6N)), were used in the present study. Mice were group-housed in standard laboratory cages measuring 42.5 (L) × 26.6 (W) × 15.5 (H) cm, with 10 animals per cage in the animal colony, at 22 ± 1 °C under a 12:12 h light/dark cycle (lights off at 19:00 h). A 2 cm layer of aspen bedding (Tapvei, Estonia) and 0.5 l of aspen nesting material (Tapvei, Estonia) were used in each cage and changed every week. Water and food pellets (R70, Lactamin AB, Sweden) were available ad libitum. Breeding and the maintenance of the mice were performed at the animal facility of the Institute of Biomedicine and Translational Medicine, University of Tartu, Estonia. The use of mice was conducted in accordance with the regulations and guidelines approved by the Laboratory Animal Centre at the Institute of Biomedicine and Translational Medicine, University of Tartu, Estonia. All animal procedures were conducted in accordance with the European Communities Directive (2010/63/EU) with permit (No. 150, 27 September 2019) from the Estonian National Board of Animal Experiments.

2.2. Diet Composition

High-fat (HF) chow (DIO-45 kJ% fat (lard)) corresponds to the D12451 diet from Ssniff Spezialiäten GmbH (Soest, Germany) and its physiological energetic value is 4.615 kcal/kg. It contains 45 kJ% fat, 20 kJ% proteins, and 35 kJ% carbohydrates. This diet is characterized by high fat content (lard) and high sucrose levels. It is used to induce obesity and metabolic syndrome/diabetes in rats and mice.

The caloric value of regular chow (V1534-000 rat/mice universal maintenance diet, autoclavable (10mm) from Ssniff Spezialiäten GmbH, Soest, Germany) corresponds to 3.225 kcal/kg. It contains 9 kJ% fat, 24 kJ% proteins, and 67 kJ% carbohydrates. This diet is suitable for long-term experiments.

2.3. High Fat Diet

The chronic high-fat diet experiment was performed with two different batches of mice (Figure 1). For the first batch of mice, 19 (±1)-week-old WT and *Negr1*$^{-/-}$ male and female mice were divided into two groups; one group received regular food and another group received an HF diet (Ssniff Spezialiäten, Soest, Germany) (12 males and 12 females WT mice + 12 males and 12 females *Negr1*$^{-/-}$ mice). Mice in the first batch received the HF diet (Ssniff Spezialiäten, Soest, Germany) for 7 weeks. All the mice were weighed weekly, starting from 10 weeks before the beginning of the HF diet (at the age of 10± 1 weeks); altogether, the body weight dynamics of the mice were tracked for 16 weeks. In the second batch of mice, 15 (±1)-week-old WT and *Negr1*$^{-/-}$ male and female mice were divided into two groups; one group received regular food and another group the HF diet (Ssniff Spezialiäten, Soest, Germany) (10 males and 10 females WT mice + 10 males and 10 females *Negr1*$^{-/-}$ mice). In the second batch, all mice received the HF diet (Ssniff Spezialiäten, Soest, Germany) for 6 weeks. As female mice showed no genotype effect in the glucose tolerance test, females received the HF diet (Ssniff Spezialiäten, Soest, Germany) for another 7 weeks (total of 13 weeks for females). In the second batch of mice, the food consumed was also weighed to evaluate their food consumption. At the end of both experimental periods, brain tissue, liver, and plasma were collected from all mice.

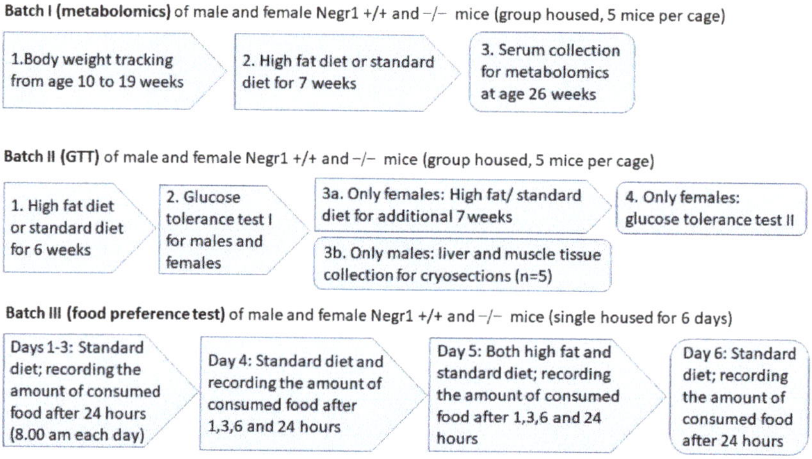

Figure 1. Schematic overview of the batches of mice and tests conducted in the current study. The age of Batch II and Batch III mice was around 12 weeks in the beginning of the experiments.

2.4. Food Preference Test (Batch III)

The food preference test was performed with minor modifications as described earlier in Leidmaa et al. [34]. All the mice were housed in single cages and, on days 1–3, mice received regular food. Both food and mice were weighed every morning at the same time (8.00 a.m.). On the morning of the 4th day, the mice and food (Ssniff Spezialiäten, Soest, Germany) were weighed, and the food was further weighed at specific timepoints (1 h, 3 h, 6 h, and 24 h) throughout the day. On the 24 h timepoint the next morning (5th day), mice were weighed, and the HF food (Ssniff Spezialiäten, Soest, Germany) was added for the food preference experiment. Both regular food and HF food were weighed at the same timepoints (1 h, 3 h, 6 h and 24 h) as the previous day. On the 24 h weighing (6th day), mice were weighed again, and HF food was removed; only regular food was retained. On the 7th day, the food was weighed for the last time to see if *Negr1*$^{-/-}$ mice showed any withdrawal effects.

2.5. Sample Collection

Mice were euthanized by decapitation, and trunk blood was collected into EDTA-coated microcentrifuge tubes and stored at 4 °C. All the tubes were centrifuged at 2000× *g* for 15 min at 4 °C. Plasma supernatant was separated and stored at −80 °C until further analysis.

2.6. Measurement of Metabolites

AbsoluteIDQ™ p180 kit (BIOCRATES Life Sciences AG, Innsbruck, Austria) was used to determine plasma levels of metabolites according to the manufacturer's protocol. Amino acids and biogenic amines in the samples were measured using the liquid chromatography–mass spectrometry techniques. Acylcarnitines (Cx:y), hexoses, sphingolipids (SMx:y or SM (OH)x:y), glycerophospholipids (lysophosphatidylcholines (lysoPCx:y)), and phosphatidylcholines (PCaa x:y and PC ae x:y) were measured using flow injection mass spectrometry. For both modes of analyzing, multiple reaction monitoring was used. Concentrations of the metabolites were calculated automatically by the MetIDQ™software (BIOCRATES Life Sciences AG, Innsbruck, Austria) in μM. The analytical system was QTRAP 4500 (Sciex, Framingham, MA, USA) in combination with Agilent 1260 series high-performance liquid chromatography (HPLC) (Agilent Technologies, Waldbronn, Germany).

Citric acid cycle intermediates were analyzed on the same instrument with an in-house protocol. In total, 50 μL serum was treated with 20 μL 100 μM [2H4] succinate (internal standard) and 750 μL ice-cold methanol for 10 min for protein precipitation. After centrifugation for 10 min at 21,000× *g*, the supernatant was dried under a stream of nitrogen and dissolved in 100 μL of methanol with 0.2% formic acid. The multiple reaction monitoring transitions in negative ionization mode were as follows: malate 133/115, succinate 117/73, citrate 191/87, pyruvate 87/43, alpha-ketoglutarate 145/101, lactate 89/43, oxaloacetate 131/87, beta-hydroxybutyrate 103/59, and the internal standard 121/77.

2.7. Glucose Tolerance Test (GTT, Batch II)

Animals were deprived of food for 3 h before and during the experiment; water was available throughout the experiment. After measuring the basal glucose levels, the mice were intraperitoneally administered a glucose (Sigma-Aldrich, Burlington, MA, USA) solution in 0.9% saline (20% w/vol) at a dose of 2 g/kg of body weight. Blood glucose values were subsequently measured after 30, 60, 90, 120, and 180 min from the tail vein using a hand-held glucometer (Accu-Check Go, Roche, Mannheim, Germany). The bioavailability of glucose was estimated by calculating the under-the-curve area of plasma concentration at measured timepoints (AUC). For males, GTT was performed on the 6th week, and, for females, on 6th and on the 13th week of a high-fat diet.

2.8. Neutral Lipid and Actin Staining on the Tissue Cryosections

Male mice (n = 5 per group) from batch II were euthanized by decapitation, and dissected quadriceps femoris muscle and left lobe of the liver were immersed in 2-methylbutane (ACROS Organics™, Cat# 10511754, Carlsbad, CA, USA), precooled to −40 °C, and kept at −80 °C until further use. Cryosections (15 μm) were prepared using Leica low-profile disposable blades (DB80LS, Leica, Wetzlar, Germany) mounted to Leica CM1850-Cryostat (Leica, Wetzlar, Germany). Sections were collected onto Thermo Scientific™ SuperFrost Plus™ slides (Thermo Scientific, Cat# 10149870) and kept at −80 °C. In order to minimize experimental errors caused by washes, staining incubations, etc., tissues from animals of each experimental group were collected on the same slide (four animals per slide).

Sections were immersion-fixed in 4% paraformaldehyde (PFA, Acros Organics™, Cat# 11924801, USA)/PBS for 15 min at 37 °C in a Coplin jar and washed thrice in PBS for 5 min each. Subsequently, sections were incubated with Alexa Fluor® 555 Phalloidin (1:500, Invitrogen, Cat# A34055, Waltham, MA, USA) and BODIPY 493/503 dye (1 μM, Invitrogen, Cat# D3922, Waltham, MA, USA) in PBS over 10 min at 37 °C in a Coplin jar protected from light. Staining solution was obtained by diluting 5 mM BODIPY 493/503 in DMSO (Sigma Aldrich, Cat# D8418, Burlington, MA, USA) stock. Sections were washed thrice with PBS over 5 min each and stained with Hoechst 33,258 (5 μg/mL, Invitrogen, Cat# H1398, Waltham, MA, USA) in PBS over 5 min. Sections were subsequently rinsed with ddH_2O and mounted as described above. To visualize filamentous actin fibers, sections were immersed in phalloidin conjugates. Images were acquired with a DP71 CCD camera (Olympus, Tokyo, Japan) mounted on a BX51 microscope (Olympus, Japan). Morphometric measurements of muscle fibers were performed manually using ImageJ software version 1.53c [35], and at least 70 muscle fibers were measured for each mouse.

2.9. Statistical Analysis

Results are expressed as mean values ± SEM. Statistical analyses for metabolomic data, body weight, GTT, and food preference test were performed using GraphPad Prism 6 software (GraphPad, San Diego, CA, USA). Normal distribution of data was evaluated by the Shapiro–Wilk test. Comparison of metabolomic data between groups was performed using two-way ANOVA (diet × genotype), followed by a Bonferroni post hoc test. Comparison of GTT data between groups was performed using two-way ANOVA, followed by a Tukey post hoc test. Statistical analysis of the food preference test and cross-sectional area of muscle fibers was performed by using Mann–Whitney U-test. All differences were considered statistically significant at $p < 0.05$.

3. Results

3.1. Negr1 Deficiency Induces Lower Intake of HF Food but Higher Body Weight Gain in Male Mice

In the first batch of mice, the mice were weighed weekly from age 10 ± 1 weeks during the 9 weeks before the beginning of an HF diet; altogether, the body weight was tracked for 16 weeks. The variation in body weight was relatively high; therefore, no statistical genotype differences in body weight dynamics were detectable in the current study. In general, both male and female *Negr1*$^{-/-}$ mice tended to have slightly lower body weight when on a standard diet (Figure 2b). When consuming a high-fat (HF) diet, however, male *Negr1*$^{-/-}$ mice tended to gain more body weight compared to their WT littermates (Figure 2a). When mice consumed regular chow, there were no significant weight differences between genotypes (Figure 2b). Male *Negr1*$^{-/-}$ mice consumed less HF food in the short-term food preference test, in which the food was individually measured for 24 h (Figure 2d). Correspondingly, a tendency for a lower HF food intake in male *Negr1*$^{-/-}$ mice was also observed in group-housing settings 2 weeks before the glucose tolerance test (Figure 2e). Additionally, male *Negr1*$^{-/-}$ mice also consumed smaller amounts of standard food when the consumption of food was individually measured for 96 h (Figure 2c). More detailed information about food intake on individual days and during the 1, 3, and 6 h food intake measurements have been shown in Supplementary Figure S1.

3.2. HFD Leads to Higher Levels of Blood Glucose in Negr1$^{-/-}$ Mice, Whereas Phenotype Difference in Glucose Tolerance Test Was Apparent Only in Males

The HF diet elevated the level of basal blood glucose in both sexes; for males, there was a diet effect (p = 0.0029, F = 10.23), whereas females showed a genotype effect (p = 0.0001, F = 19.11). However, the basal level of glucose was statistically higher in the HF-diet-fed *Negr1*$^{-/-}$ mice compared to the HF-diet-fed WT group (p = 0.0360) in both sexes (Figure 3a). The HF diet increased the basal level of blood glucose in *Negr1*$^{-/-}$ male mice (p = 0.0117) (Figure 3b). In the female group, there was no diet effect (F = 0.01256); on the other hand, the genotype effect was observed (F = 19.11). The basal level of glucose was significantly higher in the HF-fed female *Negr1*$^{-/-}$ group compared to the HF-diet-fed WT group (p = 0.0015) (Figure 3c).

For male mice, GTT was performed on the 6th week of the HF diet, and, for females, on the 6th and 13th week of the diet. In GTT, AUC was calculated for every mouse, and two-way ANOVA and Tukey post hoc tests were used.

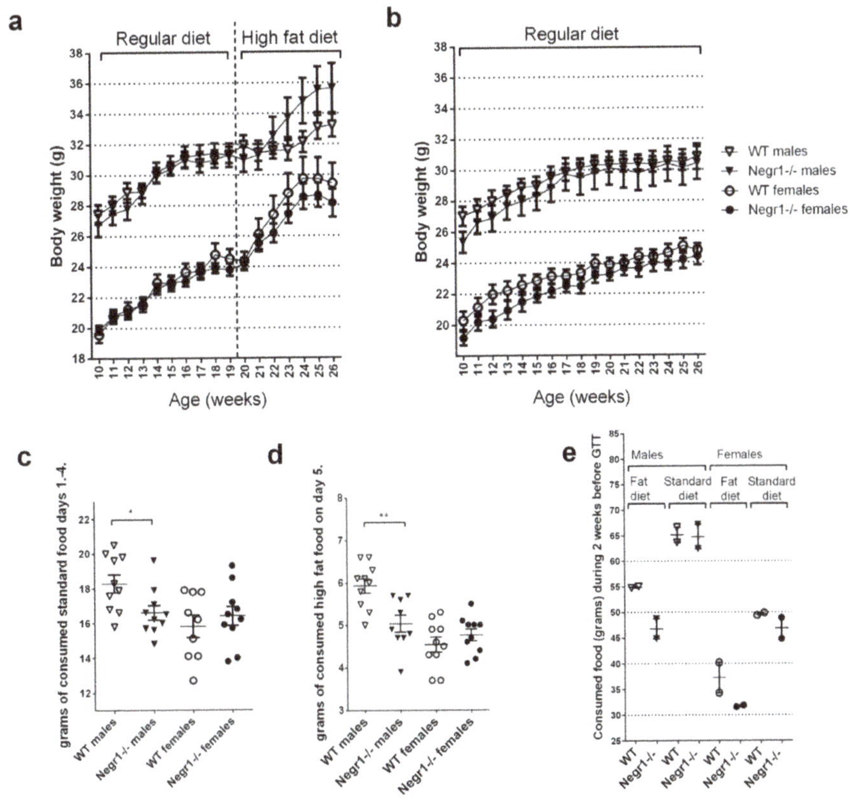

Figure 2. Body weight dynamics and food intake measurements. The body weight of the first batch of mice was measured 10 weeks before the beginning of the high-fat (HF) diet. Mice received the HF diet for 7 weeks. Body weight dynamics of (**a**) HF diet groups and (**b**) standard chow group. Daily food intake measurements in single-housed mice prior to the food preference test showed that (**c**) WT males consumed significantly more standard food (days 1–4). (**d**) WT males also consumed significantly more HF food during the food preference test (day 5) compared to the *Negr1*$^{-/-}$ mice. (**e**) Food consumed (grams) for 2 weeks (14 days) before the glucose tolerance test in the second batch of mice. The data points were calculated as food consumed per group of mice (n = 5 per group). Data represent mean ± SEM. * $p \geq 0.05$, ** $p \geq 0.01$ (Mann–Whitney U-test).

In male mice, the AUC of HF-fed *Negr1*$^{-/-}$ mice was statistically significantly higher compared to *Negr1*$^{-/-}$ mice fed regular chow (p = 0.0008) (Figure 3d). In female mice, the AUC of the HF-diet-fed mice on the 6th week of diet was statistically significantly higher compared to the regular chow group in both WT (p = 0.0026) and *Negr1*$^{-/-}$ (p = 0.018) mice (Figure 3e). In the 13th week, the results were similar to the 6th week; the AUC of female HF-diet-fed mice was statistically significantly higher than regular-chow fed mice in both genotypes: WT (p = 0.0015), *Negr1*$^{-/-}$ (p = 0.0003) (Figure 3f). When the different timepoints were analyzed separately, the blood sugar levels of HF-diet-fed male mice were significantly higher compared to HF-diet-fed WT mice at 30 min (p = 0.0017) and 60 min (p = 0.0035) timepoints (Figure 3g). The blood sugar levels of the HF-diet-fed *Negr1*$^{-/-}$ were significantly higher at 30 min ($p < 0.0001$), 60 min ($p < 0.0001$), 90 min (p = 0.0003), and 120 min (p = 0.0130) timepoints compared to regular-chow-fed *Negr1*$^{-/-}$ mice (Figure 3g). In female mice, on the 6th week of diet, the blood sugar levels of the HF-diet-fed *Negr1*$^{-/-}$ mice were significantly higher at 30 min (p = 0.0004) and 60 min (p = 0.0171) timepoints compared to regular-chow-fed *Negr1*$^{-/-}$ mice, and, for the HF-diet-fed WT mice, the blood sugar levels were significantly higher at 30 min ($p < 0.0001$)

and 60 min (p = 0.0003) timepoints (Figure 3h). On the 13th week of the diet, the blood sugar levels of the HF-diet-fed *Negr1*$^{-/-}$ female mice were significantly higher at 30 min (p = 0.0116), 60 min (p = 0.0030), 90 min (p = 0.0002), and 120 min (p = 0.0049) timepoints compared to regular-chow-fed *Negr1*$^{-/-}$ mice (Figure 3i). In the WT group the blood sugar levels of the HF-diet-fed mice were significantly higher at 30 min (p < 0.0001), 60 min (p = 0.0025), and 120 min (p = 0.0417) timepoints (Figure 3i).

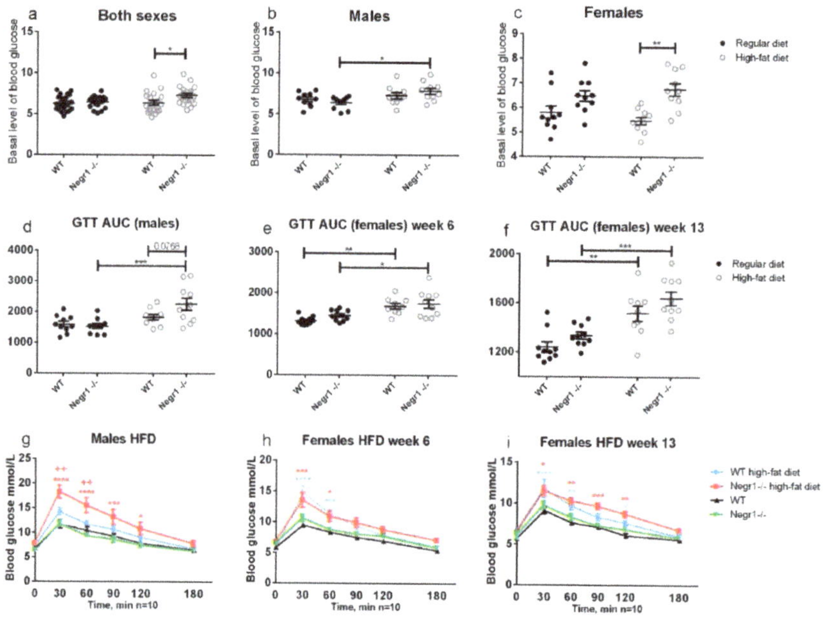

Figure 3. The basal level of glucose and glucose tolerance test. Basal level of glucose after 6 weeks of HF/standard diet when (**a**) both sexes were pooled together, (**b**) in the male group, and (**c**) in the female group. (**d**,**g**) Glucose tolerance test was performed for males after 6 weeks on the HF diet, and for females (**e**,**h**) after 6 and (**f**,**i**) 13 weeks of HF diet. (**d**–**f**) The bioavailability of glucose was estimated by calculating the area under the curve of plasma concentration (AUC) over the measured timepoints. (**g**–**i**) Mean values of blood sugar at different timepoints. Data represent mean ± SEM, * $p \geq 0.05$, ** $p \geq 0.01$, *** $p \geq 0.001$, **** $p \geq 0.0001$ (diet effect), ++ $p \geq 0.01$ (genotype effect), two-way ANOVA (Bonferroni post hoc test (basal level of glucose), Tukey post hoc test (GTT)).

For the batch I of mice, the level of hexoses was measured using the AbsoluteIDQ™ p180 kit. Although the profile of those results was slightly different from the results of basal glucose level, the HF diet also elevated the level of hexoses similarly to the basal level of glucoses. When the data of both sexes were pooled together, the HF diet significantly increased the level of hexoses (p = 0.027) in the *Negr1*$^{-/-}$ group. If the sexes were analyzed separately, the increase in hexoses remained statistically significant only in the female group (Supplementary Table S1). In female mice, the HF diet increased the level of hexoses, both in the WT (p = 0.044) and *Negr1*$^{-/-}$ (p = 0.018) groups (Supplementary Table S1).

3.3. *HFD Induces an Altered Profile of Circulating Lipids Sex-Specifically in Negr1*$^{-/-}$ *Mice*

The level of saturated fatty acids (SFA) was markedly increased in the HF-diet-fed *Negr1*$^{-/-}$ male mice group (p < 0.0001) (Figure 4a). The level of SFAs were statistically significantly higher in the HF-diet-fed *Negr1*$^{-/-}$ group compared to the regular-chow-fed *Negr1*$^{-/-}$ mice (p = 0.0173) and HF-diet-fed WT mice (p = 0.0086) (Figure 4a). In the female mice group, the HF diet increased the level of SFAs similarly for both genotypes (Figure 4b).

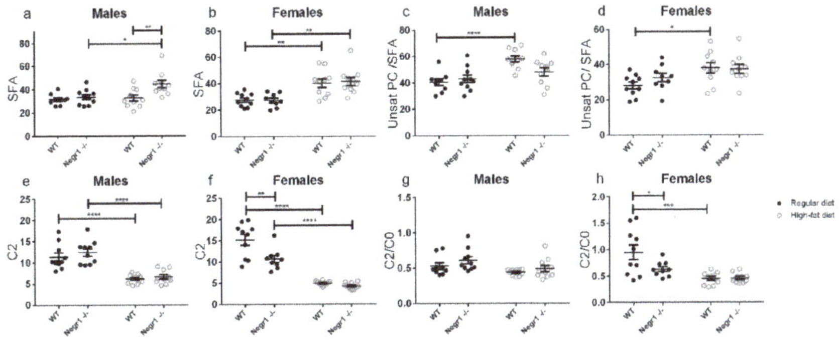

Figure 4. Effect of HF diet on the level of selected lipids and related ratios. The level of SFA for (**a**) males and (**b**) females, the ratio of unsaturated PC/SFA for (**c**) males and (**d**) females, the level of C2 for (**e**) males and (**f**) females, the ratio of C2/C0 for (**g**) males and (**h**) females. Data represent mean ± SEM, * $p \geq 0.05$, ** $p \geq 0.01$, *** $p \geq 0.001$, **** $p \geq 0.0001$, two-way ANOVA (Bonferroni post hoc test).

Serum levels of phosphatidylcholines (PC), acylcarnitines, and sphingomyelins (SM) were measured. Out of 100 quantifiable PC and SM species 87 were significantly altered due to HF diet. Among acylcarnitines, C18 and C18:1 were significantly increased by HF diet, but C0, C2, C3, C4, C14, and C18:2 decreased significantly. If C2 level was looked at in both sexes separately, there were genotype effects only for females ($p = 0.0016$) (Figure 4f). The level of C2 was statistically significantly lower in the regular-chow-fed *Negr1*$^{-/-}$ female mice group compared to the regular-chow-fed WT female mice group ($p = 0.0012$) (Figure 4f).

There were not any statistically significant changes in the ratio of C2/C0 in male mice groups (Figure 4g). In female mice groups, the HF diet decreased the ratio of C2/C0 in the WT mice group ($p < 0.0001$) and, in the regular-chow-fed mice group, the ratio of C2/C0 was statistically lower in the *Negr1*$^{-/-}$ group ($p = 0.047$) (Figure 4h). The HF diet increased the ratio of unsaturated PC/SFA in the WT group for both males ($p = 0.0005$) (Figure 4c) and females ($p = 0.046$) (Figure 4d). Although acylcarnitines with hydroxyacyl or dicarboxylic residues were frequently below the limit of quantification, or even below the limit of detection, their relative cumulative abundance among all acylcarnitines was higher in the HF diet ($p < 0.0001$).

3.4. HFD Induced an Increase in Circulating Amino Acids in Negr1$^{-/-}$ *Mice, More Prominently in Males*

To identify the differences caused by the HF diet between WT and *Negr1*$^{-/-}$, two-way ANOVA (diet (regular or HF) × genotype (WT or *Negr1*$^{-/-}$)) and Bonferroni post hoc test were used. In WT animals, the HF diet had a limited effect on serum amino acid levels.

When both sexes were analyzed separately, the total level of amino acids was statistically significantly increased only in the HF-diet-fed *Negr1*$^{-/-}$ male mice group ($p = 0.023$) (Figure 5a); in females, there were no statistically significant changes (Figure 5b). The level of branched-chain amino acids (BCAA) was also statistically significantly increased only in the HF-diet-fed *Negr1*$^{-/-}$ male mice group ($p = 0.05$) (Figure 5g). If different BCAAs were looked at separately, the levels of Leu ($p = 0.036$) (Figure 5m) and Val ($p = 0.031$) were statistically significantly increased in the HF diet *Negr1*$^{-/-}$ group. When we analyzed both sexes separately, the levels of Leu were significantly increased only in the HF-diet-fed *Negr1*$^{-/-}$ male mice group ($p = 0.0077$) (Figure 5m). The HF diet increased the level of Val in both the WT ($p = 0.023$) and *Negr1*$^{-/-}$ ($p = 0.0093$) male mice groups (Supplementary Table S1).

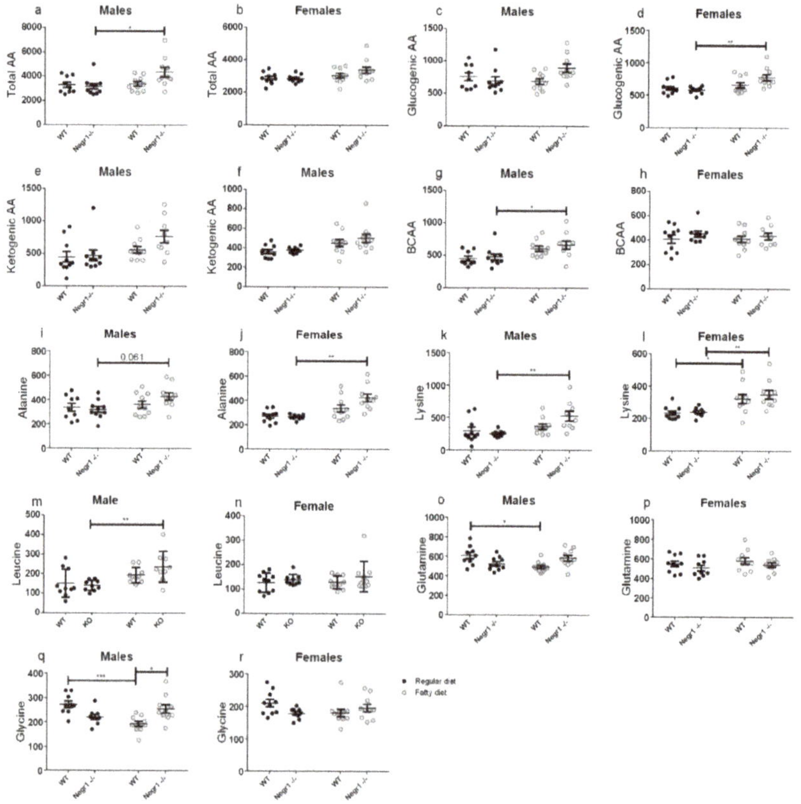

Figure 5. Effect of HF diet on the level of amino acids. In WT animals the HF diet had limited effect on serum amino acid levels. In *Negr1*$^{-/-}$ animals the HF diet increased the total pool of amino acids in blood. The level of (**a**,**b**) Total AA, (**c**,**d**) Glucogenic AA, (**e**,**f**) Ketogenic AA and (**g**,**h**) BCAA. (**i**–**r**) The level of different amino acids. Data represent mean ± SEM, * $p \geq 0.05$, ** $p \geq 0.01$, *** $p \geq 0.001$, two-way ANOVA (Bonferroni post hoc test).

In males, the HF diet increased the level of His (p = 0.0028), Ser (p = 0.0377), Thr (p = 0.0003), Pro (p = 0.0058), Asn (p = 0.0084) (Supplementary Table S1), and Lys (p = 0.0059) (Figure 5k) in the *Negr1*$^{-/-}$ group. The level of Ala was not statistically significantly increased by HF diet, but, in the *Negr1*$^{-/-}$ group, it showed a tendency towards increase (p = 0.061) (Figure 5i). In the male group, there were genotype effects for Ser (p = 0.042), Arg (p = 0.037) (Supplementary Table S1), and Gly (p = 0.011) (Figure 5q), all three were statistically significantly higher in the HF-diet-fed *Negr1*$^{-/-}$ mice group compared to the HF-diet-fed WT mice group. In male WT groups, the HF diet decreased the level of Gln (p = 0.020) (Supplementary Table S1) and Gly (p = 0.0010) (Figure 5q).

In females, HF diet increased the level of Ala (p = 0.0003) (Figure 5j), Lys (p = 0.0064) (Figure 5l), and Thr (p = 0.0200) (Supplementary Table S1) in the *Negr1*$^{-/-}$ group. The level of Lys was also increased in the WT group (p = 0.036). In other amino acids, there were no statistically significant changes caused by the HF diet.

3.5. Altered Profile of Circulating Organic Acids in Negr1$^{-/-}$ *Mice*

A few organic acids, including the citric acid cycle intermediates, were quantified in the blood serum in order to better identify the flux of metabolites (for the detailed information see Supplementary Table S2 and Supplementary Figure S2). Male *Negr1*$^{-/-}$

mice had significantly higher beta-hydroxybutyrate (F = 13.1, p = 0.0012), lactate (F = 14.7, p = 0.0007), pyruvate (F = 4.9, p = 0.035), citrate (F = 5.5, p = 0.026), and oxaloacetate (F = 5.0, p = 0.034) than WT male animals. Female mice had a weak oxaloacetate decrease due to diet (F = 4.7, p = 0.04), but, other than that, all diet and genotypes were similar. When both genders were combined, citrate (F = 8.7, p = 0.005) and lactate (F = 4.3, p = 0.05) remained significantly elevated in *Negr1*$^{-/-}$ animals. Lactate ratio to glucose was lowered by HF diet in female animals of both genotypes (F = 9.7, p = 0.005). In males, on the other hand, genotype had a significant effect, with *Negr1*$^{-/-}$ having more lactate per glucose (F = 11.1, p = 0.003).

3.6. Negr1 Decifiency Induces Hepatic Fat Accumulation in Both Male and Female Mice and Reduced Skeletal Muscle Volume in Males

As abnormal hepatic fat accumulation has been previously demonstrated in alternatively created *Negr1*$^{-/-}$ mice [33], we studied the liver cryosections of *Negr1*$^{-/-}$ male mice by using BODIPY dye, which is a fluorescent conjugate of fatty acids. In mice receiving a standard diet, markedly higher fatty-acid-specific staining could be detected in the hepatocytes from the liver of *Negr1*$^{-/-}$ mice (Figure 6b) compared to the hepatocytes from WT mice (Figure 6a). The hepatocytes from WT mice receiving a high-fat diet (Figure 6c) were similar to the hepatocytes from *Negr1*$^{-/-}$ mice receiving standard food. High-fat food did not markedly change the appearance of the liver in the *Negr1*$^{-/-}$ mice (Figure 6d). The stainings from all individual mice can be seen in Supplementary Figure S3. The supplementary section (Supplementary Figures S4 and S5) of the current study also provides histology results of a small population of middle-aged females (8–9 months old).

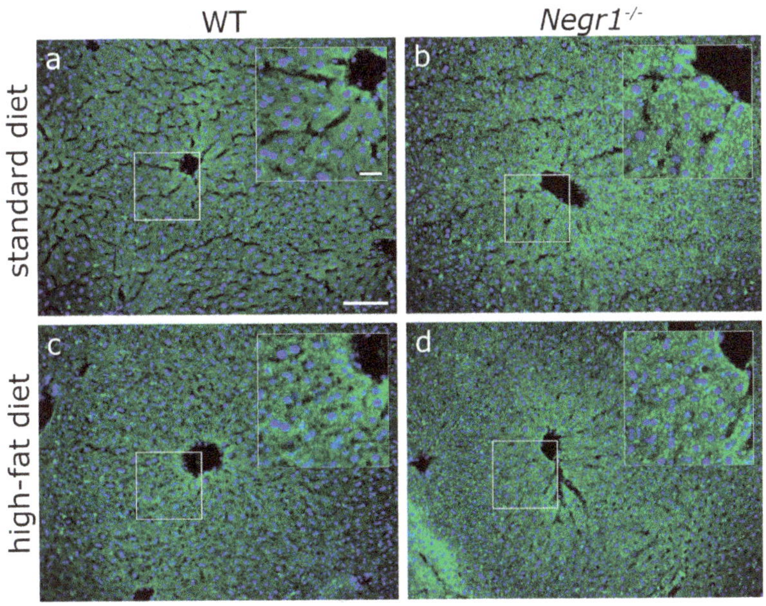

Figure 6. Hepatic lipid content. Representative images of BODIPY neutral lipid staining (green) from (**a**,**c**) WT and (**b**,**d**) *Negr1*$^{-/-}$ mouse liver sections that underwent (**a**,**b**) standard and (**c**,**d**) high-fat diets. Compared to (**a**) WT mice, the hepatocytes from (**b**) *Negr1*$^{-/-}$ mice in the standard diet display increased lipid droplet accumulation around the portal vein. This genotype-dependent difference is diminished with (**c**,**d**) high-fat diet treatment. Nuclei (blue) were stained using H33258 stain. Scale bars: 100 μm, 25 μm (inserts).

As skeletal muscle atrophy has been shown in an alternative *Negr1*$^{-/-}$ mouse strain [33], we studied the muscle cross-sectional area of quadriceps femoris muscle in *Negr1*$^{-/-}$ males in comparison with their WT littermates in both standard and high-fat diet groups. We confirmed significantly lower average muscle fiber size in the *Negr1*$^{-/-}$ mice (Figure 7b) compared to that of WT mice on a standard diet (Figure 7a). Fat food diet did not induce significant difference in the average muscle fiber size in either the *Negr1*$^{-/-}$ or WT mice (Figure 7d–f). There were no statistically significant changes in the average muscle fiber size between female *Negr1*$^{-/-}$ and WT mice (Supplementary Figure S6c,f).

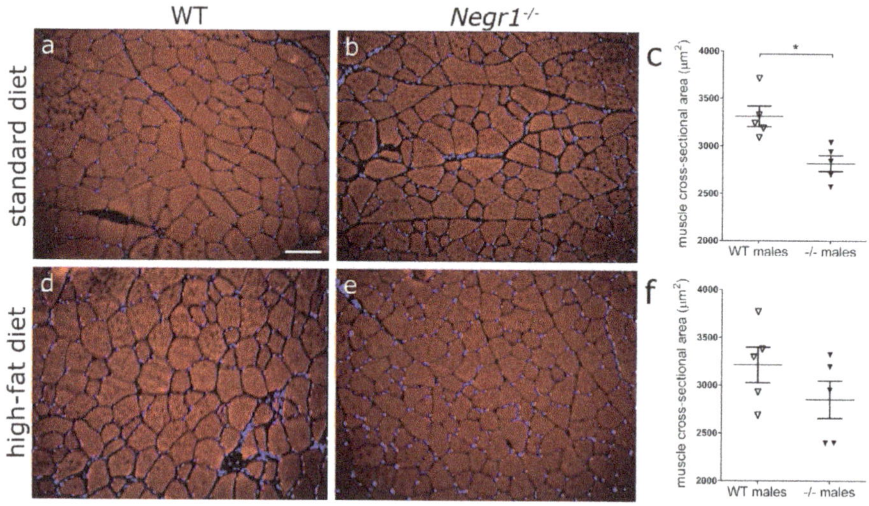

Figure 7. Reduced cross-sectional area of muscle fibers in *Negr1*$^{-/-}$ mice. Phalloidin (red) stained cryosections from the quadriceps femoris muscle of *Negr1*$^{-/-}$ (**b**,**e**) and wild-type (**a**,**d**) mice receiving standard (**a**,**b**) or high-fat (**d**,**e**) diets. (**c**) Morphometric measurements of muscle fiber cross-sectional area revealed significant decrease in *Negr1*$^{-/-}$ mice in the standard diet group, * $p \geq 0.05$ (Mann–Whitney U-test). (**f**) High-fat diet treatment resulted in no difference in muscle fiber cross-sectional area between genotypes. Nuclei (blue) were stained using H33258 stain. Scale bar: 100 µm.

4. Discussion

In the large GWAS studies, NEGR1 gene has been one of the most significant gene loci for both body mass phenotype [7–9] and depression [17,18]. Depression and obesity are leading public health concerns worldwide. Shared genetic risk factors between depression and obesity have been reported, which could be mediated through shared etiological pathways, such as dysfunction of the hypothalamic–pituitary axis [36]. In the current study, we aimed to shed light on the pleiotropic nature of the *NEGR1* gene. Our purpose was to add evidence that would enable us to determine whether the impact of NEGR1 is established mainly through its function as a cell adhesion molecule in the hypothalamus, or whether it also has a distinct role in the systemic metabolism, which could, in turn, contribute to the etiology of psychiatric disorders.

In the current study, both male and female *Negr1*$^{-/-}$ mice tended to have slightly lower body weight when on a standard diet, in accordance with an earlier study in *Negr1*$^{-/-}$ mice [21]. A HF diet leads to body weight gain in both genotypes and both genders. Surprisingly, genotype difference occurred only in male mice. The wild-type (WT) males tended to gain less body weight when on HF diet compared to *Negr1*$^{-/-}$ male mice; however, this effect was not related to higher amounts of consumed food. On the contrary, the WT males consumed higher amounts of HF food in the food preference test, in which the consumed food was individually measured for 24 h. Furthermore, a tendency to eat less of the HF food in male *Negr1*$^{-/-}$ mice and, nevertheless, gain more weight was also observed in group-housing settings over 2 weeks. Male *Negr1*$^{-/-}$ mice also consumed smaller amounts of standard food when measured individually for 96 h.

Our data indicate that *Negr1* deficiency induces alterations in the efficiency of energy storage; surprisingly, lower intake of HF food is accompanied with higher body weight gain in male *Negr1*$^{-/-}$ mice. Previously, it has been shown that mice with NEGR1 loss-of-function mutation (Negr1-I87N mice) exhibited decreased food intake but normal energy expenditure, thus fostering the positive association between NEGR1 expression and obesity [21].

Interestingly, the initial approach to and consumption of the HF food during the first hours of the food preference test was not altered in *Negr1*$^{-/-}$ mice (Figure S1). Measuring the acute consumption of energy-dense and palatable HF food allows us to estimate the changes in the reward processing (hedonic liking/wanting), as well as the acute homeostatic mechanisms that regulate the control of food intake [34]. *Negr1*$^{-/-}$ mice seem to have a normal hedonic appetite and the corresponding satiety induction in the beginning of the food preference test. During a longer exposure to HF food (24 h), however, the *Negr1*$^{-/-}$ males reduce their energy intake more than the WT. This could be explained by the impaired glucose tolerance and corresponding metabolic alterations in these mice. Reduced food intake could be an attempt to compensate for the metabolic challenge present in the *Negr1*$^{-/-}$ males, particularly during the HF diet exposure. Previous studies have also shown that a restricted feeding schedule increases *Negr1* (22%) in the arcuate nucleus/ventromedial hypothalamus of rats [25], and that NEGR1 protein is increased in the lateral hypothalamus of fasted chicks [26]. NEGR1 in certain hypothalamic nuclei might, therefore, lead to increased appetite, which would be in line with our findings of decreased food consumption in *Negr1*$^{-/-}$ mice. These effects may be site-specific, however, as administration of NEGR1 ectodomains into the paraventricular nucleus of the hypothalamus is shown to induce an opposite effect: a 20% decrease in food intake in rats [23].

The role of *NEGR1* in glucose homeostasis has been previously demonstrated in many instances. Schlauch et al. [11] have shown by genome-wide association studies (GWAS) that, besides obesity in the general population, *NEGR1* gene also associates with BMI in type 2 diabetes patients, with abnormal glucose levels and impaired fasting glucose. A direct impact on serum glucose levels has been demonstrated in *Negr1*$^{-/-}$ mice with a >1.3-fold increase in serum glucose and insulin levels [33]. While the level of leptin was substantially higher, the level of insulin-sensitizing adipokine adiponectin was lower in the *Negr1*$^{-/-}$ mice [33].

Our experiments in this study support the notion that, when on the standard diet, males and females from both genotypes have similar basal glucose levels. HF food, however, leads to higher levels of blood sugar in *Negr1*$^{-/-}$ mice, as reported by earlier studies. In male mice, HF diet resulted in altered glucose tolerance only in *Negr1*$^{-/-}$ mice. In females, HF diet altered glucose tolerance in both genotypes, and no genotype effect appeared after either 6 weeks or 13 weeks. Although males are more likely to develop insulin resistance and hyperglycemia in response to nutritional challenges [37], impaired glucose tolerance is more prevalent in women [38]. The gender dependence of glucose metabolism is in fact a complex outcome of many factors, such as muscle mass, muscle-to-fat ratio, and the nature of dysfunction in insulin signaling [38]. It should also be noted that the animals in our experiment did not develop diabetes, although the glucose tolerance test indicated a significant impairment in glucose homeostasis under some experimental conditions. With WT male mice having a slightly elevated basal glucose, and females performing worse in the glucose tolerance test, current results are in accordance with previous reports.

Our results thus far suggest that the presence of *Negr1* allows male mice to consume more HF food and hinders the development of glucose intolerance and excessive weight gain.

The HF diet is expected to overload fatty acid metabolism in one way or another. If ketogenic, HF would upregulate beta-oxidation, ketogenesis, and gluconeogenesis. If the

HF diet has enough carbohydrates, a large portion of dietary fats would be stored as fat and none of the previously mentioned metabolic pathways need to be activated.

We observed nearly unanimous increases in serum PC and SM lipids due to the HF diet. Acylcarnitines at the same time decreased, with the exception of stearyl- and oleyl-carnitine, which increased. According to the manufacturer, palmitoyl, stearyl, and oleyl residues are the most dominant lipids in the HF diet formula. A high load of long-chain acyl residues increased various species of lipids, while the amount of free carnitine and short-chain acylcarnitines decreased due to activity and low substrate specificity of carnitine-acyl transferases. Hence, the pattern of changes due to the HF diet was as expected.

Hydroxylated acylcarnitines and acyl residues with two carboxylic acids could not be properly quantified in most samples. Therefore, the observed relative increase in the total amount of all hydroxylated and dicarboxylic acyl residues may be erroneous. Even more, an overload of acyl residues is expected to activate omega oxidation, which generates dicarboxylic acids and, thereby, alleviates the overload of lipid catabolism pathways. Ketone bodies appeared not to be increased by HF diet, and the individual acylcarnitines, as well as their ratios, did not imply that beta-oxidation and ketogenesis intensified because of our dietary intervention.

The fact that serum lipoproteins have gender-dependent reference values is common knowledge in clinical chemistry. Variations in male and female serum PC species, besides cholesterol and triglycerides, have also been shown previously [39]. Higher beta-hydroxybutyrate levels in females on a standard diet have been reported before as well [40]. Thus, the gender differences in lipids found here are in accordance with previously published data. Somewhat surprisingly, the genotype effect on lipid profile appeared to be marginal in our experiment.

In WT animals, the diet had a limited effect on serum amino acid levels. In *Negr1*$^{-/-}$ animals, particularly in males, the HF diet increased the total pool of amino acids in blood. It did not seem to be related to the essentiality of glucogenicity of the amino acids. The most significant was the increase for Lys, Thr, Ala, Ser, and His. With the reduced relative abundance of proteins in the HF diet, the increased level of essential amino acids (Lys, Thr, and His, in particular) implies increased protein breakdown in the body. Indeed, decreased muscle mass has been demonstrated in another strain of *Negr1*$^{-/-}$ mice [33] and the same result was replicated in the current study, suggesting that *Negr1*$^{-/-}$ mice might also be prone to the protein breakdown in the case of standard feeding.

If the increased protein breakdown is accompanied by increased amino acid usage in peripheral tissues, Ala and Gln should increase in serum because of shuttling of the amino group to the urea production in the liver. Female mice had a relative increase in Ala, but males, on the contrary, showed a weakly significant decrease in Gln. Additionally, the urea cycle intermediates (Arg, Cit, Orn) and their ratios did not indicate an overload of the urea cycle. The ratio of short-chain acylcarnitines and BCAA decreased with HF diet for both sexes, which indicates that, although BCAA levels increase, they are not catabolized into short-chain acyl radicals, which would be a necessary step in their oxidation.

The decrease in Glu and Gln on HF while nearly all other amino acids are either unchanged or increased might be a meaningful anomaly. Most logically, Glu could be transaminated to alpha-ketoglutarate, enter the tricarboxylic acid cycle, and be used for energy, either directly or indirectly via gluconeogenesis. Indeed, there was a tendency of reduced levels of alpha-ketoglutarate and oxaloacetate in animals consuming the HF diet.

Interestingly, Ala and Ser have elevated concentrations in *Negr1*$^{-/-}$ on HF, although they are closely related to pyruvate and 3-phosphoglycerate, serving as potential substrates for gluconeogenesis. Ala generation in the periphery/muscles may simply exceed its use in the liver. This notion is supported by the fact that lactate, another gluconeogenesis substrate closely related to Ala and pyruvate, is not increased by the HF diet. However, comparing Ala and Gln levels and Ser- and alpha-ketoglutarate-related amino acids, there seems to be a preference to use carbons in alpha-ketoglutarate form rather than any gluconeogenic substrate. Alpha-ketoglutarate would be converted to oxaloacetate, which, besides being a starting point for gluconeogenesis, can be converted to aspartate by transamination. The latter is needed for the urea cycle and elimination of excessive nitrogen from increased amino acid catabolism. Gluconeogenesis from pyruvate also goes over oxaloacetate, but requires energy investment for pyruvate carboxylation, and may, therefore, be less economic.

In male *Negr1*$^{-/-}$ animals, Leu and Val are increased. BCAAs are generally upregulated in glucose intolerance [41] and, particularly, Leu is known to regulate glucose and protein metabolism [42]. Whether the BCAA increase stems from protein breakdown and contributes to glucose intolerance or whether their higher level is maintained to counter peripheral glucose resistance cannot be answered from this study. Interestingly, high Leu should inhibit protein breakdown and enhance protein synthesis [43].

Altogether, *Negr1*$^{-/-}$ mice, particularly males, break down proteins in the body in order to be able to use some amino acids, while others accumulate. This reprograming of metabolism does not seem to overload amino acid catabolic pathways, but causes reduced muscle fiber size phenotype. This reprograming also reduces metabolic flexibility and paves the way to glucose intolerance.

The citric acid cycle is a central mitochondrial pathway, which is closely related to catabolic and anabolic pathways of many biomolecules, including fatty acids, amino acids, and glucose. Above, we discussed how certain amino acids in *Negr1*$^{-/-}$ mice are a more important gluconeogenic source than in WT mice. An important note and maybe the root of all metabolic alterations are the increased citrate levels in *Negr1*$^{-/-}$ mice. High citrate is a signal for energy excess which activates fatty acid synthesizing enzymes allosterically [44,45]; on the other hand, citrate inhibits glycolytic enzymes. Thus, even on a normal diet, *Negr1*$^{-/-}$ mice are biased towards fatty acid synthesis and have reduced glycolytic efficiency. Alpha-ketoglutarate, succinate, malate, and oxaloacetate did follow citrate's pattern only in male mice, though. In *Negr1*$^{-/-}$, a part of citrate is diverted away from the citrate cycle into lipid synthesis. The latter is intense in hepatocytes and might cause the fatty liver phenotype if upregulated.

A GWAS suggests that insulin suppresses NEGR1 in adipocytes and the synthetic glucocorticoid dexamethasone induces *NEGR1* expression [29]. Insulin is known to enhance fatty acid synthesis from citrate; whether it does so via NEGR1 or independently, we cannot answer, but our results are in good accordance with these findings. Another study on monozygotic twins discordant for type 2 diabetes found *NEGR1* being upregulated in adipocytes of type 2 diabetic twins [46]. Here, the insulin level is expected to be high, but adipocytes do not recognize it properly. Upregulation of *NEGR1* in relative insulin deficiency suggests that NEGR1 has a functional role in how insulin regulates the activities of enzymes of glycolysis, citric acid cycle, and/or lipogenesis. Even more, Joo et al. [33] have shown that *Negr1* deficiency induces abnormal fat deposition in various peripheral cells, especially fat and liver tissue cells.

Still, *Negr1*$^{-/-}$ did not display increased levels of acyl residues that would be expected if high citrate diverges from glycolysis into lipid synthesis. As the synthesis also demands high amounts of ATP and reductive NADPH, it may be ineffective. Therefore, on a normal diet, the mice do not gain excessive fat or body weight, and display a normal serum lipid profile. If glycolysis is inhibited, more amino acids are catabolized for energy production, leading to a risk of muscle wasting. HF diet impairs glucose utilization even further [47]; thus, the *Negr1*$^{-/-}$, who already have trouble with glucose usage, are forced to utilize amino acids as a source of energy (Figure 8). Metabolism of muscle proteins, which are the main protein reserve, is highly dependent on sex hormones.

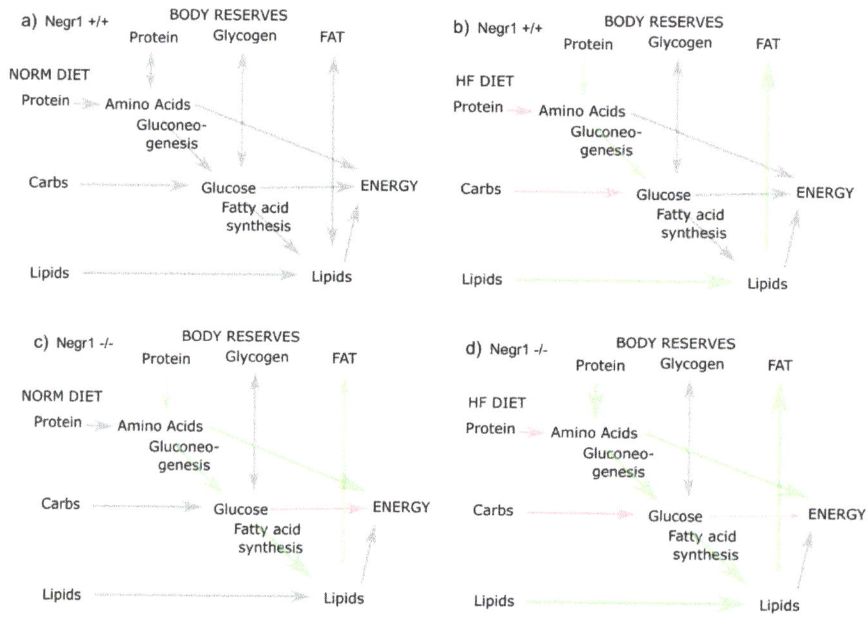

Figure 8. A hypothetical and simplified scheme of the metabolic differences in wild-type (*Negr1*$^{+/+}$) (**a**,**b**) and *Negr1*$^{-/-}$ (**c**,**d**) mice on a normal diet (**a**,**c**) and high-fat diet (**b**,**d**). Increased size and green color indicate increased metabolic flux, reduced arrow size and red color indicate inhibited process.

Gender-specific effects were most notable in some amino acid levels, which were increased in *Negr1*$^{-/-}$ animals on an HF diet, and glucose intolerance and body weight gain, in which male WT mice performed the best. Among the genders and genotypes, the WT males have the highest muscle mass, which helps to absorb and utilize glucose in its excess, and also more amino acids for gluconeogenesis to cope with time in glucose deficiency. The gluconeogenetic activity of *Negr1*$^{-/-}$ mice might be higher than in WT and, therefore, they are less capable of coping with additional requirements from an unbalanced diet. High demand for amino acids and their higher availability in males than in females might cause the most pronounced gender-specific effects on amino acid levels due to the HF diet. The Supplementary Section (Figures S4–S6) of the current study also provides histology results of a small population of middle-aged females (8–9-month-old females versus 5-month-old males). These images suggest that the reduced muscle fiber size phenotype might be sex-specific and present only in male *Negr1*$^{-/-}$ animals. The steatosis-prone phenotype seems to be present in both sexes of *Negr1*-deficient mice. These findings support our theory that the lipid metabolism of *Negr1*-deficient mice is prone to lipid synthesis and accumulation. Altogether, the results of the current study emphasize that the future metabolic studies in *Negr1*-deficient mice should be performed comparatively in males and females.

5. Conclusions

In summary, *Negr1*$^{-/-}$ mice have higher protein catabolism than WT mice. With higher usage of amino acids, the energy from carbohydrates and lipids can be diverted into reserves, e.g., formation of fat stores. Dependence of protein metabolism on sex is also a putative reason why the genotype effect is more pronounced in male mice. The HF diet does promote prediabetes and fat accumulation, which amplifies the metabolic pathways already activated by the absence of *Negr1*. Our data show that *Negr1* is one of the genetic factors that, together with other signaling molecules (e.g., sex hormones) and consumed diet, contributes to the balance of systemic metabolism, including glucose homeostasis.

Supplementary Materials: The following are available online at https://www.mdpi.com/article/10.3390/biomedicines9091148/s1, Figure S1: The time graphs of food preference test, Figure S2: The levels of selected organic acids, Figure S3: BODIPY stainings of male liver, Figure S4: BODIPY staining of female liver, Figure S5: BODIPY stainings of female liver, Figure S6: Cross-sectional area of muscle fibers from female mice, Table S1: Metabolites, Table S2: Organic acids.

Author Contributions: Conceptualization, M.K., K.K., E.L., E.V. and M.-A.P.; Data curation, M.K., K.M., K.K. and E.V.; Formal analysis, M.K., K.K., T.J., R.P., R.R., E.T. and M.P. (Mario Plaas); Funding acquisition, M.P. (Mario Plaas) and E.V.; Investigation, M.K., K.M., K.L., K.K., R.P. and K.S.; Methodology, K.K. and T.J.; Project administration, E.V. and M.-A.P.; Resources, M.K.E.S., M.P. (Mario Plaas), and E.V.; Software, M.K., K.K., T.J. and M.P. (Mario Plaas); Supervision, K.L., E.V. and M.-A.P.; Validation, K.K. and E.L.; Visualization, M.K., K.K. and T.J.; Writing—original draft, M.K., K.K., E.L. and M.-A.P.; Writing—review and editing, M.K., K.M., K.L., M.P. (Maria Piirsalu), R.P., K.S., M.K.E.S., M.P. (Mario Plaas), and E.V. All authors have read and agreed to the published version of the manuscript.

Funding: This research was supported by the European Union through the European Regional Development Fund (Project No. 2014-2020.4.01.15-0012), and by the team grant from the Estonian Research Foundation (PRG 685).

Institutional Review Board Statement: The use of mice was conducted in accordance with the regulations and guidelines approved by the Laboratory Animal Centre at the Institute of Biomedicine and Translational Medicine, University of Tartu, Estonia. All animal procedures were conducted in accordance with the European Communities Directive (2010/63/EU) with permit (No. 150, 27 September 2019) from the Estonian National Board of Animal Experiments.

Informed Consent Statement: Not applicable.

Data Availability Statement: The data presented in this study are available on request from the corresponding author.

Conflicts of Interest: The authors declare no conflict of interest.

References

1. Marg, A.; Sirim, P.; Spaltmann, F.; Plagge, A.; Kauselmann, G.; Buck, F.; Rathjen, F.G.; Brümmendorf, T. Neurotractin, A Novel Neurite Outgrowth-promoting Ig-like Protein that Interacts with CEPU-1 and LAMP. *J. Cell Biol.* **1999**, *145*, 865–876. [CrossRef]
2. Noh, K.; Lee, H.; Choi, T.-Y.; Joo, Y.; Kim, S.-J.; Kim, H.; Kim, J.Y.; Jahng, J.W.; Lee, S.; Choi, S.-Y.; et al. Negr1 controls adult hippocampal neurogenesis and affective behaviors. *Mol. Psychiatry* **2019**, *24*, 1189–1205. [CrossRef]
3. Singh, K.; Loreth, D.; Pöttker, B.; Hefti, K.; Innos, J.; Schwald, K.; Hengstler, H.; Menzel, L.; Sommer, C.J.; Radyushkin, K.; et al. Neuronal Growth and Behavioral Alterations in Mice Deficient for the Psychiatric Disease-Associated Negr1 Gene. *Front. Mol. Neurosci.* **2018**, *11*, 1662–5099. [CrossRef] [PubMed]
4. Singh, K.; Jayaram, M.; Kaare, M.; Leidmaa, E.; Jagomäe, T.; Heinla, I.; Hickey, M.; Kaasik, A.; Schäfer, M.K.; Innos, J.; et al. Neural cell adhesion molecule Negr1 deficiency in mouse results in structural brain endophenotypes and behavioral deviations related to psychiatric disorders. *Sci. Rep.* **2019**, *9*, 5457. [CrossRef]
5. Ranaivoson, F.M.; Turk, L.S.; Ozgul, S.; Kakehi, S.; von Daake, S.; Lopez, N.; Trobiani, L.; DE Jaco, A.; Denissova, N.; Demeler, B.; et al. A Proteomic Screen of Neuronal Cell-Surface Molecules Reveals IgLONs as Structurally Conserved Interaction Modules at the Synapse. *Structure* **2019**, *27*, 1055–1056. [CrossRef]
6. Willer, C.J.; Speliotes, E.K.; Loos, R.J.F.; Li, S.; Lindgren, C.M.; Heid, I.M.; Berndt, S.I.; Elliott, A.L.; Jackson, A.U.; Lamina, C.; et al. Six new loci associated with body mass index highlight a neuronal influence on body weight regulation. *Nat. Genet.* **2009**, *41*, 25–34. [CrossRef] [PubMed]
7. Thorleifsson, G.; Walters, G.B.; Gudbjartsson, D.F.; Steinthorsdottir, V.; Sulem, P.; Helgadottir, A.; Styrkarsdottir, U.; Gretarsdottir, S.; Thorlacius, S.; Jonsdottir, I.; et al. Genome-wide association yields new sequence variants at seven loci that associate with measures of obesity. *Nat. Genet.* **2009**, *41*, 18–24. [CrossRef] [PubMed]
8. Speliotes, E.K.; Willer, C.J.; Berndt, S.I.; Monda, K.L.; Thorleifsson, G.; Jackson, A.U.; Allen, H.L.; Lindgren, C.M.; Mägi, R.; Randall, J.C.; et al. Association analyses of 249,796 individuals reveal 18 new loci associated with body mass index. *Nat. Genet.* **2010**, *42*, 937–948. [CrossRef] [PubMed]
9. Locke, A.E.; Kahali, B.; Berndt, S.I.; Justice, A.E.; Pers, T.H.; Day, F.R.; Powell, C.; Vedantam, S.; Buchkovich, M.L.; Yang, J.; et al. Genetic studies of body mass index yield new insights for obesity biology. *Nature* **2015**, *518*, 197–206. [CrossRef]
10. Winkler, T.W.; Justice, A.E.; Graff, M.; Barata, L.; Feitosa, M.F.; Chu, S.; Czajkowski, J.; Esko, T.; Fall, T.; Kilpeläinen, T.O.; et al. The Influence of Age and Sex on Genetic Associations with Adult Body Size and Shape: A Large-Scale Genome-Wide Interaction Study. *PLoS Genet.* **2016**, *11*, e1005378. [CrossRef]

11. Schlauch, K.A.; Read, R.W.; Lombardi, V.C.; Elhanan, G.; Metcalf, W.J.; Slonim, A.D.; Twenty Three & Me Research Team; Grzymski, J.J. A Comprehensive Genome-Wide and Phenome-Wide Examination of BMI and Obesity in a Northern Nevadan Cohort. *G3 Genes Genomes Genet.* **2020**, *10*, 645–664. [CrossRef]
12. Wheeler, E.; Huang, N.; Bochukova, E.; Keogh, J.M.; Lindsay, S.; Garg, S.; Henning, E.; Blackburn, H.; Loos, R.; Wareham, N.J.; et al. Genome-wide SNP and CNV analysis identifies common and low-frequency variants associated with severe early-onset obesity. *Nat. Genet.* **2013**, *45*, 513–517. [CrossRef] [PubMed]
13. Merino, J.; Dashti, H.S.; Sarnowski, C.; Lane, J.M.; Udler, M.S.; Todorov, P.V.; Song, Y.; Wang, H.; Kim, J.; Tucker, C.; et al. Multi-trait genome-wide association meta-analysis of dietary intake identifies new loci and genetic and functional links with metabolic traits. *bioRxiv* **2019**. [CrossRef]
14. Niarchou, M.; Byrne, E.M.; Trzaskowski, M.; Sidorenko, J.; Kemper, K.; McGrath, J.J.; Donovan, M.C.O.; Owen, M.J.; Wray, N.R. Genome-wide association study of dietary intake in the UK biobank study and its associations with schizophrenia and other traits. *Transl. Psychiatry* **2020**, *10*, 51. [CrossRef]
15. Gamero-Villarroel, C.; González, L.M.; Gordillo, I.; Carrillo, J.A.; García-Herráiz, A.; Flores, I.; Rodríguez-López, R.; Gervasini, G. Impact of NEGR1 genetic variability on psychological traits of patients with eating disorders. *Pharm. J.* **2015**, *15*, 278–283. [CrossRef] [PubMed]
16. Lee, P.H.; Anttila, V.; Won, H.; Feng, Y.-C.A.; Rosenthal, J.; Zhu, Z.; Tucker-Drob, E.M.; Nivard, M.; Grotzinger, A.D.; Posthuma, D.; et al. Genomic Relationships, Novel Loci, and Pleiotropic Mechanisms across Eight Psychiatric Disorders. *Cell* **2019**, *179*, 1469–1482.e11. [CrossRef]
17. Hyde, C.L.; Nagle, M.; Tian, C.; Chen, X.; Paciga, S.A.; Wendland, J.R.; Tung, J.Y.; Hinds, A.; Perlis, R.H.; Winslow, A.R. Identification of 15 genetic loci associated with risk of major depression in individuals of European descent. *Nat. Genet.* **2016**, *48*, 1031–1036. [CrossRef]
18. Howard, D.M.; Adams, M.J.; Shirali, M.; Clarke, T.-K.; Marioni, R.E.; Davies, G.; Coleman, J.R.I.; Alloza, C.; Shen, X.; Barbu, M.C.; et al. Genome-wide association study of depression phenotypes in UK Biobank identifies variants in excitatory synaptic pathways. *Nat. Commun.* **2018**, *9*, 1470. [CrossRef] [PubMed]
19. Levey, D.F.; Stein, M.B.; Wendt, F.R.; Pathak, G.A.; Zhou, H.; Aslan, M.; Quaden, R.; Harrington, K.M.; Nuñez, Y.Z.; Overstreet, C.; et al. Bi-ancestral depression GWAS in the Million Veteran Program and meta-analysis in >1.2 million individuals highlight new therapeutic directions. *Nat. Neurosci.* **2021**, 954–963. [CrossRef]
20. Dall'Aglio, L.; Lewis, C.M.; Pain, O. Delineating the Genetic Component of Gene Expression in Major Depression. *Biol. Psychiatry* **2021**, *89*, 627–636. [CrossRef] [PubMed]
21. Lee, A.W.S.; Hengstler, H.; Schwald, K.; Diaz, M.B.; Loreth, D.; Kirsch, M.; Kretz, O.; Haas, C.A.; de Angelis, M.H.; Herzig, S.; et al. Functional Inactivation of the Genome-Wide Association Study Obesity Gene Neuronal Growth Regulator 1 in Mice Causes a Body Mass Phenotype. *PLoS ONE* **2012**, *7*, e41537. [CrossRef]
22. Kim, H.; Chun, Y.; Che, L.; Kim, J.; Lee, S.; Lee, S. The new obesity-associated protein, neuronal growth regulator 1 (NEGR1), is implicated in Niemann-Pick disease Type C (NPC2)-mediated cholesterol trafficking. *Biochem. Biophys. Res. Commun.* **2017**, *482*, 1367–1374. [CrossRef] [PubMed]
23. Venkannagari, H.; Kasper, J.; Misra, A.; Rush, S.; Fan, S.; Lee, H.; Sun, H.; Seshadrinathan, S.; Machius, M.; Hommel, J.D.; et al. Highly Conserved Molecular Features in IgLONs Contrast Their Distinct Structural and Biological Outcomes. *J. Mol. Biol.* **2020**, *432*, 5287–5303. [CrossRef] [PubMed]
24. Boender, A.J.; van Gestel, M.A.; Garner, K.M.; Luijendijk, M.C.M.; Adan, R.A.H. The obesity-associated gene Negr1 regulates aspects of energy balance in rat hypothalamic areas. *Physiol. Rep.* **2012**, *2*, e12083. [CrossRef] [PubMed]
25. Boender, A.J.; Van Rozen, A.J.; Adan, R.A. Nutritional State Affects the Expression of the Obesity-Associated GenesEtv5, Faim2, Fto, and Negr1. *Obesity* **2019**, *20*, 2420–2425. [CrossRef] [PubMed]
26. Liu, L.; Yi, J.; Ray, W.K.; Vu, L.; Helm, R.F.; Siegel, P.B.; Cline, M.A.; Gilbert, E.R. Fasting differentially alters the hypothalamic proteome of chickens from lines with the propensity to be anorexic or obese. *Nutr. Diabetes* **2011**, *9*, 13. [CrossRef] [PubMed]
27. Sandholt, C.H.; Vestmar, M.A.; Bille, D.S.; Borglykke, A.; Almind, K.; Hansen, L.; Sandbæk, A.; Lauritzen, T.; Witte, D.; Jørgensen, T.; et al. Studies of Metabolic Phenotypic Correlates of 15 Obesity Associated Gene Variants. *PLoS ONE* **2005**, *6*, e23531. [CrossRef]
28. Schäfer, M.; Bräuer, A.U.; Savaskan, N.E.; Rathjen, F.G.; Brümmendorf, T. Neurotractin/kilon promotes neurite outgrowth and is expressed on reactive astrocytes after entorhinal cortex lesion. *Mol. Cell. Neurosci.* **2012**, *29*, 580–590. [CrossRef]
29. Walley, A.J.; Jacobson, P.; Falchi, M.; Bottolo, L.; Andersson-Assarsson, J.; Petretto, E.; Bonnefond, A.; Vaillant, E.; Lecoeur, C.; Vatin, V.; et al. Differential coexpression analysis of obesity-associated networks in human subcutaneous adipose tissue. *Int. J. Obes.* **2017**, *36*, 137–147. [CrossRef]
30. Vanaveski, T.; Singh, K.; Narvik, J.; Eskla, K.-L.; Visnapuu, T.; Heinla, I.; Jayaram, M.; Innos, J.; Lilleväli, K.; Philips, M.-A.; et al. Promoter-Specific Expression and Genomic Structure of IgLON Family Genes in Mouse. *Front. Neurosci.* **2012**, *11*, 38. [CrossRef]
31. Bernhard, F.; Landgraf, K.; Klöting, N.; Berthold, A.; Büttner, P.; Friebe, D.; Kiess, W.; Kovacs, P.; Blüher, M.; Körner, A. Functional relevance of genes implicated by obesity genome-wide association study signals for human adipocyte biology. *Diabetol.* **2020**, *56*, 311–322. [CrossRef]
32. An, D.; Joo, Y.; Kim, H. The role of NEGR1 in the formation of lipid droplets. *FASEB J.* **2019**, *34*, 1. [CrossRef]
33. Joo, Y.; Kim, H.; Lee, S.; Lee, S. Neuronal growth regulator 1-deficient mice show increased adiposity and decreased muscle mass. *Int. J. Obes.* **2020**, *43*, 1769–1782. [CrossRef]

34. Leidmaa, E.; Gazea, M.; Patchev, A.V.; Pissioti, A.; Gassen, N.C.; Kimura, M.; Liposits, Z.; Kallo, I.; Almeida, O.F.X. Blunted leptin sensitivity during hedonic overeating can be reinstated by activating galanin 2 receptors (Gal2R) in the lateral hypothalamus. *Acta Physiol.* **2012**, *228*, e13345. [CrossRef]
35. Schindelin, J.; Arganda-Carreras, I.; Frise, E.; Kaynig, V.; Longair, M.; Pietzsch, T.; Preibisch, S.; Rueden, C.; Saalfeld, S.; Schmid, B.; et al. Fiji: An open-source platform for biological-image analysis. *Nat. Methods* **2008**, *9*, 676–682. [CrossRef] [PubMed]
36. Farmer, A.; Korszun, A.; Owen, M.J.; Craddock, N.; Jones, L.; Jones, I.; Gray, J.; Williamson, R.J.; McGuffin, P. Medical disorders in people with recurrent depression. *Br. J. Psychiatry* **2020**, *192*, 351–355. [CrossRef]
37. Tramunt, B.; Smati, S.; Grandgeorge, N.; Lenfant, F.; Arnal, J.-F.; Montagner, A.; Gourdy, P. Sex differences in metabolic regulation and diabetes susceptibility. *Diabetologia* **2018**, *63*, 453–461. [CrossRef] [PubMed]
38. Mauvais-Jarvis, F. Gender differences in glucose homeostasis and diabetes. *Physiol. Behav.* **2017**, *187*, 20–23. [CrossRef]
39. Rauschert, S.; Uhl, O.; Koletzko, B.; Mori, T.A.; Beilin, L.J.; Oddy, W.H.; Hellmuth, C. Sex differences in the association of phospholipids with components of the metabolic syndrome in young adults. *Biol. Sex Differ.* **2018**, *8*, 10. [CrossRef] [PubMed]
40. Cochran, J.; Taufalele, P.V.; Lin, K.D.; Zhang, Y.; Abel, E.D. Sex Differences in the Response of C57BL/6 Mice to Ketogenic Diets. *Diabetes* **2011**, *67*, 1884. [CrossRef]
41. Wang, T.J.; Larson, M.; Vasan, R.S.; Cheng, S.; Rhee, E.P.; McCabe, E.; Lewis, G.D.; Fox, C.S.; Jacques, P.F.; Fernandez, C.; et al. Metabolite profiles and the risk of developing diabetes. *Nat. Med.* **2014**, *17*, 448–453. [CrossRef]
42. Liu, H.; Liu, R.; Xiong, Y.; Li, X.; Wang, X.; Ma, Y.; Guo, H.; Hao, L.; Yao, P.; Liu, L.; et al. Leucine facilitates the insulin-stimulated glucose uptake and insulin signaling in skeletal muscle cells: Involving mTORC1 and mTORC2. *Amino Acids* **2014**, *46*, 1971–1979. [CrossRef] [PubMed]
43. Lynch, C.J.; Adams, S. Branched-chain amino acids in metabolic signalling and insulin resistance. *Nat. Rev. Endocrinol.* **1965**, *10*, 723–736. [CrossRef] [PubMed]
44. Kornacker, M.S.; Lowenstein, J.M. Citrate and the conversion of carbohydrate into fat. The activities of citrate-cleavage enzyme and acetate thiokinase in livers of starved and re-fed rats. *Biochem. J.* **2002**, *94*, 209–215. [CrossRef]
45. Munday, M.R. Regulation of mammalian acetyl-CoA carboxylase. *Biochem. Soc. Trans.* **2014**, *30*, 1059–1064. [CrossRef] [PubMed]
46. Nilsson, E.; Jansson, P.A.; Perfilyev, A.; Volkov, P.; Pedersen, M.; Svensson, M.K.; Poulsen, P.; Ribel-Madsen, R.; Pedersen, N.L.; Almgren, P.; et al. Altered DNA Methylation and Differential Expression of Genes Influencing Metabolism and Inflammation in Adipose Tissue from Subjects with Type 2 Diabetes. *Diabetes* **2019**, *63*, 2962–2976. [CrossRef]
47. Jana, B.A.; Chintamaneni, P.K.; Krishnamurthy, P.T.; Wadhwani, A.; Mohankumar, S.K. Cytosolic lipid excess-induced mitochondrial dysfunction is the cause or effect of high fat diet-induced skeletal muscle insulin resistance: A molecular insight. *Mol. Biol. Rep.* **2019**, *46*, 957–963. [CrossRef] [PubMed]

biomedicines

Article

Raloxifene Ameliorates Glucosamine-Induced Insulin Resistance in Ovariectomized Rats

Chung-Hwan Chen [1,2,3,4,5,6,7,8,9,†], Tsung-Lin Cheng [1,3,10,†], Chi-Fen Chang [11], Hsuan-Ti Huang [1,2,3,4,5], Sung-Yen Lin [1,2,3,4,5,6], Meng-Hsing Wu [12,13] and Lin Kang [12,13,*]

1 Orthopaedic Research Center, College of Medicine, Kaohsiung Medical University, Kaohsiung 80701, Taiwan; hwan@kmu.edu.tw (C.-H.C.); junglecc@kmu.edu.tw (T.-L.C.); hthuang@kmu.edu.tw (H.-T.H.); sungyenlin@kmu.edu.tw (S.-Y.L.)
2 Department of Orthopedics, Kaohsiung Medical University Hospital, Kaohsiung Medical University, Kaohsiung 80701, Taiwan
3 Regeneration Medicine and Cell Therapy Research Center, Kaohsiung Medical University, Kaohsiung 80701, Taiwan
4 Departments of Orthopedics, College of Medicine, Kaohsiung Medical University, Kaohsiung 80701, Taiwan
5 Department of Orthopedics, Kaohsiung Municipal Ta-Tung Hospital, Kaohsiung Medical University, Kaohsiung 80145, Taiwan
6 Department of Healthcare Administration and Medical Informatics, Kaohsiung Medical University, Kaohsiung 80701, Taiwan
7 Institute of Medical Science and Technology, National Sun Yat-Sen University, Kaohsiung 80420, Taiwan
8 Graduate Institute of Animal Vaccine Technology, College of Veterinary Medicine, National Pingtung University of Science and Technology, Pingtung 912301, Taiwan
9 Graduate Institute of Materials Engineering, College of Engineering, National Pingtung University of Science and Technology, Pingtung 912301, Taiwan
10 Department of Physiology, College of Medicine, Kaohsiung Medical University, Kaohsiung 80701, Taiwan
11 Department of Anatomy, School of Medicine, China Medical University, Taichung 40402, Taiwan; cfchang@mail.cmu.edu.tw
12 Department of Obstetrics & Gynecology, College of Medicine, National Cheng Kung University, Tainan 70101, Taiwan; mhwu68@mail.ncku.edu.tw
13 Department of Obstetrics and Gynecology, National Cheng Kung University Hospital, College of Medicine, National Cheng Kung University, Tainan 70101, Taiwan
* Correspondence: kanglin@mail.ncku.edu.tw; Tel.: +886-6-276-6685
† These authors contributed equally to this work.

Citation: Chen, C.-H.; Cheng, T.-L.; Chang, C.-F.; Huang, H.-T.; Lin, S.-Y.; Wu, M.-H.; Kang, L. Raloxifene Ameliorates Glucosamine-Induced Insulin Resistance in Ovariectomized Rats. *Biomedicines* **2021**, *9*, 1114. https://doi.org/10.3390/biomedicines9091114

Academic Editor: Susan J. Burke

Received: 4 August 2021
Accepted: 24 August 2021
Published: 30 August 2021

Abstract: Osteoarthritis (OA) and osteoporosis (OP) are common among older women, especially postmenopausal women. Glucosamine (GlcN) is a common medication for OA, but it may induce insulin resistance and β-cell dysfunction, especially if ovarian hormones are lacking. Raloxifene (RLX) is a selective estrogen receptor modulator and also an OP drug. Previously, we found that estrogen could improve GlcN-induced insulin resistance in ovariectomized (OVX) rats. Here, we further hypothesized that RLX, similarly to estrogen, can ameliorate GlcN-induced insulin resistance in OVX rats. We used GlcN to induce insulin resistance in OVX rats as a model for evaluating the protective effects of RLX in vivo. We used a pancreatic β-cell line, MIN-6, to study the mechanisms underlying the effect of RLX in GlcN-induced β-cell dysfunction in vitro. Increases in fasting plasma glucose, insulin, and homeostasis model assessments of insulin resistance in OVX Sprague Dawley rats treated with GlcN were reversed by RLX treatment ($n = 8$ in each group). Skeletal muscle GLUT-4 increased, liver PEPCK decreased, pancreatic islet hypertrophy, and β-cell apoptosis in OVX rats treated with GlcN was ameliorated by RLX. The negative effects of GlcN on insulin secretion and cell viability in MIN-6 cells were related to the upregulation of reticulum (ER) stress-associated proteins (C/EBP homologous protein, phospho-extracellular signal-regulated kinase, phospho-c-JunN-terminal kinase), the expression of which was reduced by RLX. Pretreatment with estrogen receptor antagonists reversed the protective effects of RLX. GlcN can induce insulin resistance, β-cell dysfunction, and apoptosis in OVX rats and increase ER stress-related proteins in β-cells, whereas RLX can reverse these adverse effects. The effects of RLX act mainly through estrogen receptor α; therefore, RLX may be a candidate drug for postmenopausal women with OA and OP.

Keywords: apoptosis; endoplasmic reticulum stress; glucosamine; pancreatic β-cell dysfunction; ovariectomy; raloxifene

1. Introduction

Metabolic syndrome and type 2 diabetes mellitus (T2DM) are prevalent health problems [1]. Insulin resistance (IR) is a hallmark of T2DM and metabolic syndrome, which have higher prevalence in postmenopausal women than in premenopausal women. Osteoarthritis (OA) is also prevalent in postmenopausal women [2]. T2DM, with its related chronic hyperglycemia and IR, induces possibly pathogenic effects in OA through oxidative stress and chronic low-grade inflammation [3]. Glucosamine (GlcN) is a nutritional supplement widely used for OA [4,5]. However, studies have reported that GlcN affects glucose tolerance and IR [3,6]. GlcN inhibits the insulin production of pancreatic β-cells [7,8]. The implications of clinical data regarding GlcN and glucose metabolism are conflicting. Some clinical studies have observed harmful effects of GlcN on glucose metabolism [9,10], whereas others have reported no effects [11,12].

Ovarian estrogen, 17 β-estradiol (E_2), has ameliorated both insulin sensitivity and insulin production in animal and human studies [13,14]. The risk of T2DM increased after ovariectomy in an animal study, whereas estrogen application ameliorated T2DM and enhanced insulin sensitivity [15]. In pancreatic β-cells, E_2 can ameliorate glucolipotoxicity, oxidative stress, and apoptosis [14,16]. Estrogen receptors can regulate the function and survival of β-cells [17], although estrogen use is associated with an increased risk of various cancers [18].

Selective estrogen receptor modulators (SERMs) act on estrogen receptors with agonist or antagonist activity, depending on the tissue type. Raloxifene (RLX) is an SERM used for the treatment of postmenopausal osteoporosis (OP) because of its activity as an estrogen receptor agonist in bone [19–28]. The effect of RLX can be tissue- or species-specific [29]. One study using a pancreatic β-cell line (INS-1 cells) demonstrated that RLX behaves as both an estrogen receptor antagonist through nuclear estrogen response element-dependent actions and as an estrogen receptor agonist by suppressing triglyceride accumulation [30]. The endoplasmic reticulum (ER) regulates intracellular calcium concentrations and protein folding and trafficking [31–34]. ER stress, the disruption of these ER functions, is related to T2DM in humans [35–37]. ER stress can be induced by GlcN and cause cell death [38,39]. In our previous study, we demonstrated that GlcN-induced IR in ovariectomized (OVX) rats was related to increased pancreatic islet size [40]. We also observed the protective effects of E_2 in GlcN-induced pancreatic β-cell dysfunction [41]. The rescue effects of RLX in GlcN-induced IR and pancreatic β-cell dysfunction have not been reported. In the current study, we used OVX rats treated with GlcN to induce IR and studied the protective effects of RLX in vivo. We also used pancreatic β-cell lines, MIN-6 cells, to study the protection and underlying mechanisms of RLX in GlcN-induced β-cell dysfunction in vitro.

2. Materials and Methods

2.1. Ethics Statement

All procedures were performed in accordance with the Institutional Guidelines for Animal Care of National Cheng Kung University, the Use of Laboratory Animals of the National Institutes of Health, and the guidelines of the Animal Welfare Act.

2.2. Experimental Animals

Twelve-week-old Female Sprague Dawley rats were purchased from the Animal Center of National Cheng Kung University Medical College and housed under standard laboratory conditions with free access to food and water. After acclimation, the rats were randomly allocated to one of 5 treatments: (1) sham operation (Sham group); (2) sham with 750 mg/kg/d GlcN (Sigma-Aldrich, St. Louis, MO, USA) intraperitoneally (ip) injected for 14 days (Sham + GlcN group); (3) ovariectomy (OVX group); (4) ovariectomy with 750 mg/kg/d GlcN ip treated for 14 days (OVX + GlcN group); or (5) ovariectomy with 750 mg/kg/d GlcN ip treated for 14 days with subcutaneous RLX at 0.5 mg/kg/d (Sigma Chemical Co., St. Louis, MO, USA; OVX + GlcN + RLX group; n = 8 in each group). Surgeries were performed under ip administered sodium pentobarbital (Sigma-Aldrich) anesthesia through bilateral lower back skin incisions [40–45]. Twelve weeks after the surgery, GlcN was administered for two weeks [40,41,45].

2.3. *Intraperitoneal Glucose Tolerance Test*

An intraperitoneal glucose tolerance test (IPGTT) was administered with the rats fasted for 6 h after all of the treatments were completed. Blood samples for the measurement of plasma glucose and insulin were drawn from the femoral vein before glucose loading (1 mg/kg, ip) at baseline (time 0). Blood samples were obtained at 30, 60, 90, and 120 min after glucose loading [40,41,45].

2.4. *Plasma Glucose and Insulin Concentrations*

Plasma glucose levels were evaluated by a commercial kit reagent for glucose (Biosystems SS, Barcelona, Spain) by an analyzer (Quik-Lab, Elkhart, IN, USA). Insulin concentration was evaluated by an insulin enzyme-linked immunosorbent assay (ELISA) kit (Mercodia AB, Uppsala, Sweden), as described previously [40,45].

2.5. *Determination of IR in Rats*

IR and β-cell function were assayed using homeostasis model assessments of IR (HOMA-IR). The glucose-insulin index and clinical HOMA-IR were determined to evaluate IR and compare groups after the concentrations of plasma glucose and insulin were measured. The glucose-insulin index was calculated as the product of the glucose and insulin areas under the curve (AUCs). HOMA-IR = [fasting glucose (mmol/L)] × [fasting insulin (pmol/mL)]/22.5 [40,45].

2.6. *Measurement of Islet Size*

All pancreases were immersed in phosphate-buffered saline (PBS) containing 10% formaldehyde (v/v) and kept at 4 °C for 2 days. After dehydration, the specimens were fixed in paraffin. The specimens were sliced into 5 μm thick sections with 50 μm distances; then, hematoxylin and eosin staining was performed. For each staining, more than three serial sections were used. The area of each islet was decided by Image-Pro Plus (Media Cybernetics, Inc., Rockville, MD, USA) with a total of 210–230 islets in each section [40,45].

2.7. *Immunofluorescence Stains for Insulin and Terminal Deoxynucleotidyl Transferase dUTP Nick End-Labeling in Pancreatic Islets*

Sequential 5 μm thick pancreas sections around 2 × 1 cm^2 were immunostained for transferase dUTP nick end-labeling (TUNEL), insulin, and DAPI. The immunofluorescence staining intensity was quantified using Image-Pro Plus (Media Cybernetics, Inc., Rockville, MD, USA) [41,46–48].

2.8. *Western Blot Analysis for PEPCK in the Liver and GLUT-4 in the Soleus Muscle*

The livers and soleus muscles were harvested immediately after the rats were killed, as described previously [40,45]. In brief, the tissues were washed with cold PBS and cut into 200–300 mg portions. After homogenization of the liver and soleus muscle, the homogenates (50 μg) were separated through sodium dodecyl sulfide–polyacrylamide gel electrophoresis, and Western blot analysis was performed using either an anti-rat glucose transport protein subtype 4 (GLUT-4) antibody (R&D system, Inc., Minneapolis, MN, USA) (1:1000) in the soleus muscle or an anti-rat phosphoenolpyruvate carboxykinase (PEPCK) antibody (R&D system, Inc., Minneapolis, MN, USA) (1:1000) in liver tissue.

2.9. Cell Culture and Compound Stimulation

MIN-6 cells were kept in a monolayer culture at 37 °C and 5% (*v*/*v*) CO_2 in Dulbecco's modified Eagle's medium (DMEM) supplemented with 10% (*v*/*v*) fetal bovine serum (FBS), 1.0×10^5 U/L penicillin, and 100 g/L streptomycin. MIN-6 cells were incubated with RLX 1 at μmol/L (Sigma Chemical Co.) [30], ICI 182,780 (antagonist to both estrogen receptor α and estrogen receptor β, Tocris, Ballwin, MO, USA) at 1 μmol/L, and methyl-piperidino-pyrazole (MPP, antagonist specific to estrogen receptor α, Tocris Cookson, Ellisville, MO, USA) at 1 μmol/L, as indicated, for 72 h. After estrogen receptor ligand treatment, the cells were treated with GlcN (Sigma Chemical Co.) at 10 mmol/L for 6 h before assessment in the following experiments.

2.10. Extracellular Insulin Levels

The MIN-6 cells were incubated in 6-well plates (1.0×10^4 cells/well) with 1 μmol/L ICI 182,780 or 1 μmol/L MMP for 1 h in high-glucose (4.5 g/L) DMEM if needed, and then exposed to 1 μmol/L RLX for 24 h. The cells were transferred to low-glucose (1 g/L) DMEM with 1 μmol/L RLX for another 24 h. After the supernatant was removed, the cells were washed twice with PBS and treated with 10 mmol/L GlcN for half an hour, and then exposed to glucose (5.5 mmol/L) for another half an hour. The insulin in the supernatant was assayed by an insulin ELISA kit (Mercodia AB, Uppsala, Sweden).

2.11. Cell Viability Analysis

MIN-6 cells were plated in 96-well plates (1.0×10^4 cells/well). The effects of RLX on the viability of GlcN -treated MIN-6 cells were evaluated by MTT assay. After a 1-day culture, the cells were treated with RLX, GlcN, ICI 182,780, and MPP as mentioned, and an MTT solution was subsequently used. The precipitates were dissolved in DMSO, and the absorbance was evaluated by an ELISA reader (Thermo Molecular Devices Co., Union City, CA, USA) at 570 nm after a 4 h culture. The cell viability ratio was calculated as follows:

$$\text{Inhibitory ratio (\%)} = [(\text{OD control} - \text{OD treated})/\text{OD control}] \times 100$$

2.12. Western Blot Analysis for Protein Expression Related to ER Stress

MIN-6 cells were treated as described in the preceding sections, and then proteins were harvested from the cell lysates after being treated with lysis buffer. We used 10% (*w*/*v*) sodium dodecyl sulfate–polyacrylamide gel electrophoresis to separate the protein lysates (50 μg). Western blot analysis was performed by antibodies against C/EBP homologous protein (CHOP), phospho-extracellular signal-regulated kinase (p-ERK), phospho-c-JunN-terminal kinase (p-JNK), and β-actin antibodies (Santa Cruz Biotechnology, Santa Cruz, CA, USA). The blots were treated with secondary antibodies. After washing, the blots were developed using the ECL Western blotting system (R&D system, Inc., Minneapolis, MN, USA) and quantified through laser densitometry.

2.13. Statistical Analysis

The in vivo data are expressed as the mean ± standard error of mean (SEM) for the number (*n*) of animals in each group as indicated in the Methods. Each in vitro experiment was repeated (*n*) 3 or more times, and the data are expressed as mean ± SEM. Statistical differences among groups were determined using the Friedman test in IPGTT and one-way analysis of variance in GLUT-4, *PEPCK*, TUNEL and the sizes of pancreatic islets. Dunnett range post hoc comparisons were used to determine the source of significant differences where appropriate [40,45]. A p value < 0.05 was considered significant.

3. Results

3.1. RLX Ameliorated Fasting Glucose, Insulin, and HOMA-IR in the OVX + GlcN Rats

The OVX + GlcN group (144 ± 7.8 mg/dL) presented higher fasting glucose levels than that in the Sham (120 ± 4.2 mg/dL), Sham + GlcN (121 ± 1.2 mg/dL), OVX

(117 ± 1.7 mg/dL), and OVX + GlcN + RLX (117 ± 4.9 mg/dL) groups ($p < 0.01$). RLX significantly reduced fasting glucose level in the OVX rats (Figure 1A). In addition, the OVX + GlcN group (649 ± 117 pmol/L) exhibited higher fasting plasma insulin level than that in all other groups ($p < 0.001$; Figure 1B). The OVX + GlcN group also exhibited higher fasting HOMA-IR (33.5 ± 6.6) than that in all other groups ($p < 0.001$; Figure 1C). RLX significantly reduced fasting plasma glucose, insulin levels (245 ± 57 pmol/L), and HOMA-IR (10.6 ± 2.6) in the OVX + GlcN rats.

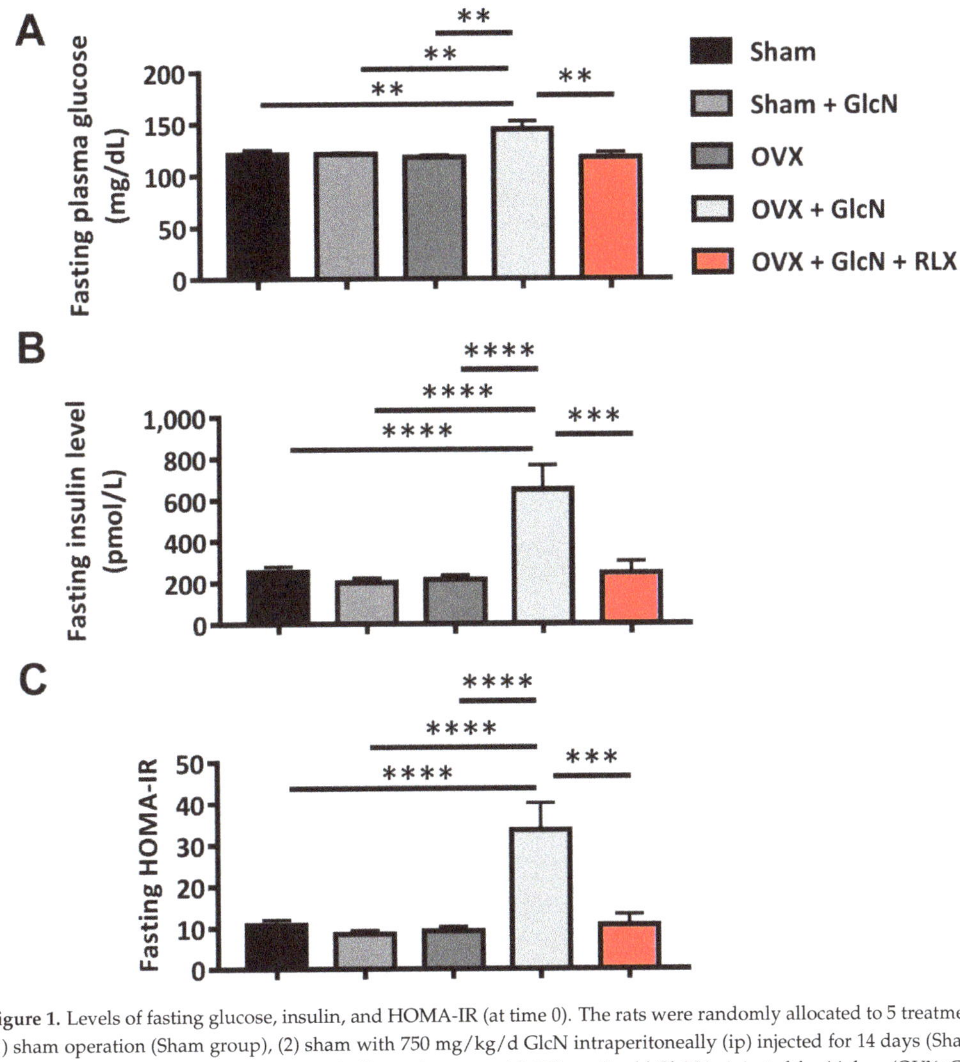

Figure 1. Levels of fasting glucose, insulin, and HOMA-IR (at time 0). The rats were randomly allocated to 5 treatments: (1) sham operation (Sham group), (2) sham with 750 mg/kg/d GlcN intraperitoneally (ip) injected for 14 days (Sham + GlcN group), (3) ovariectomy (OVX group), (4) ovariectomy with 750 mg/kg/d GlcN ip injected for 14 days (OVX+GlcN group), or (5) ovariectomy with 750 mg/kg/d GlcN ip injected for 14 days with subcutaneous RLX at 0.5 mg/kg/d. ($n = 8$ in each group). (**A**) The OVX + GlcN group exhibited higher fasting glucose than that in all other groups. RLX can significantly reduce fasting glucose levels in OVX rats. (**B**) The OVX + GlcN group exhibited higher fasting plasma insulin than that in all other groups. (**C**) RLX can significantly reduce fasting HOMA-IR in GlcN-treated OVX rats. ** $p < 0.01$; *** $p < 0.001$; **** $p < 0.0001$.

3.2. RLX Ameliorated Insulin, Glucose, Glucose-Insulin Index, and HOMA-IR in Plasma during IPGTT in the OVX + GlcN Rats

The OVX + GlcN group exhibited elevated plasma glucose level (all $p < 0.01$) than that in all other groups at 30, 60, 90, and 120 min after glucose loading. Although the levels of glucose in the OVX + GlcN + RLX group were higher than those in the three other groups (Sham, Sham + GlcN, OVX), the difference did not reach statistical significance (Figure 2A). In addition, the OVX + GlcN group also exhibited significantly higher AUC for the plasma glucose concentrations in the IPGTT than that in all other groups (Figure 2B). After glucose loading, the OVX + GlcN group exhibited significantly higher plasma insulin levels at 30, 60, 90, and 120 min than that in all other groups (all $p < 0.001$; Figure 2C). The OVX + GlcN group demonstrated higher AUC for plasma insulin concentration (Figure 2D; $p < 0.001$) than that in all other groups. The OVX + GlcN group also exhibited elevated HOMA-IR ($p < 0.001$; Figure 2E). The glucose and insulin AUCs determine the glucose-insulin index. Only the OVX + GlcN group exhibited higher glucose-insulin index ($p < 0.001$; Figure 2F). RLX reduced HOMA-IR and the glucose-insulin index nearly to the levels in the Sham group, as determined from the IPGTT. In addition to reducing fasting glucose, insulin, and HOMA-IR, RLX significantly reduced plasma glucose, insulin levels, and HOMA-IR to nearly the levels in the Sham group, as determined from the IPGTT.

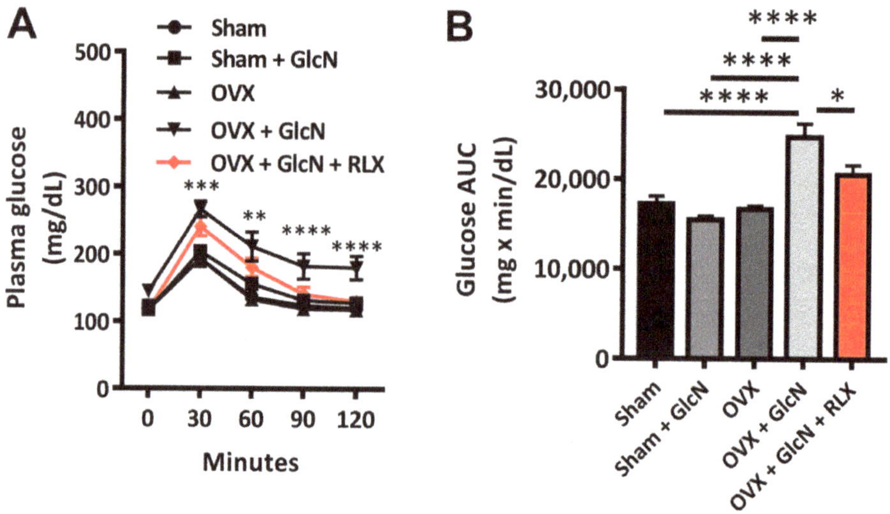

Figure 2. *Cont.*

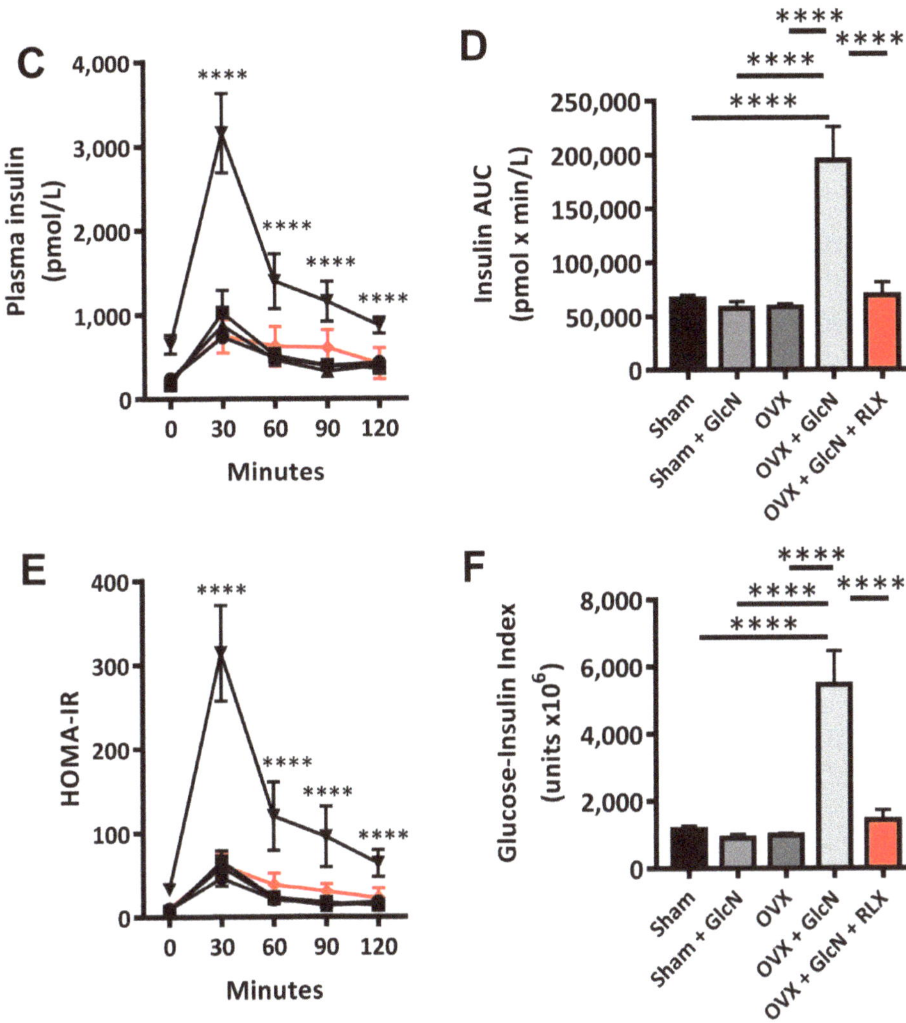

Figure 2. Levels and AUCs of plasma glucose, insulin, and HOMA-IR and glucose-insulin index in the IPGTT. The rats were randomly allocated to 5 treatments: (1) sham operation (Sham group), (2) sham with 750 mg/kg/d GlcN intraperitoneally (ip) injected for 14 days (Sham + GlcN group), (3) ovariectomy (OVX group), (4) ovariectomy with 750 mg/kg/d GlcN ip injected for 14 days (OVX+GlcN group), or (5) ovariectomy with 750 mg/kg/d GlcN ip injected for 14 days with subcutaneous RLX at 0.5 mg/kg/d. ($n = 8$ in each group). (**A**) The OVX + GlcN group exhibited elevated plasma glucose levels over all other groups at 30, 60, 90, and 120 min after glucose loading. Although the level of glucose in the OVX + GlcN + RLX group was higher than that in the other 4 groups, the difference did not reach statistical significance. (**B**) The OVX + GlcN group exhibited higher AUC for plasma glucose concentrations in the IPGTT than that in all other groups. RLX reduced the glucose AUC to nearly the level of the Sham group. (**C**) After glucose loading, the OVX + GlcN group demonstrated higher plasma insulin levels at 30, 60, 90, and 120 min than that in all other groups. (**D**) The OVX + GlcN group exhibited higher AUCs for plasma insulin concentrations than that in all other groups. RLX reduced the glucose-insulin index and the insulin AUC nearly to the levels in the Sham group. (**E**) Only the OVX + GlcN group exhibited higher HOMA-IR. The HOMA-IR of the 4 other groups exhibited no significant differences in the IPGTT. (**F**) The OVX + GlcN group demonstrated a higher glucose-insulin index. RLX reduced the glucose-insulin index to nearly the level in the Sham group. * $p < 0.05$; ** $p < 0.01$; *** $p < 0.001$; **** $p < 0.0001$.

3.3. RLX Decreased Islet Size in the OVX + GlcN Rats

The pancreatic islets in the Sham and Sham + GlcN groups were nearly the same size. The size of the pancreatic islets in the OVX group enlarged significantly and further increased in the OVX + GlcN group; these results suggest that islet hyperplasia compensated for the initial IR ($p < 0.001$). RLX treatment decreased the pancreatic islets size in the OVX + GlcN rats to the size of those in the OVX group ($p < 0.01$; Figure 3A).

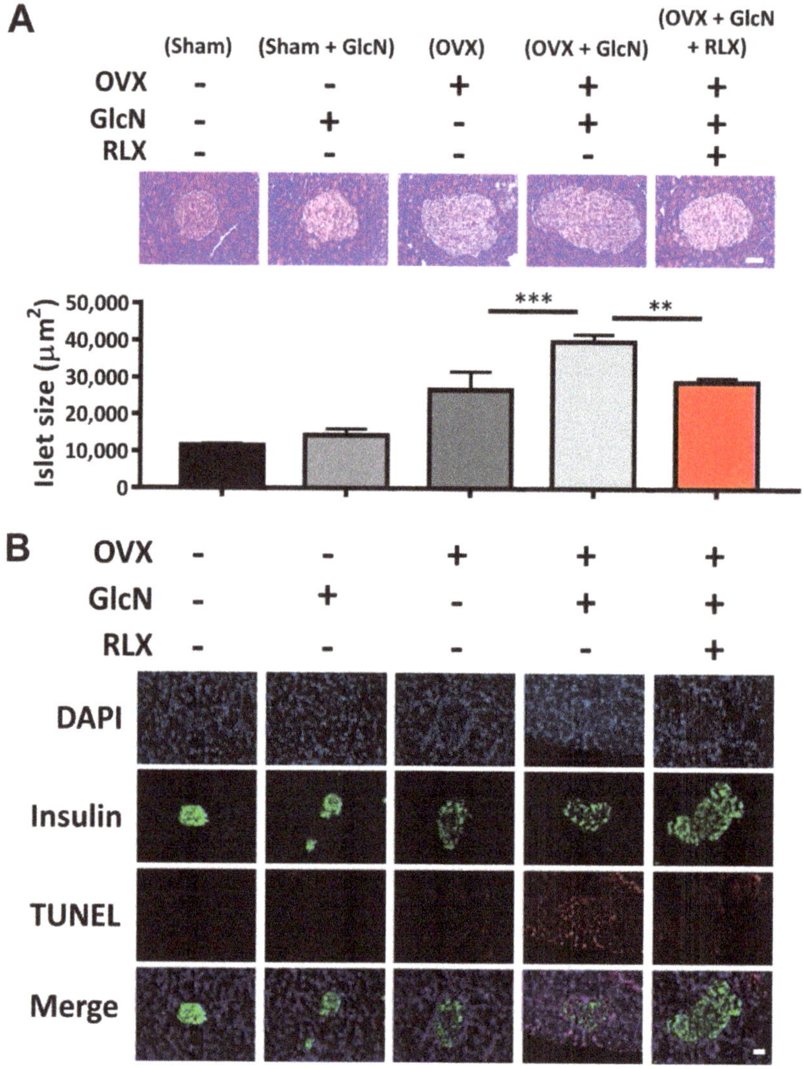

Figure 3. *Cont.*

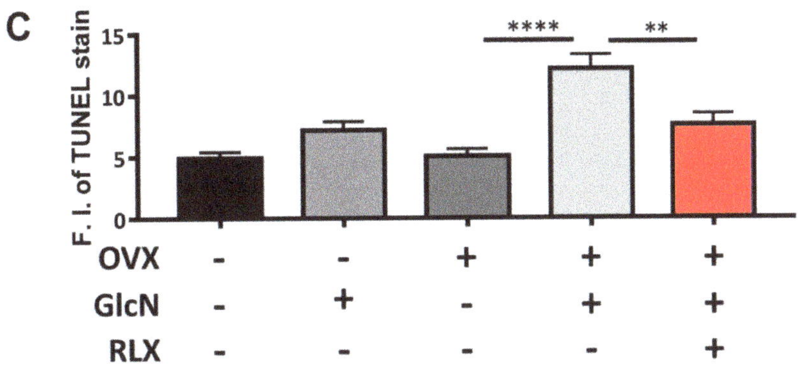

Figure 3. Islet size, immunofluorescence stains, and the quantification of insulin and TUNEL in rat pancreatic islets. The rats were randomly allocated to 5 treatments: (1) sham operation (Sham group), (2) sham with 750 mg/kg/d GlcN intraperitoneally (ip) injected for 14 days (Sham + GlcN group), (3) ovariectomy (OVX group), (4) ovariectomy with 750 mg/kg/d GlcN ip injected for 14 days (OVX+GlcN group), or (5) ovariectomy with 750 mg/kg/d GlcN ip injected for 14 days with subcutaneous RLX at 0.5 mg/kg/d. (n = 8 in each group). (**A**) The size of pancreatic islets of all groups. Pancreatic islets were nearly the same size in the Sham and Sham + GlcN groups. The size of the pancreatic islets markedly increased in the OVX group and increased further in the OVX+GlcN group, suggestive of islet hyperplasia compensating for IR. RLX treatment significantly reduced the size of the pancreatic islets in the OVX+GlcN rats to nearly the size in the OVX group. (Scale bar = 50 μm) (**B**) Immunofluorescence stains for TUNEL and insulin in pancreatic islets of all groups. Ovariectomy and GlcN reduced the immunofluorescence staining intensity for insulin in pancreatic islets; this intensity further decreased in the OVX+GlcN group. Treatment with RLX in OVX+GlcN rats increased the staining intensity of insulin. No difference was present among the OVX, GlcN, and Sham groups in apoptotic cell proportions as determined through TUNEL staining. (Scale bar = 50 μm) (**C**) Quantification of immunofluorescence stains for TUNEL in rat pancreatic islets. No difference existed among the OVX, GlcN, and Sham groups in the proportions of apoptotic cells. The proportion of apoptotic cells increased significantly in the OVX+GlcN group (12.09% ± 1.17%), whereas RLX reduced the proportion of apoptotic cells in the OVX+GlcN group (7.549% ± 0.90%). Each bar represents the mean ± SEM (n = 8–10 in each group). ** $p < 0.01$; *** $p < 0.001$; **** $p < 0.0001$.

3.4. RLX Decreased Pancreatic Islet Apoptosis in TUNEL Stain in the OVX + GlcN Rats

Immunofluorescence staining for insulin as a marker of pancreatic islets and TUNEL staining to evaluate the level of apoptosis (Figure 3B) were performed. No difference was observed in apoptotic cells among the OVX, GlcN, and Sham groups in TUNEL staining. The proportion of apoptotic cells increased significantly in the OVX + GlcN group (12.09% ± 1.17%), whereas RLX reduced the number of apoptotic cells in the OVX + GlcN group (7.549% ± 0.90%; Figure 3C).

3.5. RLX Increased the Expression of PEPCK in the Liver and Decreased the Expression of GLUT-4 in the Soleus Muscle in the OVX + GlcN Rats

OVX rats given GlcN presented with more PEPCK in the liver and less GLUT-4 expression in the soleus muscle. RLX treatment reversed these effects. No significant difference was observed in PEPCK or GLUT-4 expression among the Sham + GlcN, OVX, Sham, and OVX + GlcN + RLX groups ($p < 0.05$; Figure 4).

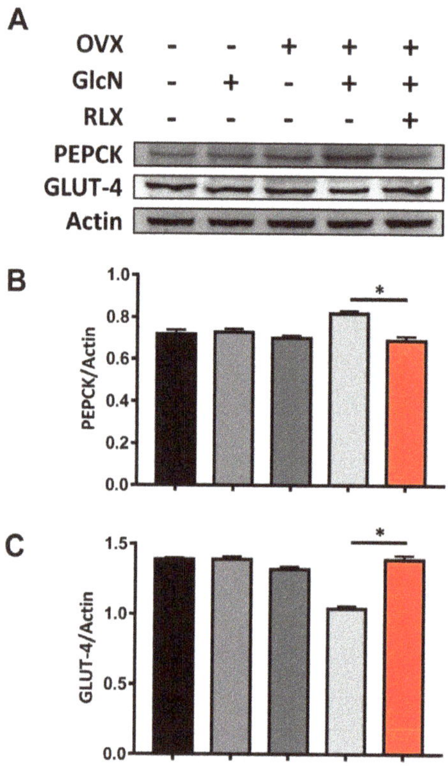

Figure 4. PEPCK expression in the liver and GLUT-4 expression in the skeletal muscle. The rats were randomly allocated to 5 treatments: (1) sham operation (Sham group), (2) sham with 750 mg/kg/d GlcN intraperitoneally (ip) injected for 14 days (Sham + GlcN group), (3) ovariectomy (OVX group), (4) ovariectomy with 750 mg/kg/d GlcN ip injected for 14 days (OVX+GlcN group), or (5) ovariectomy with 750 mg/kg/d GlcN ip injected for 14 days with subcutaneous RLX at 0.5 mg/kg/d. (n = 8 in each group). (**A**) Increased PEPCK expression in the liver and decreased GLUT-4 expression in soleus muscle were only observed in OVX rats given GlcN. RLX treatment reversed these effects. (**B**) The OVX + GlcN group exhibited increased PEPCK expression. RLX significantly reduced PEPCK expression in OVX rats with GlcN treatment to nearly the same level as the Sham group. (**C**) The OVX + GlcN group demonstrated decreased GLUT-4 expression. RLX preserved the expression of GLUT-4 in OVX rats with GlcN, maintaining it at nearly the level of the Sham group. No significant difference was present in the PEPCK or GLUT-4 expression among the Sham + GlcN, OVX, Sham, and OVX + GlcN + RLX groups. * $p < 0.05$.

3.6. RXL Increased Extracellular Insulin Secretion and Cell Viability in MIN-6 Cells Treated with GlcN

In order to clarify the roles of estrogen receptor α and estrogen receptor β in the effects of RLX, the antagonists of estrogen receptor α and estrogen receptor β were used. ICI 182,780 (antagonist to both estrogen receptor α and estrogen receptor β) and MPP (antagonist specific to estrogen receptor α) were treated as indicated. Glucose increased extracellular insulin secretion, whereas GlcN reduced it. RLX reversed the GlcN-induced decrease in extracellular insulin secretion, whereas ICI 182,780 and MPP counteracted the effects of RLX on extracellular insulin secretion (Figure 5A). GlcN treatment reduced the optical density, as determined through the MTT assay, whereas RLX reversed this effect and increased optical density in the MIN-6 cells. Pretreatment with ICI 182,780 and MPP reduced the reversal effect of RLX, and the cell viability was similar to that after GlcN treatment (Figure 5B).

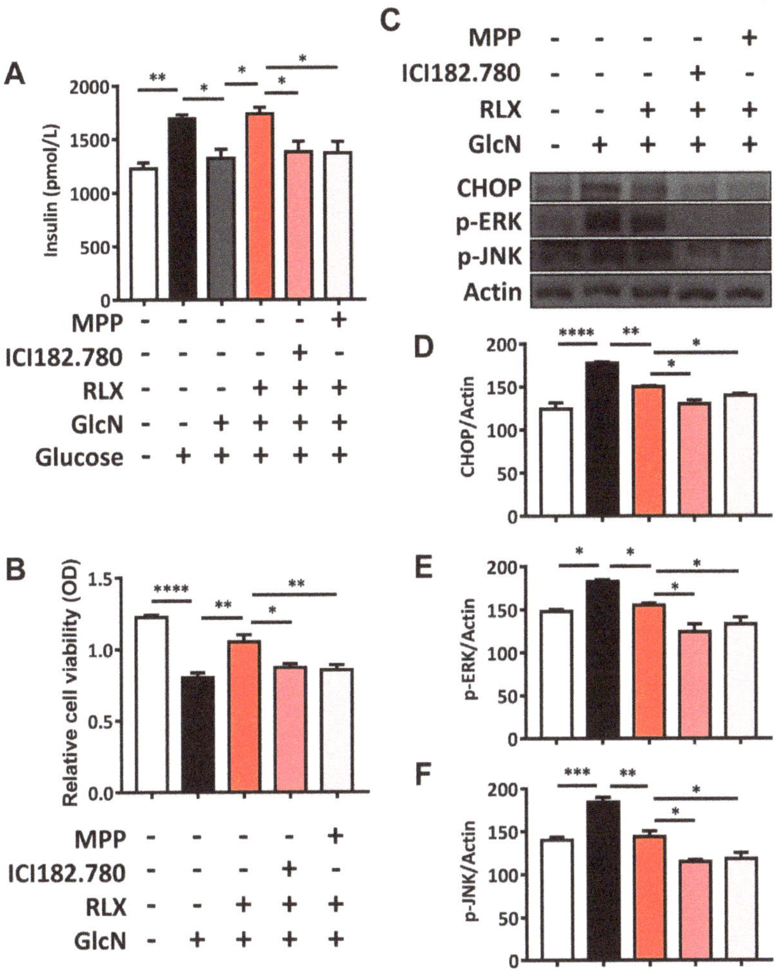

Figure 5. Extracellular insulin secretion, MTT assay, and Western blot analysis for the expression of ER stress-[associated proteins in MIN-6 cells. (**A**) Extracellular insulin secretion. Glucose increased extracellular insulin secretion, whereas GlcN reduced extracellular insulin secretion. RLX reversed GlcN-induced extracellular insulin secretion, whereas ICI 182,780 and MPP counteracted the effects of RLX. (**B**) MTT assay for cell viability. GlcN treatment reduced optical density in the MTT assay, whereas RLX counteracted this effect and even enhanced optical density. With combined ICI 182,780 and MMP treatment, the protective effect of RLX on cell viability was negated. (**C**) Western blot analysis for the expression of ER stress-associated proteins in MIN-6 cells: (**D**) CHOP, (**E**) p-ERK, and (**F**) p-JUK. GlcN treatment enhanced CHOP, p-ERK, and p-JNK expression, whereas RLX reversed these effects. Pretreatment with ICI 182,780 and MPP counteracted the reversal effect of RLX to protect against GlcN-induced CHOP, p-ERK, and p-JNK changes. Each bar represents the mean ± SEM. * $p < 0.05$; ** $p < 0.01$; *** $p < 0.001$; **** $p < 0.0001$.

3.7. RXL Decreased the Expression of ER Stress-Associated Proteins CHOP, p-ERK, and p-JUN in MIN-6 Cells Treated with GlcN in Western Blot Analysis

GlcN treatment enhanced CHOP, p-ERK, and p-JNK expression, whereas RLX reversed these effects. Pretreatment with ICI 182,780 and MPP (estrogen receptor antagonists) counteracted the reversal effect of RLX to protect against GlcN-induced CHOP, p-ERK, and p-JNK increases (Figure 5C–F). These results indicate that GlcN induces ER stress, but RLX can relieve it. The effects of RLX are evident mainly through estrogen receptor α.

4. Discussion

In previous studies, we have concluded that GlcN-induced IR in OVX rats was related to reduced insulin secretion, pancreatic β-cell apoptosis, and enlarged pancreatic islets, and that E_2 could counteract these adverse effects of in OVX rats [40,41]. In the present study, we further demonstrated that RLX, an SERM rather than an estrogen analog, can also ameliorate the deleterious effects of GlcN in OVX rats. Of the mechanisms underlying the protective effects of RLX which may be related to the combined effects of increasing GLUT-4 expression in skeletal muscle, decreasing PEPCK expression in the liver, inhibiting the growth of the pancreatic islets, and ameliorating ER stress in β-cells were shown to improve β-cell survival and function.

OP increases fracture risk and requires treatment to prevent subsequent comorbidities and mortality after fracture [19,22,25,26,49–76]. Knee OA leads to an increased risk of fall, which therefore increases fracture risk and required treatment [77–79]. OP and OA are common in older adults, especially in postmenopausal women, and they have negative impacts on quality of life [25,62,80,81]. OP is a clinically encountered comorbidity in older adults with OA [59,62,82,83]. Many patients (66%) with end-stage OA have OP or osteopenia [84]. Simultaneous lower extremity varus malalignment and OP in postmenopausal women result in more rapid OA development [85]. Hand OA and low hand and arm bone mineral density are related and can increase the risk of wrist fracture [86]. Therefore, the combined use of RLX and GlcN clinically is likely because of the frequent comorbidity of OA and OP in postmenopausal women. RLX ameliorated OA in an OVX rat model, indicating that it is a potential candidate for treating postmenopausal women with OA and OP [87]. RLX improved plasma fasting blood glucose levels in OVX rats with DM induced by a high-fat diet and ip administered streptozotocin [88]. In a clinical trial, RLX elevated 2 h insulin levels and the insulin AUC during the oral glucose tolerance test (OGTT) through reduced hepatic extraction. The insulin-retaining effect of RLX may be helpful in postmenopausal women with decreased insulin reserves or those predisposed to T2DM [89]. Compared with a placebo group, RLX resulted in significantly lower HOMA-IR levels in postmenopausal women with IR in a double-blind randomized trial [90]. RLX may improve glucose levels in postmenopausal women taking GlcN.

E_2 reduces hepatic gluconeogenesis-related genes including phosphoenolpyruvate carboxykinase 1 (*Pck-1*) and glucose 6-phosphatase (*G6Pase*). The effects of E_2 were inhibited in mice lacking liver estrogen receptor α. Hepatic estrogen receptor signaling is related to gluconeogenesis maintenance in men [91]. In the current study, RLX reduced liver PEPCK expression in OVX rats treated with GlcN. The effects may, at least partially, be exerted through liver estrogen receptor α.

Skeletal muscle estrogen receptor α plays a key protective role in the regulation of insulin action and metabolic homeostasis. Compared with estrogen receptor β and GPR30, estrogen receptor α is expressed much more in both rodent and human muscle [92,93]. Total muscular GLUT-4 level determined the IR phenotype in estrogen receptor α–knockout mice [94]. RLX can increase the mRNA expression of GLUT-4 in human skeletal muscle cells [95]. In the present study, RLX enhanced GLUT-4 expression in the soleus muscle. The effect of RLX on GLUT-4 may be through estrogen receptor α.

In a previous study, estrogen receptor α knockout reduced glucose-stimulated insulin secretion in a murine model. In addition, estrogen receptor α knockout in MIN-6 cells enhanced ER stress and apoptosis, and overexpression of estrogen receptor α reduced oxidative stress-induced CHOP expression in MIN-6 cells. Thus, estrogen receptor α targets β-cell apoptosis susceptibility and insulin secretion capacity [96]. We have previously reported that GlcN impairs the insulin secretion of β-cells and enhances β-cell apoptosis through increases in ER stress-related proteins including CHOP, p-ERK, p-EIF2α, and p-JNK in MIN-6 cells, whereas E_2 ameliorates the ER stress caused by GlcN [41]. In the current study, RLX, similarly to E_2, reversed these adverse effects of GlcN. RLX counteracted GlcN-mediated reductions in extracellular insulin secretion, cell viability, and the expression of ER stress-associated proteins. Both ICI 182,780 and MMP inhibited these benefits of RLX, with no statistical differences in extracellular insulin secretion or cell viability, but differences in the expression of ER stress-associated proteins. Our results indicate that RLX ameliorates GlcN-induced dysfunction in pancreatic β-cells through estrogen receptors, mainly estrogen receptor α for cell viability and extracellular insulin secretion, and from increases in ER stress-associated proteins. Further studies are required to explore the mechanism of RLX in ER stress-associated protein suppression.

There was one limitation to our study: we did not stain Ki67 to clarify the roles of proliferation in pancreatic β-cells. Further studies may be required to clarify the roles of proliferation in pancreatic β-cells.

5. Conclusions

GlcN may lead to β-cell dysfunction and apoptosis via diminishing pancreatic β-cell viability and insulin secretion and increasing ER stress-associated protein levels. RLX can reverse the effects of GlcN, but pretreatment with ICI 182,780 and MPP can inhibit the reversal effects of RLX. The effects of RLX are exerted mainly through estrogen receptor α. Uncovering more about the underlying mechanisms requires further study.

Author Contributions: Conceptualization, C.-H.C., H.-T.H., S.-Y.L. and L.K.; data curation, T.-L.C., C.-F.C. and S.-Y.L.; formal analysis, T.-L.C., M.-H.W. and L.K.; funding acquisition, C.-H.C. and L.K.; Investigation, C.-H.C., T.-L.C., H.-T.H. and L.K.; methodology, T.-L.C., C.-F.C. and S.-Y.L.; project administration, C.-H.C. and L.K.; Software, T.-L.C., H.-T.H., C.-F.C. and M.-H.W.; supervision, L.K. and C.-H.C.; validation, H.-T.H., S.-Y.L. and L.K.; writing—original draft, C.-H.C. and T.-L.C.; writing—review and editing, C.-F.C., S.-Y.L., M.-H.W. and L.K. All authors have read and agreed to the published version of the manuscript.

Funding: This study was supported in part by the National Health Research Institute (NHRI-EX101-9935EI) of Taiwan, National Cheng Kung University (NCKUH- 10406014), Kaohsiung Municipal Ta-Tung Hospital (KMTTH-109-R014 and KMTTH-DK(B) 110002-1), Kaohsiung Medical University (NPUST KMU-109-P002, NCTUKMU108-BIO-04, KMU-TC108A02-1, KMU-DK(A)110003 and KMU-DK(B) 110002) and the Minister of Science and Technology, Taiwan (MOST 104-2314-B-006-071-MY2, 108-2314-B-037-059-MY3 and 110-2314-B-037-029-MY3) of Taiwan. The funders had no role in the study design, data collection, and analysis; decision to publish; or preparation of the manuscript.

Institutional Review Board Statement: Not applicable.

Informed Consent Statement: Not applicable.

Data Availability Statement: Data are available on request.

Acknowledgments: We are grateful to W.T. Chang and Y.S. Lin for their assistance in this study. This manuscript was edited by Wallace Academic Editing.

Conflicts of Interest: The authors declare no conflict of interest.

References

1. Jiajue, R.; Qi, X.; Jiang, Y.; Wang, Q.; Wang, W.; Pei, Y.; Wang, X.; Huang, W.; Zheng, X.; Ning, Z.; et al. Incident Fracture Risk in Type 2 Diabetic Postmenopausal Women in Mainland China: Peking Vertebral Fracture Study. *Calcif. Tissue Int.* **2019**, *105*, 466–475. [CrossRef]
2. Cross, M.; Smith, E.; Hoy, D.; Nolte, S.; Ackerman, I.; Fransen, M.; Bridgett, L.; Williams, S.; Guillemin, F.; Hill, C.L.; et al. The global burden of hip and knee osteoarthritis: Estimates from the global burden of disease 2010 study. *Ann. Rheum. Dis.* **2014**, *73*, 1323–1330. [CrossRef]

3. Veronese, N.; Cooper, C.; Reginster, J.Y.; Hochberg, M.; Branco, J.; Bruyère, O.; Chapurlat, R.; Al-Daghri, N.; Dennison, E.; Herrero-Beaumont, G.; et al. Type 2 diabetes mellitus and osteoarthritis. *Semin. Arthritis Rheum.* **2019**, *49*, 9–19. [CrossRef] [PubMed]
4. Hsu, C.H.; Hsu, N.C.; Shih, C.L.; Huang, H.T.; Chen, C.H.; Chou, P.H. Medication-Taking Habit and Outcome of Glucosamine Sulfate for Osteoarthritis Patients Influenced by National Health Insurance Regulations in Taiwan. *J. Clin. Med.* **2019**, *8*, 1734. [CrossRef]
5. Ghouri, A.; Conaghan, P.G. Prospects for Therapies in Osteoarthritis. *Calcif. Tissue Int.* **2020**, 1–12. [CrossRef] [PubMed]
6. Guo, Q.; Hu, H.; Zhou, Y.; Yan, Y.; Wei, X.; Fan, X.; Yang, D.; He, H.; Oh, Y.; Chen, K.; et al. Glucosamine induces increased musclin gene expression through endoplasmic reticulum stress-induced unfolding protein response signaling pathways in mouse skeletal muscle cells. *Food Chem. Toxicol.* **2019**, *125*, 95–105. [CrossRef]
7. Ciaraldi, T.P.; Carter, L.; Nikoulina, S.; Mudaliar, S.; McClain, D.A.; Henry, R.R. Glucosamine regulation of glucose metabolism in cultured human skeletal muscle cells: Divergent effects on glucose transport/phosphorylation and glycogen synthase in non-diabetic and type 2 diabetic subjects. *Endocrinology* **1999**, *140*, 3971–3980. [CrossRef] [PubMed]
8. D'Alessandris, C.; Andreozzi, F.; Federici, M.; Cardellini, M.; Brunetti, A.; Ranalli, M.; Del Guerra, S.; Lauro, D.; Del Prato, S.; Marchetti, P.; et al. Increased O-glycosylation of insulin signaling proteins results in their impaired activation and enhanced susceptibility to apoptosis in pancreatic beta-cells. *FASEB J.* **2004**, *18*, 959–961. [CrossRef] [PubMed]
9. Biggee, B.A.; Blinn, C.M.; Nuite, M.; Silbert, J.E.; McAlindon, T.E. Effects of oral glucosamine sulphate on serum glucose and insulin during an oral glucose tolerance test of subjects with osteoarthritis. *Ann. Rheum. Dis.* **2007**, *66*, 260–262. [CrossRef]
10. Monauni, T.; Zenti, M.G.; Cretti, A.; Daniels, M.C.; Targher, G.; Caruso, B.; Caputo, M.; McClain, D.; Del Prato, S.; Giaccari, A.; et al. Effects of glucosamine infusion on insulin secretion and insulin action in humans. *Diabetes* **2000**, *49*, 926–935. [CrossRef]
11. Muniyappa, R.; Karne, R.J.; Hall, G.; Crandon, S.K.; Bronstein, J.A.; Ver, M.R.; Hortin, G.L.; Quon, M.J. Oral glucosamine for 6 weeks at standard doses does not cause or worsen insulin resistance or endothelial dysfunction in lean or obese subjects. *Diabetes* **2006**, *55*, 3142–3150. [CrossRef] [PubMed]
12. Scroggie, D.A.; Albright, A.; Harris, M.D. The effect of glucosamine-chondroitin supplementation on glycosylated hemoglobin levels in patients with type 2 diabetes mellitus: A placebo-controlled, double-blinded, randomized clinical trial. *Arch. Intern. Med.* **2003**, *163*, 1587–1590. [CrossRef]
13. Mauvais-Jarvis, F.; Clegg, D.J.; Hevener, A.L. The role of estrogens in control of energy balance and glucose homeostasis. *Endocr. Rev.* **2013**, *34*, 309–338. [CrossRef] [PubMed]
14. Le May, C.; Chu, K.; Hu, M.; Ortega, C.S.; Simpson, E.R.; Korach, K.S.; Tsai, M.J.; Mauvais-Jarvis, F. Estrogens protect pancreatic beta-cells from apoptosis and prevent insulin-deficient diabetes mellitus in mice. *Proc. Natl. Acad. Sci. USA* **2006**, *103*, 9232–9237. [CrossRef] [PubMed]
15. Riant, E.; Waget, A.; Cogo, H.; Arnal, J.F.; Burcelin, R.; Gourdy, P. Estrogens protect against high-fat diet-induced insulin resistance and glucose intolerance in mice. *Endocrinology* **2009**, *150*, 2109–2117. [CrossRef]
16. Tiano, J.P.; Delghingaro-Augusto, V.; Le May, C.; Liu, S.; Kaw, M.K.; Khuder, S.S.; Latour, M.G.; Bhatt, S.A.; Korach, K.S.; Najjar, S.M.; et al. Estrogen receptor activation reduces lipid synthesis in pancreatic islets and prevents beta cell failure in rodent models of type 2 diabetes. *J. Clin. Investig.* **2011**, *121*, 3331–3342. [CrossRef]
17. Liu, S.; Le May, C.; Wong, W.P.; Ward, R.D.; Clegg, D.J.; Marcelli, M.; Korach, K.S.; Mauvais-Jarvis, F. Importance of extranuclear estrogen receptor-alpha and membrane G protein-coupled estrogen receptor in pancreatic islet survival. *Diabetes* **2009**, *58*, 2292–2302. [CrossRef]
18. Rodriguez, C.; Patel, A.V.; Calle, E.E.; Jacob, E.J.; Thun, M.J. Estrogen replacement therapy and ovarian cancer mortality in a large prospective study of US women. *JAMA* **2001**, *285*, 1460–1465. [CrossRef]
19. Conley, R.B.; Adib, G.; Adler, R.A.; Akesson, K.E.; Alexander, I.M.; Amenta, K.C.; Blank, R.D.; Brox, W.T.; Carmody, E.E.; Chapman-Novakofski, K.; et al. Secondary Fracture Prevention: Consensus Clinical Recommendations from a Multistakeholder Coalition. *J. Bone Miner. Res.* **2020**, *35*, 36–52. [CrossRef]
20. Tatangelo, G.; Watts, J.; Lim, K.; Connaughton, C.; Abimanyi-Ochom, J.; Borgstrom, F.; Nicholson, G.C.; Shore-Lorenti, C.; Stuart, A.L.; Iuliano-Burns, S.; et al. The Cost of Osteoporosis, Osteopenia, and Associated Fractures in Australia in 2017. *J. Bone Miner. Res.* **2019**, *34*, 616–625. [CrossRef]
21. Shieh, A.; Greendale, G.A.; Cauley, J.A.; Karvonen-Gutierrez, C.; Crandall, C.J.; Karlamangla, A.S. Estradiol and Follicle-Stimulating Hormone as Predictors of Onset of Menopause Transition-Related Bone Loss in Pre- and Perimenopausal Women. *J. Bone Miner. Res.* **2019**, *34*, 2246–2253. [CrossRef] [PubMed]
22. Lewiecki, E.M.; Binkley, N.; Bilezikian, J.P. Treated Osteoporosis Is Still Osteoporosis. *J. Bone Miner. Res.* **2019**, *34*, 605–606. [CrossRef]
23. Crandall, C.J.; Larson, J.; Manson, J.E.; Cauley, J.A.; LaCroix, A.Z.; Wactawski-Wende, J.; Datta, M.; Sattari, M.; Schousboe, J.T.; Leslie, W.D.; et al. A Comparison of US and Canadian Osteoporosis Screening and Treatment Strategies in Postmenopausal Women. *J. Bone Miner. Res.* **2019**, *34*, 607–615. [CrossRef]
24. Chang, P.Y.; Feldman, D.; Stefanick, M.L.; McDonnell, D.P.; Thompson, B.M.; McDonald, J.G.; Lee, J.S. 27-Hydroxycholesterol, an Endogenous SERM, and Risk of Fracture in Postmenopausal Women: A Nested Case-Cohort Study in the Women's Health Initiative. *J. Bone Miner. Res.* **2019**, *34*, 59–66. [CrossRef]

25. Wu, C.H.; Hung, W.C.; Chang, I.L.; Tsai, T.T.; Chang, Y.F.; McCloskey, E.V.; Watts, N.B.; McClung, M.R.; Huang, C.F.; Chen, C.H.; et al. Pharmacologic intervention for prevention of fractures in osteopenic and osteoporotic postmenopausal women: Systemic review and meta-analysis. *Bone Rep.* **2020**, *13*, 100729. [CrossRef]
26. Wu, C.H.; Chang, Y.F.; Chen, C.H.; Lewiecki, E.M.; Wuster, C.; Reid, I.; Tsai, K.S.; Matsumoto, T.; Mercado-Asis, L.B.; Chan, D.C.; et al. Consensus Statement on the Use of Bone Turnover Markers for Short-Term Monitoring of Osteoporosis Treatment in the Asia-Pacific Region. *J. Clin. Densitom.* **2021**, *24*, 3–13. [CrossRef] [PubMed]
27. Centofanti, F.; Santoro, M.; Marini, M.; Visconti, V.V.; Rinaldi, A.M.; Celi, M.; D'Arcangelo, G.; Novelli, G.; Orlandi, A.; Tancredi, V.; et al. Identification of Aberrantly-Expressed Long Non-Coding RNAs in Osteoblastic Cells from Osteoporotic Patients. *Biomedicines* **2020**, *8*, 65. [CrossRef]
28. Taylor, E.A.; Donnelly, E.; Yao, X.; Johnson, M.L.; Amugongo, S.K.; Kimmel, D.B.; Lane, N.E. Sequential Treatment of Estrogen Deficient, Osteopenic Rats with Alendronate, Parathyroid Hormone (1-34), or Raloxifene Alters Cortical Bone Mineral and Matrix Composition. *Calcif. Tissue Int.* **2020**, *106*, 303–314. [CrossRef]
29. Hozumi, Y.; Kawano, M.; Jordan, V.C. In vitro study of the effect of raloxifene on lipid metabolism compared with tamoxifen. *Eur. J. Endocrinol.* **2000**, *143*, 427–430. [CrossRef]
30. Tiano, J.; Mauvais-Jarvis, F. Selective estrogen receptor modulation in pancreatic beta-cells and the prevention of type 2 diabetes. *Islets* **2012**, *4*, 173–176. [CrossRef] [PubMed]
31. Bernales, S.; Papa, F.R.; Walter, P. Intracellular signaling by the unfolded protein response. *Annu. Rev. Cell Dev. Biol.* **2006**, *22*, 487–508. [CrossRef] [PubMed]
32. Chiu, C.F.; Lai, G.Y.; Chen, C.H.; Chiu, C.C.; Hung, S.W.; Chang, C.F. 6,7-Dihydroxy-2-(4′-hydroxyphenyl)naphthalene induces HCT116 cell apoptosis through activation of endoplasmic reticulum stress and the extrinsic apoptotic pathway. *Drug Des. Dev. Ther.* **2019**, *13*, 1609–1621. [CrossRef]
33. Gundamaraju, R.; Lu, W.; Azimi, I.; Eri, R.; Sohal, S.S. Endogenous Anti-Cancer Candidates in GPCR, ER Stress, and EMT. *Biomedicines* **2020**, *8*, 402. [CrossRef] [PubMed]
34. Osman, A.; Benameur, T.; Korashy, H.M.; Zeidan, A.; Agouni, A. Interplay between Endoplasmic Reticulum Stress and Large Extracellular Vesicles (Microparticles) in Endothelial Cell Dysfunction. *Biomedicines* **2020**, *8*, 409. [CrossRef] [PubMed]
35. Marchetti, P.; Bugliani, M.; Lupi, R.; Marselli, L.; Masini, M.; Boggi, U.; Filipponi, F.; Weir, G.C.; Eizirik, D.L.; Cnop, M. The endoplasmic reticulum in pancreatic beta cells of type 2 diabetes patients. *Diabetologia* **2007**, *50*, 2486–2494. [CrossRef] [PubMed]
36. Laybutt, D.R.; Preston, A.M.; Akerfeldt, M.C.; Kench, J.G.; Busch, A.K.; Biankin, A.V.; Biden, T.J. Endoplasmic reticulum stress contributes to beta cell apoptosis in type 2 diabetes. *Diabetologia* **2007**, *50*, 752–763. [CrossRef] [PubMed]
37. Galli, A.; Marciani, P.; Marku, A.; Ghislanzoni, S.; Bertuzzi, F.; Rossi, R.; Di Giancamillo, A.; Castagna, M.; Perego, C. Verbascoside Protects Pancreatic beta-Cells against ER-Stress. *Biomedicines* **2020**, *8*, 582. [CrossRef] [PubMed]
38. Hwang, M.S.; Baek, W.K. Glucosamine induces autophagic cell death through the stimulation of ER stress in human glioma cancer cells. *Biochem. Biophys. Res. Commun.* **2010**, *399*, 111–116. [CrossRef]
39. Morin, M.J.; Porter, C.W.; McKernan, P.; Bernacki, R.J. The biochemical and ultrastructural effects of tunicamycin and D-glucosamine in L1210 leukemic cells. *J. Cell Physiol.* **1983**, *114*, 162–172. [CrossRef]
40. Kang, L.; Chen, C.H.; Cheng, Y.C.; Chang, C.H.; Lee, C.T.; Chang, J.K.; Cheng, J.T.; Chang, F.M. Glucosamine-induced insulin resistance in ovariectomized rats is relevant to decreasing the expression of glucose transport protein subtype 4 in the skeletal muscle and in increasing the size of pancreatic islets. *Menopause* **2012**, *19*, 496–502. [CrossRef]
41. Kang, L.; Chen, C.H.; Wu, M.H.; Chang, J.K.; Chang, F.M.; Cheng, J.T. 17beta-estradiol protects against glucosamine-induced pancreatic beta-cell dysfunction. *Menopause* **2014**, *21*, 1239–1248. [CrossRef] [PubMed]
42. Ho, M.L.; Chen, Y.H.; Liao, H.J.; Chen, C.H.; Hung, S.H.; Lee, M.J.; Fu, Y.C.; Wang, Y.H.; Wang, G.J.; Chang, J.K. Simvastatin increases osteoblasts and osteogenic proteins in ovariectomized rats. *Eur. J. Clin. Investig.* **2009**, *39*, 296–303. [CrossRef]
43. Juan, Y.S.; Chuang, S.M.; Long, C.Y.; Chen, C.H.; Levin, R.M.; Liu, K.M.; Huang, C.H. Neuroprotection of green tea catechins on surgical menopause-induced overactive bladder in a rat model. *Menopause* **2012**, *19*, 346–354. [CrossRef] [PubMed]
44. Chen, C.H.; Kang, L.; Lin, R.W.; Fu, Y.C.; Lin, Y.S.; Chang, J.K.; Chen, H.T.; Chen, C.H.; Lin, S.Y.; Wang, G.J.; et al. (-)-Epigallocatechin-3-gallate improves bone microarchitecture in ovariectomized rats. *Menopause* **2013**, *20*, 687–694. [CrossRef]
45. Chen, C.H.; Huang, T.H.; Cheng, T.L.; Chang, C.F.; Wang, C.Z.; Wu, M.H.; Kang, L. Exercise training ameliorates glucosamine-induced insulin resistance in ovariectomized rats. *Menopause* **2017**, *24*, 617–623. [CrossRef]
46. Chen, C.H.; Ho, M.L.; Chang, L.H.; Kang, L.; Lin, Y.S.; Lin, S.Y.; Wu, S.C.; Chang, J.K. Parathyroid hormone-(1-34) ameliorated knee osteoarthritis in rats via autophagy. *J. Appl. Physiol.* **2018**, *124*, 1177–1185. [CrossRef] [PubMed]
47. Chou, H.C.; Chen, C.H.; Chou, L.Y.; Cheng, T.L.; Kang, L.; Chuang, S.C.; Lin, Y.S.; Ho, M.L.; Wang, Y.H.; Lin, S.Y.; et al. Discoidin Domain Receptors 1 Inhibition Alleviates Osteoarthritis via Enhancing Autophagy. *Int. J. Mol. Sci.* **2020**, *21*, 6991. [CrossRef] [PubMed]
48. Chen, C.H.; Kang, L.; Chang, L.H.; Cheng, T.L.; Lin, S.Y.; Wu, S.C.; Lin, Y.S.; Chuang, S.C.; Lee, T.C.; Chang, J.K.; et al. Intra-articular low-dose parathyroid hormone (1-34) improves mobility and articular cartilage quality in a preclinical age-related knee osteoarthritis model. *Bone Jt. Res.* **2021**, *10*, 514–525. [CrossRef] [PubMed]
49. Lee, T.C.; Lee, Y.L.; Chen, J.C.; Chen, C.H.; Ho, P.S. Impact of type 2 diabetes on postoperative outcome after hip fracture: Nationwide population-based study in Taiwan. *BMJ Open Diabetes Res. Care* **2020**, *8*, e000843. [CrossRef] [PubMed]

50. Chen, C.H.; Huang, P.J.; Huang, H.T.; Lin, S.Y.; Wang, H.Y.; Fang, T.J.; Lin, Y.C.; Ho, C.J.; Lee, T.C.; Lu, Y.M.; et al. Impact of orthogeriatric care, comorbidity, and complication on 1-year mortality in surgical hip fracture patients: An observational study. *Medicine* **2019**, *98*, e17912. [CrossRef] [PubMed]
51. Chen, C.L.; Chen, C.M.; Wang, C.Y.; Ko, P.W.; Chen, C.H.; Hsieh, C.P.; Chiu, H.C. Frailty is Associated with an Increased Risk of Major Adverse Outcomes in Elderly Patients Following Surgical Treatment of Hip Fracture. *Sci. Rep.* **2019**, *9*, 19135. [CrossRef] [PubMed]
52. Chen, C.H.; Elsalmawy, A.H.; Ish-Shalom, S.; Lim, S.J.; Al-Ali, N.S.; Cunha-Borges, J.L.; Yang, H.; Casas, N.; Altan, L.; Moll, T.; et al. Study description and baseline characteristics of the population enrolled in a multinational, observational study of teriparatide in postmenopausal women with osteoporosis: The Asia and Latin America Fracture Observational Study (ALAFOS). *Curr. Med. Res. Opin.* **2019**, *35*, 1041–1049. [CrossRef] [PubMed]
53. Borhan, S.; Papaioannou, A.; Gajic-Veljanoski, O.; Kennedy, C.; Ioannidis, G.; Berger, C.; Goltzman, D.; Josse, R.; Kovacs, C.S.; Hanley, D.A.; et al. Incident Fragility Fractures Have a Long-Term Negative Impact on Health-Related Quality of Life of Older People: The Canadian Multicentre Osteoporosis Study. *J. Bone Miner. Res.* **2019**, *34*, 838–848. [CrossRef] [PubMed]
54. Bliuc, D.; Tran, T.; van Geel, T.; Adachi, J.D.; Berger, C.; van den Bergh, J.; Eisman, J.A.; Geusens, P.; Goltzman, D.; Hanley, D.A.; et al. Reduced Bone Loss Is Associated With Reduced Mortality Risk in Subjects Exposed to Nitrogen Bisphosphonates: A Mediation Analysis. *J. Bone Miner. Res.* **2019**, *34*, 2001–2011. [CrossRef] [PubMed]
55. Anastasilakis, A.D.; Papapoulos, S.E.; Polyzos, S.A.; Appelman-Dijkstra, N.M.; Makras, P. Zoledronate for the Prevention of Bone Loss in Women Discontinuing Denosumab Treatment. A Prospective 2-Year Clinical Trial. *J. Bone Miner. Res.* **2019**, *34*, 2220–2228. [CrossRef] [PubMed]
56. Osagie-Clouard, L.; Sanghani-Kerai, A.; Coathup, M.; Meeson, R.; Briggs, T.; Blunn, G. The influence of parathyroid hormone 1-34 on the osteogenic characteristics of adipose- and bone-marrow-derived mesenchymal stem cells from juvenile and ovarectomized rats. *Bone Jt. Res.* **2019**, *8*, 397–404. [CrossRef] [PubMed]
57. Wu, Y.Z.; Huang, H.T.; Cheng, T.L.; Lu, Y.M.; Lin, S.Y.; Ho, C.J.; Lee, T.C.; Hsu, C.H.; Huang, P.J.; Huang, H.H.; et al. Application of microrna in human osteoporosis and fragility fracture: A systemic review of literatures. *Int. J. Mol. Sci.* **2021**, *22*, 5232. [CrossRef] [PubMed]
58. Saad, R.K.; Harb, H.; Bou-Orm, I.R.; Ammar, W.; El-Hajj Fuleihan, G. Secular Trends of Hip Fractures in Lebanon, 2006 to 2017: Implications for Clinical Practice and Public Health Policy in the Middle East Region. *J. Bone Miner. Res.* **2020**, *35*, 71–80. [CrossRef]
59. Li, S.; Mao, Y.; Zhou, F.; Yang, H.; Shi, Q.; Meng, B. Gut microbiome and osteoporosis: A review. *Bone Jt. Res.* **2020**, *9*, 524–530. [CrossRef]
60. Huang, H.T.; Cheng, T.L.; Lin, S.Y.; Ho, C.J.; Chyu, J.Y.; Yang, R.S.; Chen, C.H.; Shen, C.L. Osteoprotective Roles of Green Tea Catechins. *Antioxidants* **2020**, *9*, 1136. [CrossRef]
61. Chou, Y.S.; Jiang, H.J.; Chen, C.H.; Ho, P.S.; Lee, T.C. Proton pump inhibitor use and risk of hip fracture in patients with type 2 diabetes. *Sci. Rep.* **2020**, *10*, 1–8. [CrossRef] [PubMed]
62. Dahl, C.; Holvik, K.; Meyer, H.E.; Stigum, H.; Solbakken, S.M.; Schei, B.; Sogaard, A.J.; Omsland, T.K. Increased Mortality in Hip Fracture Patients Living Alone: A NOREPOS Study. *J. Bone Miner. Res.* **2021**, *36*, 480–488. [CrossRef]
63. Chen, C.H.; Lim, S.J.; Oh, J.K.; Huang, T.W.; Zeng, Y.H.; Wu, M.T.; Yang, H.L.; Cheung, J.P.; Kim, J.W.; Han, J.H.; et al. Teriparatide in East Asian Postmenopausal Women with Osteoporosis in a Real-World Setting: A Baseline Analysis of the Asia and Latin America Fracture Observational Study (ALAFOS). *Clin. Interv. Aging* **2020**, *15*, 111–121. [CrossRef]
64. Lin, S.Y.; Huang, H.T.; Chou, S.H.; Ho, C.J.; Liu, Z.M.; Chen, C.H.; Lu, C.C. The Safety of Continuing Antiplatelet Medication Among Elderly Patients Undergoing Urgent Hip Fracture Surgery. *Orthopedics* **2019**, *42*, 268–274. [CrossRef] [PubMed]
65. Chan, D.D.; Chang, L.Y.; Akesson, K.E.; Mitchell, P.; Chen, C.H.; Lewiecki, E.M.; Lee, J.K.; Lau, T.C.; Songpatanasilp, T.; Lee, K.B.; et al. Consensus on best practice standards for Fracture Liaison Service in the Asia-Pacific region. *Arch. Osteoporos.* **2018**, *13*, 1–6. [CrossRef] [PubMed]
66. Chang, L.Y.; Tsai, K.S.; Peng, J.K.; Chen, C.H.; Lin, G.T.; Lin, C.H.; Tu, S.T.; Mao, I.C.; Gau, Y.L.; Liu, H.C.; et al. The development of Taiwan Fracture Liaison Service network. *Osteoporos. Sarcopenia* **2018**, *4*, 47–52. [CrossRef]
67. Chen, C.H.; Elsalmawy, A.H.; Ish-Shalom, S.; Lim, S.J.; AlAli, N.S.; Cunha-Borges, J.L.; Yang, H.; Casas, N.; Altan, L.; Belaya, Z.; et al. The Effect of Teriparatide Treatment on the Risk of Fragility Fractures in Postmenopausal Women with Osteoporosis: Results from the Asian and Latin America Fracture Observational Study (ALAFOS). *Calcif. Tissue Int.* **2021**, in press. [CrossRef]
68. Lisk, R.; Yeong, K.; Fluck, D.; Fry, C.H.; Han, T.S. The Ability of the Nottingham Hip Fracture Score to Predict Mobility, Length of Stay and Mortality in Hospital, and Discharge Destination in Patients Admitted with a Hip Fracture. *Calcif. Tissue Int.* **2020**, *107*, 319–326. [CrossRef]
69. Mortensen, S.J.; Mohamadi, A.; Wright, C.L.; Chan, J.J.; Weaver, M.J.; von Keudell, A.; Nazarian, A. Medications as a Risk Factor for Fragility Hip Fractures: A Systematic Review and Meta-analysis. *Calcif. Tissue Int.* **2020**, *107*, 1–9. [CrossRef]
70. Tei, R.M.H.; Ramlau-Hansen, C.H.; Plana-Ripoll, O.; Brink, O.; Langdahl, B.L. OFELIA: Prevalence of Osteoporosis in Fragility Fracture Patients. *Calcif. Tissue Int.* **2019**, *104*, 102–114. [CrossRef] [PubMed]
71. Silverman, S.; Langdahl, B.L.; Fujiwara, S.; Saag, K.; Napoli, N.; Soen, S.; Enomoto, H.; Melby, T.E.; Disch, D.P.; Marin, F.; et al. Reduction of Hip and Other Fractures in Patients Receiving Teriparatide in Real-World Clinical Practice: Integrated Analysis of Four Prospective Observational Studies. *Calcif. Tissue Int.* **2019**, *104*, 193–200. [CrossRef] [PubMed]

72. Chen, L.R.; Ko, N.Y.; Chen, K.H. Medical Treatment for Osteoporosis: From Molecular to Clinical Opinions. *Int. J. Mol. Sci.* **2019**, *20*, 2213. [CrossRef]
73. Macias, I.; Alcorta-Sevillano, N.; Rodriguez, C.I.; Infante, A. Osteoporosis and the Potential of Cell-Based Therapeutic Strategies. *Int. J. Mol. Sci.* **2020**, *21*, 1653. [CrossRef] [PubMed]
74. Ukon, Y.; Makino, T.; Kodama, J.; Tsukazaki, H.; Tateiwa, D.; Yoshikawa, H.; Kaito, T. Molecular-Based Treatment Strategies for Osteoporosis: A Literature Review. *Int. J. Mol. Sci.* **2019**, *20*, 2557. [CrossRef]
75. Hsu, C.Y.; Chen, L.R.; Chen, K.H. Osteoporosis in Patients with Chronic Kidney Diseases: A Systemic Review. *Int. J. Mol. Sci.* **2020**, *21*, 6864. [CrossRef] [PubMed]
76. Chandra, A.; Rajawat, J. Skeletal Aging and Osteoporosis: Mechanisms and Therapeutics. *Int. J. Mol. Sci.* **2021**, *22*, 3553. [CrossRef]
77. Wang, J.; Zhou, L.; Zhang, Y.; Huang, L.; Shi, Q. Mesenchymal stem cells—A promising strategy for treating knee osteoarthritis. *Bone Jt. Res.* **2020**, *9*, 719–728. [CrossRef]
78. Luk, H.Y.; Appell, C.; Chyu, M.C.; Chen, C.H.; Wang, C.Y.; Yang, R.S.; Shen, C.L. Impacts of Green Tea on Joint and Skeletal Muscle Health: Prospects of Translational Nutrition. *Antioxidants* **2020**, *9*, 1050. [CrossRef] [PubMed]
79. Hain, B.A.; Jude, B.; Xu, H.; Smuin, D.M.; Fox, E.J.; Elfar, J.C.; Waning, D.L. Zoledronic Acid Improves Muscle Function in Healthy Mice Treated with Chemotherapy. *J. Bone Miner. Res.* **2020**, *35*, 368–381. [CrossRef]
80. Chakhtoura, M.; Dagher, H.; Sharara, S.; Ajjour, S.; Chamoun, N.; Cauley, J.; Mahfoud, Z.; Boudreau, R.; El Hajj Fuleihan, G. Systematic review of major osteoporotic fracture to hip fracture incidence rate ratios worldwide: Implications for Fracture Risk Assessment Tool (FRAX)-derived estimates. *J. Bone Miner. Res.* **2021**. [CrossRef]
81. Pickering, M.E.; Chapurlat, R. Where Two Common Conditions of Aging Meet: Osteoarthritis and Sarcopenia. *Calcif. Tissue Int.* **2020**, *107*, 203–211. [CrossRef] [PubMed]
82. Chu, L.; Liu, X.; He, Z.; Han, X.; Yan, M.; Qu, X.; Li, X.; Yu, Z. Articular Cartilage Degradation and Aberrant Subchondral Bone Remodeling in Patients with Osteoarthritis and Osteoporosis. *J. Bone Miner. Res.* **2020**, *35*, 505–515. [CrossRef] [PubMed]
83. Shao, L.T.; Gou, Y.; Fang, J.K.; Hu, Y.P.; Lian, Q.Q.; Zhang, Y.Y.; Wang, Y.D.; Tian, F.M.; Zhang, L. Parathyroid hormone (1-34) ameliorates cartilage degeneration and subchondral bone deterioration in collagenase-induced osteoarthritis model in mice. *Bone Jt. Res.* **2020**, *9*, 675–688. [CrossRef]
84. Lingard, E.A.; Mitchell, S.Y.; Francis, R.M.; Rawlings, D.; Peaston, R.; Birrell, F.N.; McCaskie, A.W. The prevalence of osteoporosis in patients with severe hip and knee osteoarthritis awaiting joint arthroplasty. *Age Ageing* **2010**, *39*, 234–239. [CrossRef]
85. Wang, S.P.; Wu, P.K.; Lee, C.H.; Shih, C.M.; Chiu, Y.C.; Hsu, C.E. Association of osteoporosis and varus inclination of the tibial plateau in postmenopausal women with advanced osteoarthritis of the knee. *BMC Musculoskelet. Disord.* **2021**, *22*, 1–8. [CrossRef]
86. Kasher, M.; Williams, F.M.K.; Freidin, M.B.; Cherny, S.; Livshits, G. An in-depth study of the associations between osteoarthritis- and osteoporosis-related phenotypes at different skeletal locations. *Osteoporos. Int.* **2020**, *31*, 2197–2208. [CrossRef]
87. Bei, M.J.; Tian, F.M.; Xiao, Y.P.; Cao, X.H.; Liu, N.; Zheng, Z.Y.; Dai, M.W.; Wang, W.Y.; Song, H.P.; Zhang, L. Raloxifene retards cartilage degradation and improves subchondral bone micro-architecture in ovariectomized rats with patella baja-induced-patellofemoral joint osteoarthritis. *Osteoarthr. Cartil.* **2020**, *28*, 344–355. [CrossRef] [PubMed]
88. Ebrahimi, M.N.; Khaksari, M.; Sepehri, G.; Karam, G.A.; Raji-Amirhasani, A.; Azizian, H. The effects of alone and combination tamoxifen, raloxifene and estrogen on lipid profile and atherogenic index of ovariectomized type 2 diabetic rats. *Life Sci.* **2020**, *263*, 118573. [CrossRef]
89. Nagamani, M.; Szymajda, A.; Sepilian, V.; Urban, R.J.; Gilkison, C. Effects of raloxifene on insulin sensitivity, beta-cell function, and hepatic insulin extraction in normal postmenopausal women. *Fertil. Steril.* **2008**, *89*, 614–619. [CrossRef] [PubMed]
90. Grover-Paez, F.; Zavalza-Gomez, A.B.; Anaya-Prado, R. Raloxifene modifies the insulin sensitivity and lipid profile of post-menopausal insulin resistant women. *Gynecol. Endocrinol.* **2013**, *29*, 674–677. [CrossRef]
91. Qiu, S.; Vazquez, J.T.; Boulger, E.; Liu, H.; Xue, P.; Hussain, M.A.; Wolfe, A. Hepatic estrogen receptor α is critical for regulation of gluconeogenesis and lipid metabolism in males. *Sci. Rep.* **2017**, *7*, 1–12. [CrossRef] [PubMed]
92. Hevener, A.L.; Ribas, V.; Moore, T.M.; Zhou, Z. The Impact of Skeletal Muscle ERα on Mitochondrial Function and Metabolic Health. *Endocrinology* **2020**, *161*, bqz017. [CrossRef] [PubMed]
93. Ribas, V.; Drew, B.G.; Zhou, Z.; Phun, J.; Kalajian, N.Y.; Soleymani, T.; Daraei, P.; Widjaja, K.; Wanagat, J.; de Aguiar Vallim, T.Q.; et al. Skeletal muscle action of estrogen receptor alpha is critical for the maintenance of mitochondrial function and metabolic homeostasis in females. *Sci. Transl. Med.* **2016**, *8*, 334ra54. [CrossRef] [PubMed]
94. Bryzgalova, G.; Gao, H.; Ahren, B.; Zierath, J.R.; Galuska, D.; Steiler, T.L.; Dahlman-Wright, K.; Nilsson, S.; Gustafsson, J.A.; Efendic, S.; et al. Evidence that oestrogen receptor-alpha plays an important role in the regulation of glucose homeostasis in mice: Insulin sensitivity in the liver. *Diabetologia* **2006**, *49*, 588–597. [CrossRef] [PubMed]
95. Dieli-Conwright, C.M.; Spektor, T.M.; Rice, J.C.; Todd Schroeder, E. Oestradiol and SERM treatments influence oestrogen receptor coregulator gene expression in human skeletal muscle cells. *Acta Physiol.* **2009**, *197*, 187–196. [CrossRef]
96. Zhou, Z.; Ribas, V.; Rajbhandari, P.; Drew, B.G.; Moore, T.M.; Fluitt, A.H.; Reddish, B.R.; Whitney, K.A.; Georgia, S.; Vergnes, L.; et al. Estrogen receptor alpha protects pancreatic beta-cells from apoptosis by preserving mitochondrial function and suppressing endoplasmic reticulum stress. *J. Biol. Chem.* **2018**, *293*, 4735–4751. [CrossRef]

biomedicines

MDPI

Article

High Prevalence of Insulin Resistance in Asymptomatic Patients with Acute Intermittent Porphyria and Liver-Targeted Insulin as a Novel Therapeutic Approach

Isabel Solares [1,†], Laura Izquierdo-Sánchez [2,3,†], Montserrat Morales-Conejo [1,4], Daniel Jericó [2,5], Francisco Javier Castelbón [1], Karol Marcela Córdoba [2,5], Ana Sampedro [2,5], Carlos Lumbreras [1], María Jesús Moreno-Aliaga [5,6,7], Rafael Enríquez de Salamanca [1], Pedro Berraondo [5,8,9] and Antonio Fontanellas [2,5,10,*]

1 Reference Center for Inherited Metabolic Disease-MetabERN, Department of Internal Medicine, University Hospital 12 de Octubre, UCM, 28041 Madrid, Spain; isolares@alumni.unav.es (I.S.); montserrat.morales@salud.madrid.org (M.M.-C.); fjcastelbon@yahoo.es (F.J.C.); clumbrerasb@gmail.com (C.L.); salamanca@med.ucm.es (R.E.d.S.)
2 Hepatology Program, Cima Universidad de Navarra, 31008 Pamplona, Spain; laura.izquierdo@biodonostia.org (L.I.-S.); djerico@alumni.unav.es (D.J.); kcordoba@alumni.unav.es (K.M.C.); asampedro@unav.es (A.S.)
3 Department of Liver and Gastrointestinal Diseases, Biodonostia Health Research Institute, Donostia University Hospital, University of the Basque Country (UPV/EHU), 20014 San Sebastian, Spain
4 Grupo de Enfermedades Mitocondriales y Neuromusculares, Instituto de Investigación Hospital 12 de Octubre (i+12), Centro de Investigación Biomédica en Red de Enfermedades Raras (CIBERER), Instituto de Salud Carlos III, 28029 Madrid, Spain
5 Instituto de Investigación Sanitaria de Navarra (IdiSNA), 31008 Pamplona, Spain; mjmoreno@unav.es (M.J.M.-A.); pberraondol@unav.es (P.B.)
6 Center for Nutrition Research and Department of Nutrition, Food Science and Physiology, University of Navarra, 31008 Pamplona, Spain
7 CIBERobn Physiopathology of Obesity and Nutrition, Carlos III Health Institute, 28029 Madrid, Spain
8 Program of Immunology and Immunotherapy, Cima Universidad de Navarra, 31008 Pamplona, Spain
9 Centro de Investigación Biomédica en Red de Cáncer, CIBERonc, Instituto de Salud Carlos III, 28029 Madrid, Spain
10 Centro de Investigación Biomédica en Red de Enfermedades Hepáticas y Digestivas (CIBERehd), Instituto de Salud Carlos III, 28029 Madrid, Spain
* Correspondence: afontanellas@unav.es; Tel.: +34-948194700
† These two authors contributed equally.

Citation: Solares, I.; Izquierdo-Sánchez, L.; Morales-Conejo, M.; Jericó, D.; Castelbón, F.J.; Córdoba, K.M.; Sampedro, A.; Lumbreras, C.; Moreno-Aliaga, M.J.; Enríquez de Salamanca, R.; et al. High Prevalence of Insulin Resistance in Asymptomatic Patients with Acute Intermittent Porphyria and Liver-Targeted Insulin as a Novel Therapeutic Approach. *Biomedicines* **2021**, *9*, 255. https://doi.org/10.3390/biomedicines9030255

Academic Editor: Susan J. Burke

Received: 19 January 2021
Accepted: 27 February 2021
Published: 5 March 2021

Abstract: Acute porphyria attacks are associated with the strong up-regulation of hepatic heme synthesis and over-production of neurotoxic heme precursors. First-line therapy is based on carbohydrate loading. However, altered glucose homeostasis could affect its efficacy. Our first aim was to investigate the prevalence of insulin resistance (IR) in an observational case-control study including 44 Spanish patients with acute intermittent porphyria (AIP) and 55 age-, gender- and BMI-matched control volunteers. Eight patients (18.2%) and one control (2.3%, $p = 0.01$) showed a high HOMA-IR index (cut-off ≥ 3.4). Patients with IR and hyperinsulinemia showed clinically stable disease. Thus, the second aim was to evaluate the effect of the co-administration of glucose and a fast-acting or new liver-targeted insulin (the fusion protein of insulin and apolipoprotein A-I, Ins-ApoAI) in AIP mice. The combination of glucose and the Ins-ApoAI promoted partial but sustained protection against hepatic heme synthesis up-regulation compared with glucose alone or co-injected with fast-acting insulin. In a prevention study, Ins-ApoAI improved symptoms associated with a phenobarbital-induced attack but maintained high porphyrin precursor excretion, probably due to the induction of hepatic mitochondrial biogenesis mediated by apolipoprotein A-I. In conclusion, a high prevalence of IR and hyperinsulinemia was observed in patients with AIP. The experimental data provide proof-of-concept for liver-targeted insulin as a way of enhancing glucose therapy for AIP.

Keywords: acute intermittent porphyria; carbohydrate loading therapy; insulin resistance; hyperinsulinemia; fast-acting insulin; experimental liver-targeted insulin

1. Introduction

Acute intermittent porphyria (AIP, MIM 176000) is an autosomal dominant metabolic disease caused by a partial deficiency of the hepatic porphobilinogen deaminase (PBGD, EC 4.3.1.8), the third enzyme of the heme synthesis pathway [1–3]. The prevalence of the genetic defect is considered high (~1 in 1700 individuals), although the clinical penetrance of the disease is very low (≤1%) [4]. The main clinical manifestation of AIP is acute neurovisceral attacks characterized by abdominal pain, often accompanied by severe fatigue, nausea, vomiting, constipation, and appetite loss. Patients may also have hypertension, tachycardia, trouble sleeping, and anxiety. Severe attacks can lead to hyponatremia, seizures, sensory loss, or motor neuropathy [5].

Acute attacks are associated with the high accumulation of porphyrin precursors, δ-aminolevulinic acid (ALA) and porphobilinogen (PBG), when hepatic heme synthesis is up-regulated by endogenous or exogenous factors, such as fasting, hormonal fluctuations during the menstrual cycle, infection, stress, smoking, alcohol, or exposure to porphyrinogenic drugs. All these factors strongly induce the transcription of the first and rate-limiting enzyme in the heme synthesis pathway, *ALA Synthase 1* (ALAS1, EC 2.3.1.37) in hepatocytes [1,6–8]. Substances such as endogenous intermediates and xenobiotics, metabolized by the CYP450 enzyme system in the liver, can result in the enhanced transcription of both specific CYP450 and *ALAS1* through nuclear receptors binding a cis-acting regulatory drug-responsive sequence (ALAS drug-responsive element, ADRES) located upstream of the promoter of such genes [9]. Fasting, acting through the nuclear receptor peroxisome proliferator-activated receptor-alpha (PPARα) and the PPAR-gamma coactivator-1 alpha (PGC-1α), can also induce a direct transcriptional up-regulation of hepatic *ALAS1* [10,11]. A third potential mechanism of hepatic *ALAS1* up-regulation is associated with the inducible and highly dynamic heme oxygenase-1 (HMOX1, EC 1.14.14.18), which produces an active hepatic heme turnover in response to cell stress or inflammatory and infectious diseases [12]. Finally, the exacerbated abundance of intracellular heme decreases the stability of ALAS1 mRNA, inhibits the uptake of pre-ALAS into mitochondria, and reduces the stability of the mitochondrial ALAS1 protein via Lon Peptidase 1 (LONP1)-mediated degradation [13].

The first-line therapeutic approach for acute attacks is based on both carbohydrate loading and intravenous hemin therapy to down-regulate hepatic *ALAS1* transcription. Hemin replenishes the regulatory heme pool in hepatocytes and is more effective than glucose [7,14]. Carbohydrate loading (300 to 500 g/day), based on oral or intravenous glucose infusions, has been used to alleviate relapses or milder attacks with low requirements for narcotics and the absence of hyponatremia or motor impairment [6,7]. However, recent studies have not shown a clear preventive effect. While high carbohydrate intake and the subsequent increase of insulin levels were associated with lower biochemical disease activity in a case-control trial performed in northern Norway [15], prophylactic infusion of dextrose or a carbohydrate-rich diet have yielded inconclusive clinical findings in epidemiological studies in the USA [16].

Early studies reported an abnormal oral glucose tolerance test (GTT) and hyperinsulinemia in patients with AIP [17,18], which resemble the findings in cellular insulin resistance (IR). More recently, we have reported a delayed GTT, serum hyperinsulinemia, and abnormal carbohydrate metabolism in AIP mice [19]. Since insulin promotes hepatic *ALAS1* transcription repression through the PI3-K/Akt system [10], IR could reduce the efficiency of carbohydrate loading as a treatment. Thus, the first aim of this study was to evaluate the prevalence of IR in patients with AIP in an observational case-control study. In AIP mice, we assayed the synergistic effect of glucose and two types of insulin (a fast-acting and an experimental liver-targeted insulin [20] as a treatment and prevention therapy for acute attacks induced both by fasting and by challenging the mice with multiple increasing doses of a porphyrinogenic drug. The novel liver-targeted insulin is a fusion protein formed by the insulin B chain followed by a linker QRGGGGGQR [21], the insulin A chain, the short linker GAP, and the apolipoprotein A-I. This fusion protein prolongs the insulin half-life in circulation and increases insulin activity in the liver. In the db/db model of metabolic syndrome, subcutaneous administration of the fusion protein reduced body weight and improved steatosis [20]. In this study, the activity of this novel liver-targeted insulin is evaluated in a mouse model of AIP.

2. Materials and Methods

2.1. Reagents

Fast-acting insulin (Actrapid®) was from Novo Nordisk Production SAS, Chartres, France. The recombinant fusion protein insulin fused to apolipoprotein A-I (Ins-ApoAI) and apolipoprotein A-I (Apo) were expressed in *E. coli* and purified by GenScript Corp. (Piscataway, NJ, USA).

2.2. Participants and Study Design

A case-control study was conducted in 44 patients with AIP and 55 age- and gender-matched healthy volunteers. Participants were recruited during the follow-up in the Porphyria Unit at the Hospital Universitario 12 de Octubre from May 2018 to May 2019. Among patients, there was high variability in the *PBGD* gene mutation, with the most prevalent being the 340-341insT (4 of 44 cases). Control volunteers were selected among family members of the AIP patients once the presence of the family mutation and biochemistry compatible with acute porphyria had been ruled out. The study was approved by the Hospital Ethics committee (CEIm: 19/262). A medical doctor questioned participants about the presence or absence of AIP symptoms, the time of diagnosis, the number and duration of attacks, and about triggering and relieving factors during attacks. Blood sampling was routinely performed between 08:00 and 10:00 a.m. after overnight fasting, according to hospital protocols. The data collected included a physical examination and laboratory tests (glucose, insulin, and urinary ALA and PBG levels). Factors related to the development of insulin resistance were also collected: (i) metabolic syndrome [22], (ii) overweight [23] measured according to the WHO recommendations, (iii) sedentary lifestyle [24] according to the international physical activity questionary (IPAQ, www.ipaq.ki.se., last entry March 3, 2021), (iv) the presence of polycystic ovary syndrome (PCOS) [25], (v) glucocorticoid treatment [26], (vi) human immunodeficiency virus infection [27], and (vii) genetic syndromes of severe IR [28]. Hypertension (HT) was defined as a blood pressure $\geq$130/$\geq$80 mmHg, HOMA was calculated as described by Matthews et al. [29], and insulin resistance was defined as a HOMA-IR index of >3.4, corresponding to the 90th percentile of the HOMA-IR index distribution in a Spanish adult nondiabetic population [30]. Sedentary lifestyle was measured according to the international physical activity questionary (IPAQ) and dichotomized into sedentary (category 1) or non-sedentary lifestyle (categories 2 and 3). Metabolic syndrome was defined according to the American College of Endocrinology criteria (among other parameters BMI $\geq$ 25 kg/m^2 or waist circumference $\geq$102 cm [men] or $\geq$88 cm [women] and fasting glucose $\geq$6.1 mmol/L ($\geq$110 mg/dL) [22]).

2.3. Experimental Studies in a Murine Model of AIP

AIP mice (C57BL/6-$^{pbgdtm1(neo)Uam}$/C57BL/6-$^{pbgdtm2(neo)Uam}$) exhibit hepatic PBGD activity reduced to 30% of normal and reproduce the biochemical test results, together with the presence of pain and motor neuropathy that characterize human porphyria [31]. Experimental protocols were approved by the Ethics Committee of the University of Navarra (CEEA032-13) according to European Council guidelines.

The glucose tolerance test (GTT) was performed after 15 h of fasting. Male mice were intraperitoneally administered with one dose of 20% glucose (10 µL/g mice). Glycemia was quantified 5 min later, followed by one single subcutaneous administration of Ins-ApoAI (90, 30, or 10 µg/kg; dose eq. to 2.3, 0.77 or 0.26 µg of pure insulin/mouse) or fast-acting insulin (180 µg/kg or 18 µg/kg; dose eq. to 4.55 or 0.45 µg of pure insulin/mouse). Glycemia was quantified every 30 min using the Accu-Chek® Aviva meter (Roche, Sant Cugat del Vallès, Spain) for 6 h. Supplementary doses of glucose were administered when serum glucose levels were lower than 100 mg/dl. After 6 h, the animals were euthanised and liver samples were frozen at −80 °C.

The steady-state mRNA levels of the genes were analyzed by quantitative RT-PCR using iQ SYBR Green supermix in an iQ5 real-time PCR detection system (Bio-Rad, Hercules, CA, USA) and specific primers (*alas1*, forward: 5′-CAAAGAAACCCCTCCAGCCAATGA-3′, reverse: 5′-GCTGTGTGCCGTCTGGAGTCTGTG-3′, product length: 104 bp; *pgc-1α*, forward: 5′-GAAGTGGTGTAGCGACCAATC-3′, reverse: 5′-AATGAGGGCAATCCGTCTTCA-3′, product length: 162 bp; *hmox1*, forward: 5′-CCAGAGTGTTCATTCGAGCA-3′, reverse: 5′-CTGCAGGGGCAGTATCTTGC-3′, product length: 116 bp; *g6pase*, forward 5′-AACGCCTTCTATGTCCTCTTT-3′, reverse: 5′-GTTGCTGTAGTAGCTGGTGTC-3′, product length: 168 bp; *pepck*, forward: 5′-AGCCTGCCCCAGGCAGTGAG-3′, reverse: 5′-CATGCACCCTGGGAACCTGGC-3′, product length: 339 bp and *cyp7a1*, forward: 5′-GCTGTGGTAGTGAGCTGTTGCA-3′, reverse: 5′-CACAGCCCAGGTATGGAATCA-3′, product length: 103 bp). PCR amplification was performed under the following conditions: one cycle of 3 min at 95 °C, followed by 35 cycles of 15 s at 95 °C, 30 s at 60 °C, 30 s at 72 °C, and 30 s at the detection temperature of each gen, followed by a single final extension cycle of 72 °C for 4 min. The amount of gene transcript was calculated as the n-fold difference relative to the control gene *actin* (forward: 5′-CGCGTCCACCCGCGAG-3′, reverse: 5′-CCTGGTGCCTAGGGCG-3′, product length: 125 bp). The results were expressed according to the formula $2^{Ct(Actin)-Ct(gene)}$, where Ct represents the difference in threshold cycle between the target and control genes.

The treatment after the co-administration of glucose and exogenous insulin was studied during an ongoing acute attack in male AIP mice. Acute attack was induced by an intraperitoneal administration of four increasing doses of phenobarbital (75, 80, 85, and 90 mg/kg) at 24 h intervals. On day 3 and 4, mice were treated with glucose one hour before and nine hours after the administration of phenobarbital. Co-administration with insulin was performed 5 min after the first daily dose of glucose (18 µg/kg of fast-acting insulin or 90 µg/kg of Ins-ApoAI). Mice were housed in metabolic cages (BIOSIS Biologic Systems, SL, Madrid, Spain), and urine samples were collected for 24 h. On day 5, animals were euthanised 20 min after an additional administration of 90 mg/kg of phenobarbital, and then liver tissue samples were frozen at −80 °C.

In a prevention trial, male mice were fed *ad libitum* and two glucose doses (2 h apart) were injected daily between days −5 and +4. Animals received Ins-ApoAI on days –5 and 0, two hours before the first dose of phenobarbital. Co-administration with Ins-ApoAI (90 µg/kg, sc) was performed two hours before the first dose of phenobarbital. Pain scores and motor coordination were measured four hours after the fourth dose of phenobarbital injection, as previously described [32]. After each dose of phenobarbital, mice were housed in metabolic cages, and urine samples were collected after 24 h. Urinary excretion of ALA and PBG were quantified using a quantitative ion exchange column method (BioSystems SA, Barcelona, Spain) and measured at 555nm in an Ultrospc 3000 spectrophotometry (Pharmacia Biotech, Buckinghamshire, UK).

The ratio of mitochondria per hepatocyte was scored in different groups of male mice between 8 and 10 weeks old after a 10-day protocol of glucose administration combined or not with fast-acting insulin (10 ui/mL, eq. to 18 µg/kg of crystallized insulin equivalent), 1 mg/kg of ApoAI or 1 mg/kg of Ins-ApoAI (equivalent to 90 µg/kg of crystallized insulin) on days 1, 4, 6, and 8. Animals were euthanised on day 11. An immunohistochemistry assay was performed on liver tissue samples using an antibody to detect mitochondria MTCO1 protein, 1:4000 dilution (ref: 1D6E1A8, ab14705 abcam, Cambridge, UK). Detection and counting of mitochondria in histological images was carried out using a plugin developed for Fiji/ImageJ, an open-source Java-based image processing software [33]. The plugin was developed by the Imaging Platform of the Cima Universidad de Navarra. The image processing pipeline includes automatic tissue detection, individual color-channel retrieval (hematoxylin and DAB) using a color deconvolution plugin for stain separation [34], and automatic nuclei detection and counting from the hematoxylin channel and mitochondria detection and counting from the DAB channel. The proportion of anti-MTCO1 stained mitochondria was counted from at least 1200 nuclei from four microscopic fields from each liver.

The body composition was measured in 15-h-fasted mice by quantitative magnetic resonance (QMR) technology (EchoMRI-100-700, Echo Medical Systems, Houston, TX, USA) as previously described [35]. At the end of the experimental period, mice were euthanised, and blood and tissue samples including liver, kidney, heart, soleus and gastrocnemius muscles, white adipose tissue (WAT) (gonadal, retroperitoneal, mesenteric, and subcutaneous), and brown adipose tissue depots (BAT) were collected as previously described [36]. The visceral WAT was estimated by the sum of gonadal, retroperitoneal, and mesenteric depot weights.

2.4. Statistics

The results were plotted as the mean ± s.d. The Fisher's exact test was used to compare the distribution of qualitative data. Prior to statistical analysis, quantitative data were transformed using the formula Log (1 + x) in order to normalize the variances. Comparisons between two groups were analyzed by Student's *t* test. In the case of comparisons across more than two groups, data were analyzed with the ANOVA test, and pairwise comparisons were made using Bonferroni´s multiple comparison tests. The null hypothesis was rejected when $p < 0.05$. Statistical analysis was performed using GraphPad Prism® 5 (GraphPad Software, Inc., La Jolla, CA, USA)

3. Results

3.1. High Prevalence of a Pathological HOMA-IR Index in Patients with AIP

Forty-four Spanish patients with AIP and 55 control volunteers (CV) matched by age, gender, and BMI were enrolled in an observational case-control study in order to estimate the homeostasis model assessment of insulin resistance (HOMA-IR) index. Subjects had no HIV infection, genetic syndromes associated with severe IR, or steroid treatment. Clinical and biochemical descriptions of cases and control volunteers are shown in Table 1. A significantly higher frequency of pathological levels of the HOMA-IR index (cut-off ≥ 3.4) [30] was observed in eight (18.2%) patients with AIP as compared to one single subject (1.82%) in the CV group ($p = 0.01$, Fisher's exact test) (Figure 1a). The only individual included in the CV group showing a high HOMA-IR index was also overweight and met the criteria for metabolic syndrome (see Material and Methods). Among the eight AIP cases, a high HOMA-IR index was associated with obesity (BMI ≥ 30) [37] in five participants; one had polycystic ovary syndrome (associated with insulin resistance as reported in [25]), and two showed no specific etiology related to IR other than porphyria. Finally, a sedentary lifestyle was found to be more frequent among CV (45%) than in patients with AIP (21%) (Table 1).

Among the 44 individuals carrying a mutation in the *PBGD* gene, eight (18%) were classified with active disease (AIP-AD) as defined by high urinary porphyrin precursor excretion (ALA, ≥7.5 mmol/mol creat.; PBG, ≥3.4 mmol/mol creat.) and at least one attack in the last year or were under prophylactic hemin treatment (Figure 1a). Eighteen cases (41%) showed high urinary precursor levels but clinically stable disease and these constituted the asymptomatic high excreters (AIP-ASHE) group. Furthermore, 18 individuals (41%) with clinically stable disease and low urinary precursor excretion (ALA, <7.5 mmol/mol creat.; PBG, <3.4 mmol/mol creat.) were included in the stable disease (AIP-SD) group. Of interest, among patients with high HOMA-IR (Figure 1a) and hyperinsulinemia (Figure 1b), six were classified as AIP-SD (33% of the group) and two were AIP-ASHE (11%), whereas none of the AIP-AD group showed values outside the normal range for this index. Indeed, only the HOMA-IR index (Figure 1a) and serum insulin levels (Figure 1b) from the AIP-SD group showed significant differences with the CV group. These data suggest that increased serum insulin levels are associated with a biochemical and clinical improvement in patients with AIP.

Table 1. Demographic and clinical characteristics of the study population.

	CV (n=55)	AIP (n=44)	AIP-SD (n=18)	AIP-ASHE (n=18)	AIP-AD (n=8)	*p*. CV vs AIP
Woman (n) **(%)**	43 78.2%	36 81.8%	13 72.2%	15 83.3%	8 100%	0.8
Age (y) **(range)**	39.2 ± 12.9 17–68	42 ± 14.1 17–68	44.3 ± 16.3 17–68	44.5 ± 11.2 22–63	33 ± 10.9 17–46	0.15
Body weight (kg) **(range)**	65.3 ± 11.7 40–92	63.6 ± 16.3 30–98	65.7 ± 17.3 43–97	66.8 ± 16.1 45–98	51.9 ± 9.6 30–59	0.58
HOMA index **(range)**	1.64 ± 0.74 0.56–3.53	2.22 ± 1.23 0.68–6.07	2.47 ± 1.56 0.88–6.07	2.16 ± 0.92 0.8–4.1	1.80 ± 0.82 0.68–2.93	0.008
Serum glucose (mg/dl) (range)	89.2 ± 8.57 68.0–111.5	93.7 ± 9.40 71.0–125.0	95.7 ± 9.95 71.0–113.0	92.8 ± 10.04 81.0–125.0	91.3 ± 6.16 83.0–99.0	0.015
Serum insulin (µU/dl) (range)	7.55 ± 3.23 2.63–14.9	9.48 ± 5.13 1.8–26.5	10.4 ± 7.03 1.8–26.5	9.27 ± 3.23 3.7–13.9	7.9 ± 3.38 3–11.5	0.032
Metabolic Sd. (n°) **(%)**	1/55 1.82%	3/44 6.82%	0/18 0%	1/18 5.6%	2/8 25%	0.333
Sedentary lifestyle (n°) (%)	22 of 49 * 44.89%	9/44 20.45%	3/18 16.7%	4/18 22.2%	2/8 25%	0.012
BMI (kg/m²) **(range)**	22.4 ± 3.87 16.7–32.98	24.2 ± 5.88 11.9–36.4	24.9 ± 5.70 16.2–36.4	25.8 ± 5.99 16.7–36.3	19.14 ± 3.27 11.9–22.5	0.99
ALA (µg/mg creat.) **(range)**	4.97 ± 0.24 3.2–5.0	9.54 ± 10 0.98–57	4.48 ± 1.17 1.3–5.0	9.63 ± 6.60 0.98–25	20.7 ± 17.3 3.3–57	0.003

Data represent mean ± SD (range), except counts of women and individuals with metabolic syndrome and sedentary lifestyle (percentage of total). Data were analyzed using a two-tailed Student's *t* test on total AIP cases (n = 44) versus matched control volunteers (n = 55) and differences in frequency distribution were analyzed using Fisher's exact test. CV: control volunteers; AIP: acute intermittent porphyria; AD: cases with active disease; SD: cases with stable disease; ASHE: asymptomatic high excreters patients. Normal urinary levels are <4.64 mg ALA/g creat. (<4 mmol ALA/mol creat.) and <3 mg PBG/g creat. (<1.5 mmol PBG/mol creat.). * There were missing data for six patients.

3.2. *Ins-ApoAI Induced a Fast and Sustained Normalization in the Gene Transcription Involved in the Liver Regulation of Heme Synthesis, Gluconeogenesis, and Bile Acid Synthesis in Fasted WT and AIP Mice*

The transcriptional effects of the exogenous administration of glucose and/or insulin on important genes related to hepatic heme synthesis were evaluated in fasted mice injected with both fast-acting insulin (Actrapid®) and an experimental liver-targeted insulin, Ins-ApoAI. The matched glucose/insulin doses were titrated with a GTT in 15-h-fasted AIP mice (Figure 2a). The mice received three doses of 2 mg/kg of glucose (20% solution) every 2 h throughout the 6 h of study and supplementary doses when they were close to hypoglycemic values (~70 mg/dL). Co-administration with Ins-ApoAI or fast-acting insulin was performed 5 min after the first glucose dose. Mice treated with subcutaneous Ins-ApoAI (0.11, 0.33 and 1 mg/kg of Ins-ApoAI, equivalent to 10, 30, and 90 µg/kg of crystallized insulin equivalent, respectively) showed blood glucose kinetics similar to the control group (Figure 2a) and received a total of three doses of glucose. In contrast, mice that received fast-acting insulin (10 and 100 ui/mL, equivalent to 18 µg/kg and 180 µg/kg of crystallized insulin) showed a rapid organ glucose uptake and needed up to six doses of glucose (one dose/h). At the end of the study, the administration of three doses of glucose alone did not reduce hepatic *alas1* over-expression, but the administration of the highest doses of Ins-ApoAI (30 and 90 µg/kg) and the lowest fast-acting insulin (18 µg/kg) reduced its expression to values found in fed AIP mice (Supplementary Figure S1). The doses with the best outcomes (i.e., 90 µg/kg Ins-ApoAI and 18 µg/kg fast-acting insulin) were chosen for further transcription studies of the genes of interest in AIP and WT mice.

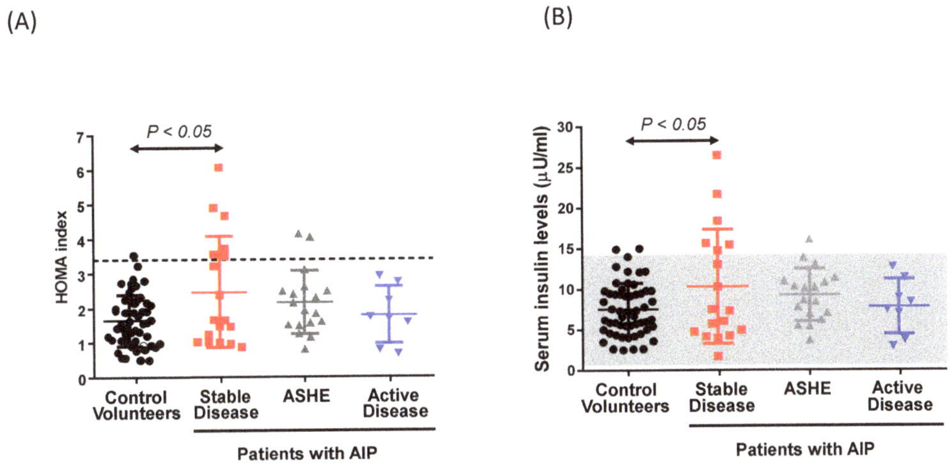

Figure 1. Range distribution of the HOMA-IR index and serum insulin levels in control volunteers and carriers of a mutation in the *PBGD* gene classified according to biochemical and clinical characteristics in patients with stable disease (SD), asymptomatic high excreters (ASHE), or those with active disease (AD). (**A**) The HOMA-IR index and (**B**) serum insulin levels significantly increased in patients with stable disease (AIP-SD). Dotted lines indicate the cut-off values of the HOMA-IR index ($\geq$3.4) and the gray rectangle corresponds to the 95% confidence interval of serum insulin levels corresponding to the group of voluntary controls. Comparisons were performed by one-way-ANOVA followed by Bonferroni post-hoc correction.

The transcription of genes involved in the regulation of hepatic heme synthesis (*alas1*, *pgc-1α*) and catabolism (*hmox1*) were strongly induced in both male WT and AIP mice after a 15-h-fasting period (Figure 2b–d). Of interest, the fold-change expression for *alas1* was significantly higher in AIP than in WT animals, both in the fed (2-fold induction vs. WT mice) and fasted conditions (3.4-fold induction vs. WT mice) (Figure 2b). Gluconeogenesis, as measured by the fold-change up-regulation of *glucose 6-phosphatase* (*g6pase*) (Figure 2e) and *phosphoenolpyruvate carboxykinase* (*pepck*) (Figure 2f) genes, was also significantly activated in fasted WT (2-fold and 4.3-fold induction, respectively) and AIP mice (3-fold and 5.3-fold induction, respectively).

The expression of genes involved in the heme synthesis pathway, *alas1*, and the *pgc1α* gene, returned to normal after co-administration of glucose with the exogenous insulins in both WT and AIP mice (Figure 2b,d) but not with glucose alone. To better understand what occurs in the regulation of hepatic *alas1*, a group of mice was euthanised 2.5 h after the end of fasting. Alas-1 tended to return to baseline levels 30 min after the second glucose administration (Supplementary Figure S2a). However, 6 h after ending the fast (2 h after the third glucose dose) its expression showed a re-induction to near fasting values while both *alas1* (Figure 2b and Supplementary Figure S2a) and *pgc-1α* (Figure 2c) gene transcription displayed pre-fasting levels in both insulin-treated groups. Regarding protein analysis, we detected a low ratio of pAkt/Akt in the liver of 15-h-fasted mice when compared with animals fed ad libitum (Supplementary Figure S3). The co-administration of glucose and Ins-ApoAI significantly increased the pAkt/Akt ratio within 10 min, whereas this was delayed until 2.5 h in mice treated with glucose alone. Based on these results, it is tempting to speculate that pAkt induces rapid Forkhead Box O1 (FOXO1) phosphorylation and breaks up the FOXO1/PGC1α complex, causing the sustained repression of alas1 transcription in the liver of AIP mice treated with Ins-ApoAI.

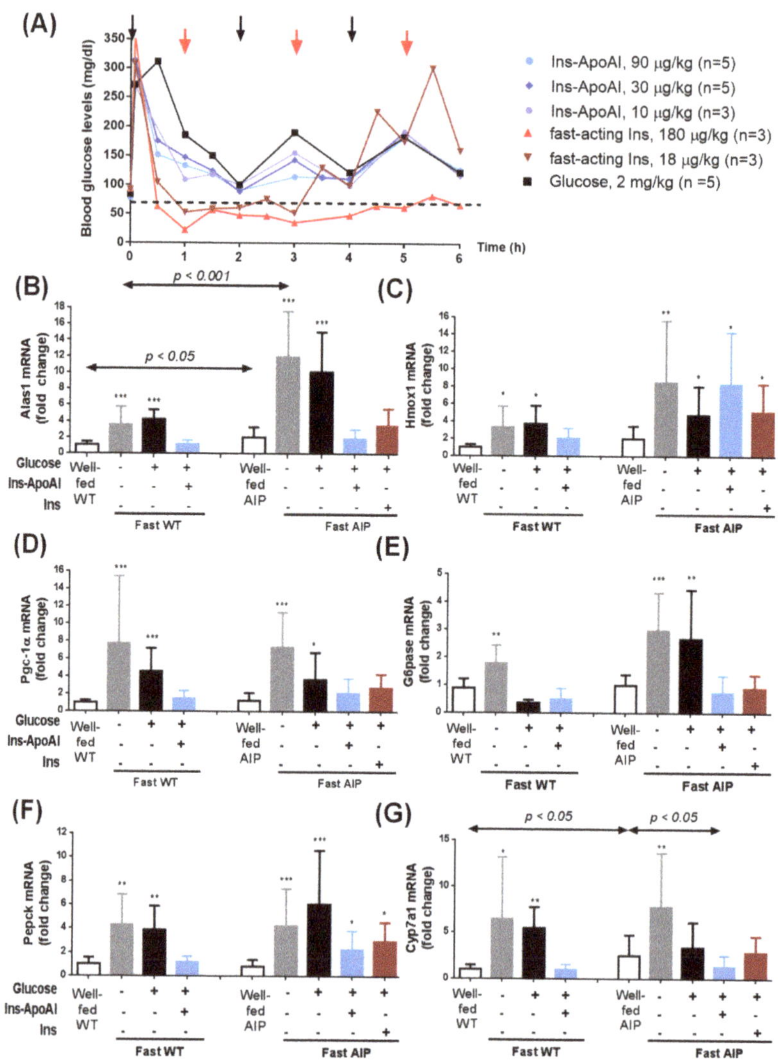

Figure 2. Transcriptional analysis of important genes in fasted animals treated with glucose, and glucose with a fast-acting insulin or an experimental liver-targeted insulin (Ins-ApoAI). (**A**) Serum glucose kinetics over 6 h measured after glucose overloads in 15-h-fasted AIP mice treated with glucose. Glycemia was measured at 30 min intervals, starting 5 min post-initial dose. Black arrows represent glucose administration for all groups and red arrows represent supplementary glucose administration for the fast-acting insulin group. Kinetics of the (**B**) *alas1*, (**C**) *hmox1*, (**D**) *pgc-1α*, (**E**) *g6pase*, (**F**) *pepck*, and (**G**) *cyp1a7* gene transcription in the liver were measured in male WT and AIP mice at baseline (well-fed condition), 15-h-post starvation and after the administration of three doses of glucose (2 mg/kg, i.p.), three doses of glucose with a single subcutaneous dose of Ins-ApoAI (eq. from 90 µg/kg of crystallized insulin equivalent) or six doses of glucose with a single dose of a commercial fast-acting insulin (10 ui/mL, eq. to 18 µg/kg). Data are mean ± s.d. of five animals per group. Comparisons were performed by one-way ANOVA followed by Bonferroni post-hoc correction. Alas1, aminolevulinate synthase 1; hmox1, heme oxygenase-1; pgc-1α, Peroxisome proliferator-activated receptor-gamma coactivator-1 alpha; g6pase, glucose 6-phosphatase; pepck, phosphoenolpyruvate carboxykinase and cyp1a7, cholesterol 7 alpha-hydroxylase. *, $p < 0.05$; **, $p < 0.01$; ***, $p < 0.001$ vs. well-fed mice and. Ins-ApoAI, the fusion protein of a single chain insulin and apolipoprotein A-I.

Another difference between WT and AIP mice was the down-regulation of the key limiting enzyme of the hepatic heme catabolism. Although, fasting induced the expression of *hmox1* in both strains (3.3-fold induction in WT and 4.2-fold induction in AIP mice), co-administration of exogenous insulin treatment only normalized its expression in WT mice (Figure 2c).

In WT mice, the administration of three doses of glucose normalized the expression of *g6pase*, whereas in AIP mice its expression was only normalized after co-administration of glucose with the exogenous insulins (Figure 2e). *Pepck* up-regulation was normalized after co-administration with glucose and Ins-ApoAI in fasted WT mice, whereas in the liver of AIP mice it remained overexpressed when compared to pre-fasted values (Figure 2f). These data suggest a differential regulation of the glucose supply from the liver to the bloodstream during fasting in the livers of WT and AIP mice. Treatment with glucose overload tended to normalize the expression of *g6pase* and *pepck* at 2.5 h (30 min after the second dose of glucose), but their expression increased again at 6 h, that is, 2 h after the third dose of glucose (Supplementary Figure S2b,c). Co-administration of glucose and Ins-ApoAI or fast-acting insulin induced better transcriptional control of both gluconeogenic enzymes throughout the study period (Supplementary Figure S2b,c); however, fast-acting insulin required double the dose of glucose to maintain the same transcriptional level of these two enzymes when compared to Ins-ApoAI.

Finally, transcription levels of the *cyp7a1* gene, an insulin-regulated gene encoding the enzyme cholesterol 7α-hydroxylase (EC 1.14.14.23) which catalyzes the initial step in bile acid synthesis [38] also showed a significant induction in fasted mice (Figure 2g). Well-fed AIP mice showed a significant overexpression of the insulin-dependent *cyp7a1* gene compared to WT mice (2.7-fold induction), and fasting strongly induced its over-expression in both strains (6.4-fold induction in WT and 2.9-fold induction in AIP mice) (Figure 2g). Of interest, the administration of glucose and glucose with rapid insulin reduced this up-regulation to values within the range observed in fed AIP mice while co-administration of glucose and Ins-ApoA1 normalized its hepatic expression when compared to well-fed WT mice (Figure 2g). These data suggest that *cyp7a1* regulation could be affected by a hepatocyte insulin resistance in AIP mice that is reversed with the administration of Ins-ApoAI.

3.3. *The Relative Contribution of Insulin to Protect against alas1 Induction Modulated by Barbiturate Challenge*

Drugs up-regulate the *alas1* gene via direct interactions with 5′-upstream response elements in the promoter region of the gene. Given that it is an enhanced transcriptional pathway different from that activated during fasting, we decided to test the co-administration of glucose and exogenous insulin during acute attacks of porphyria induced by the administration of increasing doses of phenobarbital for four consecutive days. In a treatment study, the administration of glucose or glucose and fast-acting insulin did not alter daily ALA (Figure 3a) and PBG (Figure 3b) excretion on the last day of the study (24 h after the fourth-last dose of phenobarbital). However, the administration of glucose and Ins-ApoAI halved the excretion of both precursors. On day five, mice received a supplementary dose of phenobarbital (90 mg/kg) and were euthanized 20 min later to study the hepatic expression of the genes involved in the heme synthesis and catabolism. While glucose administration with or without fast-acting insulin did not repress *alas1* transcription, co-administration of Ins-ApoAI reduced its overexpression by 60% (Figure 3c). Insulin, either free or conjugated with ApoAI, tended to induce the expression of *hmox1*, although no significant differences were observed between groups (Figure 3d).

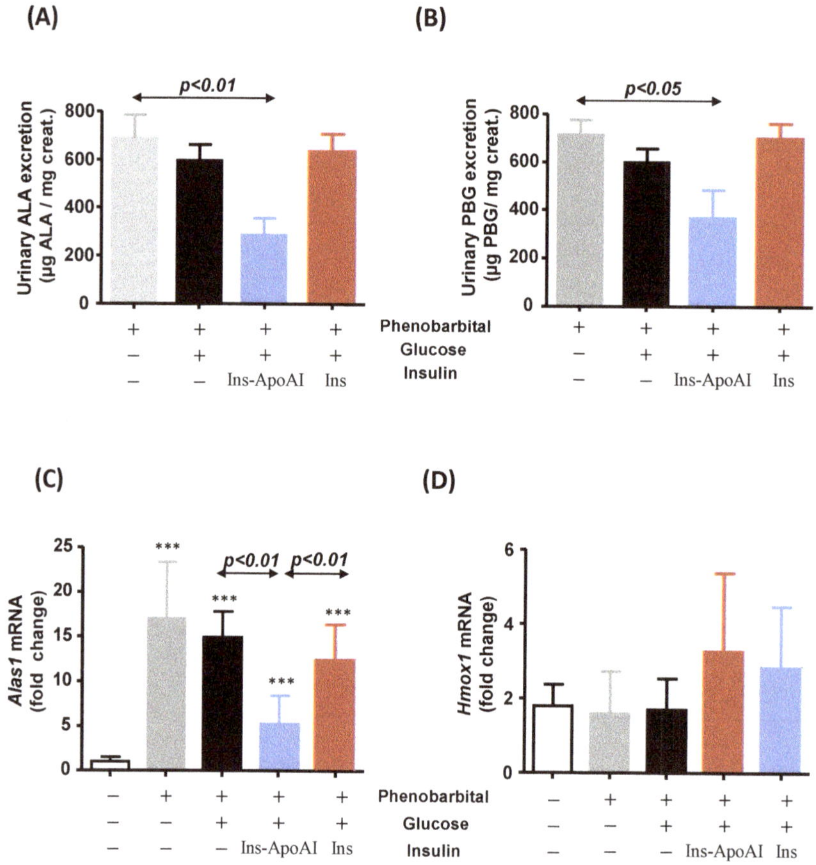

Figure 3. Therapeutic efficacy of subcutaneous administration of Ins-ApoAI and glucose during an ongoing phenobarbital-induced attack. Urinary (**A**) ALA and (**B**) PBG excretion on day 4 of the phenobarbital-induced attack. While co-administration of glucose and Ins-ApoAI halved the excretion, glucose alone or glucose co-administered with fast acting insulin failed to reduce the excretion of the neurotoxic precursors ALA and PBG. Hepatic expression of (**C**) *alas1* and (**D**) *hmox1* measured on day 5, 20 min after a supplementary phenobarbital dose of 90 mg/kg. Data are mean ± s.d. of at least four animals per group. Comparisons were performed by one-way ANOVA followed by Bonferroni post-test. ***, $p < 0.001$ vs. control untreated AIP mice. Daily urinary excretion of porphyrin precursors corresponding to baseline values are: 52 ± 12.7 µg ALA/mg creat. and 11.8 ± 4.2 µg PBG/mg creat. *Alas1, aminolevulinate synthase 1; hmox1, heme oxygenase-1.* Ins-ApoAI, the fusion protein of a single chain insulin and apolipoprotein A-I.

In a prevention study, the protective effect of the recurrent administration of Ins-ApoAI and/or glucose was assayed in AIP mice to mimic the prophylactic therapy some patients receive when beginning with prodromal symptoms associated with an acute attack (Figure 4). While PBG levels were unchanged (Figure 4b), urinary ALA quantification measured as daily excretion during the four days of phenobarbital challenge (Figure 4a, left) or area under the curve (Figure 4a, right) showed a significant reduction in the two groups treated with glucose, with or without Ins-ApoAI (62 or 67% vs. the control phenobarbital group). However, this reduction was not translated into a clinical improvement in phenobarbital–challenged AIP mice treated with glucose alone, as measured by both a pain score (Figure 4c) and motor coordination in the rotarod test (Figure 4d). Of interest, these two parameters showed a significant improvement in the group that received Ins-ApoAI together with glucose, as compared with mice treated with glucose alone (Figure 4c,d).

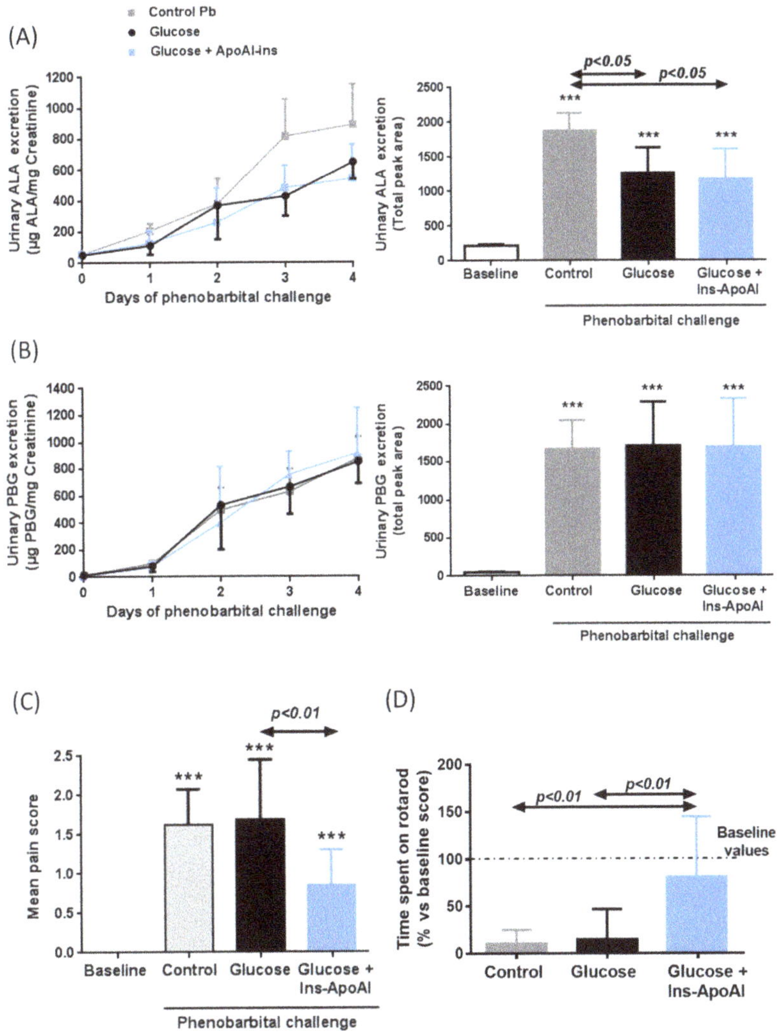

Figure 4. Therapeutic efficacy of multi-dose co-administration of glucose and liver-targeted insulin against a drug-induced attack in AIP mice. Daily urinary (**A**) ALA and (**B**) PBG excretion (left) and quantification of the area under the curve over time (right) during a prevention trial (as detailed in material and method section). (**C**) Pain scoring measured by the mouse grimace scale (MGS) and (**D**) motor coordination score assessed by the rotarod test (% respect baseline score) performed 4 h after the fourth dose of phenobarbital. Data are mean ± s.d. of six animals per group. Comparisons were performed by one-way ANOVA followed by Bonferroni post-test. ***, $p < 0.001$ vs. baseline values. Glu, Glucose; Ins-ApoAI, the fusion protein of a single chain insulin and apolipoprotein A-I.

An important finding observed in the liver of AIP mice treated with repeated doses of Ins-ApoAI was an increase in the number of mitochondria per hepatocyte (Supplementary Figure S4). We suggest that an increase in the mitochondria per hepatocyte ratio together with the glucose supplement could enhance energy synthesis. In order to better explore the effect of ApoAI, new cohorts of AIP mice were exposed to the 10d-protocol with glucose with or without exogenous insulin. These animals were not challenged with phenobarbital to avoid interference from the barbiturate effect. Of interest, the administration of ApoAI significantly increased the number of mitochondria per hepatocyte (120% vs. control AIP) (Figure 5a). Mice injected with Ins-ApoAI also showed an increased ratio (113% vs. control AIP), although differences were not statistically significant due to high intragroup dispersion. No changes were observed in the liver of mice co-injected with glucose and fast-acting insulin (Figure 5a).

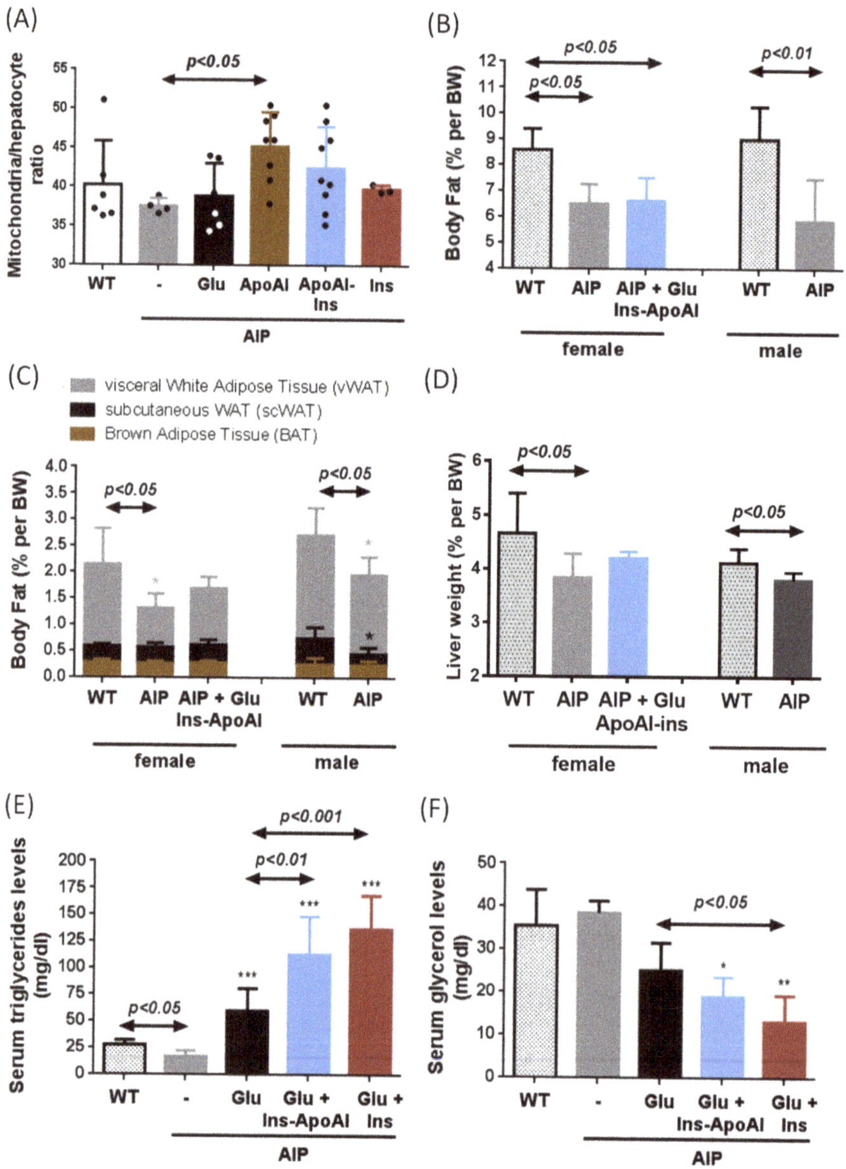

Figure 5. Changes in mitochondrial count per hepatocyte, serum levels of triglycerides and glycerol and body mass composition of WT and AIP mice after repeated administration of glucose and exogenous insulin. (**A**) The ratio of mitochondria per hepatocyte was measured in AIP mice treated with glucose combined or not with fast-acting insulin, ApoAI or Ins-ApoAI. (**B**) Body fat composition, (**C**) brown adipose tissue (BAT), visceral white adipose tissue (vWAT), and

subcutaneous white adipose tissue (scWAT) measured in fasted mice after 15 h of starvation using quantitative magnetic resonance. (**D**) Liver weight in WT and AIP mice. Serum (**E**) triglyceride and (**F**) glycerol levels measured in male mice treated with glucose for 10 days combined or not with four doses of either fast-acting insulin or Ins-ApoAI. All the assays presented in this figure were performed in mice not challenged with phenobarbital to avoid interference from the barbiturate effect. WT: wild type; AIP: acute intermittent porphyria; Glu: glucose; Ins: fast-acting insulin; Ins-ApoAI: Apolipoprotein AI conjugated with insulin. Data are mean ± s.d. of four mice per group. Comparisons were performed by one-way ANOVA followed by Bonferroni post-test. *, $p < 0.05$; **, $p < 0.01$; ***, $p < 0.001$ vs. control untreated AIP mice. Glu: glucose; Ins-ApoAI: the fusion protein of a single chain insulin and apolipoprotein A-I; Ins: fast-acting insulin.

Regarding body composition during starvation, AIP mice showed reduced fat accumulation when compared to age-matched WT mice, suggesting a predominance of the lipid catabolic processes in white adipose tissue (WAT) of AIP mice (Figure 5b). AIP mice showed reduced fat mass in both females (75% vs. WT group) and males (65% vs WT group) (Figure 5b). Within the fat tissue, a significant reduction in the percentage of visceral WAT (vWAT) depots was observed in AIP mice, with no changes in brown adipose tissue (BAT) (Figure 5c). In female AIP mice, a 10 day-protocol of glucose (two doses of 2 mg/kg, 2 h apart) together with four doses of Ins-ApoAI (every three days) did not modify the weight of vWAt (Figure 5c), or the weight of other organs such as the liver (Figure 5d), heart, kidney, or the soleus and gastrocnemius muscles (data not shown).

Fasted AIP mice showed low levels of serum triglycerides (TG, 61% as compared to WT mice) (Figure 5e). TG levels (Figure 5e) were increased in AIP mice treated with glucose (3.4-fold increase) (two doses of 2 mg/kg, 2 h apart) for 10 days and serum levels rose even higher after co-administration with exogenous insulin, both fast-acting insulin (7.8-fold increase) or Ins-ApoAI (6.5-fold increase) (four doses, three days apart). These data suggest that the co-administration of glucose and insulin increases the availability of circulating high-energy molecules in AIP mice. While serum TG levels rose, serum glycerol levels were significantly reduced in insulin-treated animals (Figure 5c), according to the lipogenic/antilipolytic effect previously described during insulin treatment [39,40].

4. Discussion

Carbohydrate loading is a well-documented treatment for mild acute attacks of porphyria [7,8]. Fasting and glucagon up-regulate the hepatic ALAS1 gene by the interaction of the complexes formed by PGC-1α with the FOXO1 and nuclear respiratory factor 1 (NRF-1) with a sequence within the ALAS1 promoter. The therapeutic effect of carbohydrates is based on their ability to induce endogenous insulin synthesis and disrupt the FOXO1-PGC1α complex through the hepatic pAkt signaling [10]. However, early reports showed abnormal oral glucose tolerance tests, as well as hyperinsulinemia in patients with AIP [17,18,41,42] which could lead to the reduced effectiveness of the glucose-treatment. The nature of IR in patients with AIP remains unknown. Chronic ALA accumulation, as occurs in patients with active disease, can increase reactive oxygen species (ROS), which could cause the dysfunction of glutathione-insulin transhydrogenase (the enzyme responsible for insulin degradation) [43]. However, the effect of ALA accumulation is ruled out in our study because patients with active disease (the AIP-AD group) showed serum insulin levels and a HOMA-IR index within the normal range. Of interest, *PBGD* mutation carriers with hyperinsulinemia report no clinical symptoms related to AIP. Similarly, Storjord et al. also found higher serum insulin levels associated with lower biochemical disease activity in their patient cohort [15]. These data suggest that IR and high-serum insulin levels are not a consequence of disease activity, but rather that sustained hyperinsulinemia can protect against acute attacks of porphyria.

Here, we assayed the efficacy of repressing *alas1* transcription after the administration of a commercial fast-acting insulin or an experimental hepatotropic Ins-ApoAI in an AIP mouse model. In these mice, abnormal GTT, hyperinsulinemia, and blockage of glycogenolysis pathways have been previously reported [19,44]. Significant repression of the hepatic *alas1* and *pgc1α* gene expressions were observed 6 h post-insulin administration with both fast-acting and Ins-ApoAI in fasted AIP mice, while treatment based exclusively on glucose did not yield a comparable decrease. Our data show that the administration of exogenous insulin quickly increased the pAkt/Akt ratio and induced a sustained repression of hepatic alas1 transcription, through the breakup of the transactivator FOXO1/PGC1α complex. In line with our results, Stein and Tschudy [18] reported clinical improvement in patients when small amounts of insulin were administered together with carbohydrates. More recently, Handschin et al. confirmed that the combination of glucose and insulin causes a more potent inhibition of ALAS1 than the administration of glucose alone [10]. Oliveri et al. (2012) also found hepatic *alas1* down-regulation concomitant with an activation of the phosphoinositide 3-kinase/Akt pathway and subsequent reduction of nuclear FOXO1-PGC-1α complex levels [45].

The advantage of our experimental Ins-ApoAI protein is that the apolipoprotein A-I moiety promotes liver targeting [20], thus promoting quicker and sustained repression of hepatic *alas1* expression. One important finding in our study is the repressive effect of the co-administration of glucose and Ins-ApoAI to counteract direct hepatic *alas1* gene induction modulated by barbiturate administration. The co-administration of glucose with fast-acting insulin was unable to modify urine porphyrin precursor excretion in an ongoing attack (treatment study). A possible drawback of the use of fast-acting insulin could be the iatrogenic induction of hypoglycemia. In these circumstances, glucagon secretion and *pgc-1α* are stimulated, and therefore, the effect obtained would be the re-induction of hepatic *alas1*. To avoid this effect in our study, AIP mice co-injected with fast-acting insulin received twice as many glucose doses as those treated with Ins-ApoAI, which would lead to an increased risk of weight gain if the protocol were repeated regularly.

In a preventive study, recurrent administration of glucose for ten days in AIP mice fed ad libitum partially protected against ALA accumulation induced after phenobarbital challenge. However, this reduction did not translate into a positive effect on the behavior of AIP mice because all the animals showed the same degree of pain and motor disability as control AIP mice. Of interest, administration of glucose with two doses of Ins-ApoAI improved pain and motor coordination, although it was unable to protect against heme precursor accumulation. Probably, ApoAI-induced mitochondrial biogenesis in the liver (as shown in Figure 5a and previously reported in [46]), together with increased metabolite supply could promote increased energy production that further improves behavioral parameters in AIP mice.

In the fasted state, the liver is a primary energy source and secretes glucose through both the breakdown of glycogen (glycogenolysis) and de novo glucose synthesis (gluconeogenesis) [47,48]. The hepatic G6Pase enzyme plays an important role in blood glucose homeostasis, and its expression is highly up-regulated during starvation. In fasted WT mice, glucose administration quickly normalized its up-regulation, while in the livers of AIP mice *g6pase* gene overexpression remained. Given that AIP mice showed reduced fat accumulation in vWAT and low levels of circulating triglycerides when compared to WT mice, all these data together suggest that the adipose-liver axis is an important source of energy supply for the rest of the organs in fasted AIP mice, but also that glucose secretion reduces the metabolite availability for the TCA cycle in the hepatocytes. Of interest, the administration of glucose and exogenous insulin in AIP mice increased circulating triglyceride levels as an energy source in non-adipose tissues while reversing hepatic *g6pase* overexpression, thus reducing the hepatic glucose supply.

During the phenobarbital-induced acute attack, there is a large loss of hepatic succinyl-CoA and glycine to produce ALA and PBG, which are rapidly excreted in the urine [49]. Thus, preventive treatment with glucose and insulin would increase hepatocyte metabolite availability that compensates for the loss of succinyl-CoA. In addition, another advantage of using a liver-targeted insulin is the increased hepatic glucose availability for its incorporation into the TCA cycle through pyruvate, enhanced mitochondrial biogenesis, and improved beta-oxidation that increases the energy status in the liver of AIP mice and reduces the porphyrinogenic effects of phenobarbital administration.

5. Conclusions

A high prevalence of hyperinsulinemia and IR was observed in Spanish patients carrying a mutation in the *PBGD* gene who had either stable (without clinical outcomes associated with acute porphyria) or asymptomatic disease (AIP-ASHE). The treatment study in AIP mice provides a proof-of-concept for a rapid and sustained repressive effect of Ins-ApoAI on hepatic *alas1* transcription during fasting and acute attacks induced by barbiturate administration. The prophylactic administration of this fusion protein, although associated with behavioral improvements, showed lower effectiveness to reduce porphyrin precursor accumulation compared with a direct *ALAS* inhibitor [50] or with therapies aimed at restoring hepatic PBGD expression [32,51]. Nevertheless, and given the anti-obesity effect associated with an increase of energy expenditure of the ApoAI moiety [52], further studies are needed to evaluate the effect of ApoAI on patients with AIP who are overweight and/or with metabolic syndrome associated with excessive carbohydrate intake or sedentary habits.

Supplementary Materials: The following are available online at https://www.mdpi.com/2227-9059/9/3/255/s1. Figure S1: Hepatic expression of the *alas1* gene in fasted animals treated with glucose, and glucose with a fast-acting insulin or an experimental liver-targeted insulin (Ins-ApoAI). To explore in the animal model the matched dose of insulin and glucose to obtain a treatment effect, in other words, to mimic the "glucose/insulin clamp", male AIP mice between 12 and 16 weeks old were injected with glucose combined with different concentrations of fast-acting insulin (180 μg/kg or 18 μg/kg; dose eq. to 4.55 or 0.45 μg of pure insulin/mouse) or Ins-ApoAI (90, 30, or 10 μg/kg; dose eq. to 2.3, 0.77 or 0.26 μg of pure insulin/mouse) after 15 h of fasting. The protocol is shown in Figure 2a. At the end of the study, the hepatic expression of the *alas1* gene was measured. The AIP mice that received higher doses of fast-acting insulin (180 μg/kg) showed very low glycemia during the 6 h of study despite being injected with repeated doses of glucose. Hepatic *alas* gene was overexpressed compared to the glucose-treated control. Given that hypoglycemia probably induced the opposite effect than was expected, the results of this group were eliminated from the graph. Data are the mean ± s.d. of five mice per group. Comparisons were performed by one-way ANOVA followed by a Bonferroni post-test. ***, $p < 0.001$ vs. control fasted AIP mice (grey column). *Alas1*: *aminolevulinate synthase 1*. Ins-ApoAI: the fusion protein of a single chain insulin and apolipoprotein A-I; Ins: fast-acting insulin. Supplementary Figure S2: Gene expression profiles of key genes involved in heme synthesis and gluconeogenesis in 15-h-fasted mice fed with glucose with or without exogenous insulin. Fold change expression of the main genes involved in the hepatic regulation of (A) *alas1*, (B) *g6pase*, and (C) *pepck* were analyzed in 15-h-fasted mice and 2.5 h and 6 h after intraperitoneal administration of glucose (2 mg/kg) combined with subcutaneous administration of fast-acting insulin (18 μg/kg, dose eq. to 0.45 μg of pure insulin/mouse) or Ins-ApoAI (eq. from 10, 30, and 90 μg/kg; dose eq. to 0.26, 0.77 or 2.3μg of pure insulin/mouse). Mice received an initial intraperitoneal dose of 2 mg/kg of glucose (20% solution) and supplementary doses when close to hypoglycemic values (<70 mg/dL). Glycemia was measured at 30 min intervals, starting 5 min post-initial dose. Mice treated with Ins-ApoAI showed blood glucose kinetics similar to the control glucose group and received a total of three glucose doses (every 2 h among the 6 h of study). Serum glucose quickly decreased in AIP mice treated with fast-acting insulin in which one dose per hour was necessary to avoid hypoglycemia. Data are mean ± s.d. of five mice per group. Comparisons were performed by one-way ANOVA followed by Bonferroni post-test. *, $p < 0.05$, **, $p < 0.01$; ***, $p < 0.001$ vs. well-fed mice. *Alas1*: *aminolevulinate synthase 1; g6pase*: *glucose 6-phosphatase; pepck*: *phosphoenolpyruvate carboxykinase;* Glu: Glucose; Ins-ApoAI: the fusion protein of a single chain insulin and apolipoprotein A-I; Ins: fast-acting insulin. Supplementary Figure S3: (A) Kinetics of total pAkt/Akt ratio in the liver of AIP mice at different times after glucose or co-administration of glucose and Ins-ApoAI. (B) Representative western blot of pAkt/Akt ratio in 15-h-fasted AIP mice at different times post-treatment. Immunodetection of relevant proteins was determined on PVDF/nitrocellulose membranes. The membranes were blocked (5% milk for Akt, and 5% BSA for GAPDH and pAkt) for 1 h at room temperature and incubated at 4 °C overnight with the corresponding primary antibody for anti-AKT (1:1000 dilution, #9272, Cell Signaling, MA, USA), anti-GAPDH (1:10.000 dilution, #2118, Cell Signaling) and anti-phosphorylated-AKT (1:1000 dilution, #9271S, Cell Signaling). Mice treated with glucose and Ins-ApoAI showed a significant increase at

10 min, with maximum Akt phosphorylation at 30 min. Quantification of bands was performed using the Image J program in a total of four different Western blots. The normalization of the values was undertaken on the same baseline group of mice. Supplementary Figure S4: Representative liver sections of (A) a non-injected AIP mouse, (B) an AIP mouse treated with glucose and Ins-ApoAI and (C) an AIP mouse co-administered with glucose and fast-acting insulin. Scale ×80.

Author Contributions: Conceptualization and design of the work, I.S.; L.I.-S., D.J., K.M.C., A.S., M.M.-C., F.J.C., C.L., M.J.M.-A., R.E.d.S., P.B. and A.F.; Methodology (performed the experiments and behavior assays in AIP mice and processed animal samples and tissues) I.S.; L.I.-S., D.J., K.M.C., A.S.; Methodology (cases and control recruitment): I.S.; M.M.-C., F.J.C., C.L., M.J.M.-A., R.E.d.S.; Software (plugin developed for mitochondria counting), D.J.; Statistical Analysis, I.S.; L.I.-S., D.J., K.M.C., P.B. and A.F.; Interpretation of data, I.S.; L.I.-S., D.J., K.M.C., M.M.-C., F.J.C., C.L., M.J.M.-A., R.E.d.S., P.B. and A.F.; Resources, I.S.; M.J.M.-A., F.J.C., C.L., M.J.M.-A., R.E.d.S., P.B. and A.F.; Writing—Original Draft Preparation, I.S.; L.I.-S., P.B. and A.F., assisted by D.J. and K.M.C. for figures and tables. All authors have read and agreed to the published version of the manuscript.

Funding: This research was supported in part by grants from Spanish Institute of Health Carlos III (FIS) cofunded by European Union (ERDF/ESF, "A way to make Europe"/"Investing in your future" [grant numbers PI15/01951, PI18/00860 and PI19/01128], the Spanish Fundación Mutua Madrileña de Investigación Médica, the Spanish Fundación Eugenio Rodríguez Pascual, the CaixaImpulse program, CIBERobn and the Spanish Fundación FEDER para la investigación de enfermedades raras. The financial sponsors had no role in the analysis or the development of conclusions. The investigators are solely responsible for the content and the decision to submit the manuscript for publication.

Institutional Review Board Statement: The case-control observational study was approved by the Hospital Ethics committee (CEIm: 19/262). The experimental study was conducted according to the European Council Guidelines, and approved by the Ethics Committee of the University of Navarra (protocol code CEEA032-13, date of approval: 08/03/2012).

Informed Consent Statement: Informed consent was obtained from all subjects involved in the study.

Data Availability Statement: The data presented in this study are available on request from the corresponding author.

Acknowledgments: Original T1 and T2 mouse strains were provided by U.A. Meyer (Biozentrum of University of Basel, Switzerland). We thank S. Arcelus, C.M. Rodríguez-Ortigosa and N. Sáinz for technical assistance.

Conflicts of Interest: The authors declare no conflict of interest. The funders had no role in the design of the study; in the collection, analyses, or interpretation of data; in the writing of the manuscript, or in the decision to publish the results.

References

1. Elder, G.; Harper, P.; Badminton, M.; Sandberg, S.; Deybach, J.-C. The incidence of inherited porphyrias in Europe. *J. Inherit. Metab. Dis.* **2013**, *36*, 849–857. [CrossRef] [PubMed]
2. Puy, H.; Gouya, L.; Deybach, J.C. Porphyrias. *Lancet* **2010**, *375*, 924–937. [CrossRef]
3. Anderson, K.; Sassa, S.; Bishop, D.; Desnick, R. Disorders of Heme Biosynthesis: X-Linked Sideroblastic Anemia and the Porphyrias. In *The Metabolic and Molecular Bases of Inherited Disease*, 8th ed.; Scriver, C.R.B.A., Sly, W.S., Valle, E., Eds.; McGraw Hill: New York, NY, USA, 2001; Volume 1, pp. 2991–3062.
4. Chen, B.; Solis-Villa, C.; Hakenberg, J.; Qiao, W.; Srinivasan, R.R.; Yasuda, M.; Balwani, M.; Doheny, D.; Peter, I.; Chen, R.; et al. Acute Intermittent Porphyria: Predicted Pathogenicity of HMBS Variants Indicates Extremely Low Penetrance of the Autosomal Dominant Disease. *Hum. Mutat.* **2016**, *37*, 1215–1222. [CrossRef]
5. Jaramillo-Calle, D.A.; Solano, J.M.; Rabinstein, A.A.; Bonkovsky, H.L. Porphyria-induced posterior reversible encephalopathy syndrome and central nervous system dysfunction. *Mol. Genet. Metab.* **2019**, *128*, 242–253. [CrossRef]
6. Anderson, K.E.; Bloomer, J.R.; Bonkovsky, H.L.; Kushner, J.P.; Pierach, C.A.; Pimstone, N.R.; Desnick, R.J. Recommendations for the Diagnosis and Treatment of the Acute Porphyrias. *Ann. Intern. Med.* **2005**, *142*, 439–450. [CrossRef] [PubMed]
7. Stein, P.; Badminton, M.; Barth, J.; Rees, D.; Stewart, M.F. Best practice guidelines on clinical management of acute attacks of porphyria and their complications. *Ann. Clin. Biochem.* **2013**, *50*, 217–223. [CrossRef]
8. Robert, T.L.; Varella, L.; Meguid, M.M. Nutrition management of acute intermittent porphyria. *Nutrition* **1994**, *10*, 551. [PubMed]
9. Podvinec, M.; Handschin, C.; Looser, R.; Meyer, U.A. Identification of the xenosensors regulating human 5-aminolevulinate synthase. *Proc. Natl. Acad. Sci. USA* **2004**, *101*, 9127–9132. [CrossRef] [PubMed]

10. Handschin, C.; Lin, J.; Rhee, J.; Peyer, A.K.; Chin, S.; Wu, P.H.; Meyer, U.A.; Spiegelman, B.M. Nutritional regulation of hepatic heme biosynthesis and porphyria through PGC-1α. *Cell* **2005**, *122*, 505–515. [CrossRef] [PubMed]
11. Degenhardt, T.; Väisänen, S.; Rakhshandehroo, M.; Kersten, S.; Carlberg, C. Peroxisome proliferator-activated receptor alpha controls hepatic heme biosynthesis through ALAS1. *J. Mol. Biol.* **2009**, *388*, 225–238. [CrossRef]
12. Duvigneau, J.C.; Esterbauer, H.; Kozlov, A.V. Role of Heme Oxygenase as a Modulator of Heme-Mediated Pathways. *Antioxidants* **2019**, *8*, 475. [CrossRef]
13. Tian, Q.; Li, T.; Hou, W.; Zheng, J.; Schrum, L.W.; Bonkovsky, H.L. Lon peptidase 1 (LONP1)-dependent breakdown of mitochondrial 5-aminolevulinic acid synthase protein by heme in human liver cells. *J. Biol. Chem.* **2011**, *286*, 26424–26430. [CrossRef]
14. Schmitt, C.; Lenglet, H.; Yu, A.; Delaby, C.; Benecke, A.; Lefebvre, T.; Letteron, P.; Paradis, V.; Wahlin, S.; Sandberg, S.; et al. Recurrent attacks of acute hepatic porphyria: Major role of the chronic inflammatory response in the liver. *J. Intern. Med.* **2018**, *284*, 78–91. [CrossRef]
15. Storjord, E.; Dahl, J.A.; Landsem, A.; Ludviksen, J.K.; Karlsen, M.B.; Karlsen, B.O.; Brekke, O.-L. Lifestyle factors including diet and biochemical biomarkers in acute intermittent porphyria: Results from a case-control study in northern Norway. *Mol. Genet. Metab.* **2019**, *128*, 254–270. [CrossRef] [PubMed]
16. Balwani, M.; Wang, B.; Anderson, K.E.; Bloomer, J.R.; Bissell, D.M.; Bonkovsky, H.L.; Phillips, J.D.; Desnick, R.J.; Porphyrias Consortium of the Rare Diseases Clinical Research Network. Acute hepatic porphyrias: Recommendations for evaluation and long-term management. *Hepatology* **2017**, *66*, 1314–1322. [CrossRef] [PubMed]
17. Sixel-Dietrich, F.; Verspohl, F.; Doss, M. Hyperinsulinemia in Acute Intermittent Porphyria. *Horm. Metab. Res.* **1985**, *17*, 375–376. [CrossRef] [PubMed]
18. Stein, J.A.; Tschudy, D.P. Acute Iintermittent Porphyria: A Clinical and Biochemical Study of 46 Patients. *Medicine* **1970**, *49*, 1–6. [CrossRef] [PubMed]
19. Collantes, M.; Serrano-Mendioroz, I.; Benito, M.; Molinet-Dronda, F.; Delgado, M.; Vinaixa, M.; Sampedro, A.; de Salamanca, R.E.; Prieto, E.; Pozo, M.A.; et al. Glucose metabolism during fasting is altered in experimental porphobilinogen deaminase deficiency. *Hum. Mol. Genet.* **2016**, *25*, 1318–1327. [CrossRef] [PubMed]
20. Ardaiz, N.; Gomar, C.; Vasquez, M.; Tenesaca, S.; Fernandez-Sendin, M.; Di Trani, C.; Belsué, V.; Escalada, J.; Werner, U.; Tennagels, N.; et al. Insulin fused to Apolipoprotein AI reduces body weigth and steatosis in db/db mice. *Front. Pharmacol.* **2021**. [CrossRef]
21. Rajpal, G.; Liu, M.; Zhang, Y.; Arvan, P. Single-Chain Insulins as Receptor Agonists. *Mol. Endocrinol.* **2009**, *23*, 679–688. [CrossRef] [PubMed]
22. Einhorn, D.; Reaven, G.M.; Cobin, R.H.; Ford, E.; Ganda, O.P.; Handelsman, Y.; Hellman, R.; Jellinger, P.S.; Kendall, D.; Krauss, R.M.; et al. American College of Endocrinology position statement on the insulin resistance syndrome. *Endocr. Pr.* **2003**, *9*, 237–252.
23. Hotamisligil, G.S.; Shargill, N.S.; Spiegelman, B.M. Adipose expression of tumor necrosis factor-alpha: Direct role in obesity-linked insulin resistance. *Science* **1993**, *259*, 87–91. [CrossRef]
24. Hamburg, N.M.; McMackin, C.J.; Huang, A.L.; Shenouda, S.M.; Widlansky, M.E.; Schulz, E.; Gokce, N.; Ruderman, N.B.; Keaney, J.F., Jr.; Vita, J.A. Physical inactivity rapidly induces insulin resistance and microvascular dysfunction in healthy volunteers. *Arterioscler. Thromb. Vasc. Biol.* **2007**, *27*, 2650–2656. [CrossRef]
25. Dunaif, A.; Segal, K.R.; Shelley, D.R.; Green, G.; Dobrjansky, A.; Licholai, T. Evidence for Distinctive and Intrinsic Defects in Insulin Action in Polycystic Ovary Syndrome. *Diabetes* **1992**, *41*, 1257–1266. [CrossRef] [PubMed]
26. Geer, E.B.; Islam, J.; Buettner, C. Mechanisms of glucocorticoid-induced insulin resistance: Focus on adipose tissue function and lipid metabolism. *Endocrinol. Metab. Clin. N. Am.* **2014**, *43*, 75–102. [CrossRef] [PubMed]
27. Hadigan, C.; Kattakuzhy, S. Diabetes mellitus type 2 and abnormal glucose metabolism in the setting of human immunodeficiency virus. *Endocrinol. Metab. Clin. N. Am.* **2014**, *43*, 685–696. [CrossRef] [PubMed]
28. Semple, R.K.; Savage, D.B.; Cochran, E.K.; Gorden, P.; O'Rahilly, S. Genetic syndromes of severe insulin resistance. *Endocr. Rev.* **2011**, *32*, 498–514. [CrossRef] [PubMed]
29. Matthews, D.R.; Hosker, J.P.; Rudenski, A.S.; Naylor, B.A.; Treacher, D.F.; Turner, R.C. Homeostasis model assessment: Insulin resistance and β-cell function from fasting plasma glucose and insulin concentrations in man. *Diabetologia* **1985**, *28*, 412–419. [CrossRef]
30. Gayoso-Diz, P.; Otero-González, A.; Rodriguez-Alvarez, M.X.; Gude, F.; García, F.; De Francisco, A.; Quintela, A.G. Insulin resistance (HOMA-IR) cut-off values and the metabolic syndrome in a general adult population: Effect of gender and age: EPIRCE cross-sectional study. *BMC Endocr. Disord.* **2013**, *13*, 47. [CrossRef]
31. Lindberg, R.L.P.; Porcher, C.; Grandchamp, B.; Ledermann, B.; Bürki, K.; Brandner, S.; Aguzzi, A.; Meyer, U.A. Porphobilinogen deaminase deficiency in mice causes a neuropathy resembling that of human hepatic porphyria. *Nat. Genet.* **1996**, *12*, 195. [CrossRef]
32. Jiang, L.; Berraondo, P.; Jericó, D.; Guey, L.T.; Sampedro, A.; Frassetto, A.; Benenato, K.E.; Burke, K.; Santamaría, E.; Alegre, M.; et al. Systemic messenger RNA as an etiological treatment for acute intermittent porphyria. *Nat. Med.* **2018**, 1. [CrossRef] [PubMed]
33. Schneider, C.A.; Rasband, W.S.; Eliceiri, K.W. NIH Image to ImageJ: 25 years of image analysis. *Nat. Methods* **2012**, *9*, 671–675. [CrossRef]

34. Ruifrok, A.; Johnston, D. Ruifrok AC, Johnston DA. Quantification of histochemical staining by color deconvolution. *Anal. Quant. Cytol. Histol.* **2001**, *23*, 291–299. [PubMed]
35. Moreno-Aliaga, M.J.; Pérez-Echarri, N.; Marcos-Gómez, B.; Larequi, E.; Gil-Bea, F.J.; Viollet, B.; Gimenez, I.; Martínez, J.A.; Prieto, J.; Bustos, M. Cardiotrophin-1 Is a Key Regulator of Glucose and Lipid Metabolism. *Cell Metab.* **2011**, *14*, 242–253. [CrossRef]
36. Prieto-Hontoria, P.L.; Pérez-Matute, P.; Fernández-Galilea, M.; Martínez, J.A.; Moreno-Aliaga, M.J. Lipoic acid inhibits leptin secretion and Sp1 activity in adipocytes. *Mol. Nutr. Food Res.* **2011**, *55*, 1059–1069. [CrossRef]
37. *Obesity: Preventing and Managing the Global Epidemic; World Health Organ Techical Report Series*. 2000, pp. 1–253. Available online: https://www.who.int/nutrition/publications/obesity/WHO_TRS_894/en/ (accessed on 4 March 2021).
38. Shin, D.-J.; Campos-Sandoval, J.; Gil, G.; Osborne, T. PGC-1 activates CYP7A1 and bile acid biosynthesis. *J. Biol. Chem.* **2004**, *278*, 50047–50052. [CrossRef] [PubMed]
39. Brown, M.S.; Goldstein, J.L. Selective versus Total Insulin Resistance: A Pathogenic Paradox. *Cell Metab.* **2008**, *7*, 95–96. [CrossRef]
40. Bódis, K.; Roden, M. Energy metabolism of white adipose tissue and insulin resistance in humans. *Eur. J. Clin. Investig.* **2018**, *48*, e13017. [CrossRef] [PubMed]
41. Sterling, K.; Silver, M.; Ricketts, H.T. Development of pophyria in Diabetes Mellitus: Report of Three Cases. *Arch. Intern. Med.* **1949**, *84*, 965–975. [CrossRef] [PubMed]
42. Andersson, C.; Bylesjö, I.; Lithner, F. Effects of diabetes mellitus on patients with acute intermittent porphyria. *J. Intern. Med.* **1999**, *245*, 193–197. [CrossRef] [PubMed]
43. Matkovic, L.B.; D'Andrea, F.; Fornes, D.; San Martín de Viale, L.C.; Mazzetti, M.B. How porphyrinogenic drugs modeling acute porphyria impair the hormonal status that regulates glucose metabolism. Their relevance in the onset of this disease. *Toxicology* **2011**, *290*, 22–30. [CrossRef] [PubMed]
44. Lelli, S.M.; De Viale, L.C.S.M.; Mazzetti, M.B. Response of glucose metabolism enzymes in an acute porphyria model: Role of reactive oxygen species. *Toxicology* **2005**, *216*, 49–58. [CrossRef] [PubMed]
45. Oliveri, L.M.; Davio, C.; Batlle, A.M.d.C.; Gerez, E.N. ALAS1 gene expression is down-regulated by Akt-mediated phosphorylation and nuclear exclusion of FOXO1 by vanadate in diabetic mice. *Biochem. J.* **2012**, *442*, 303–310. [CrossRef]
46. Song, P.; Kwon, Y.; Yea, K.; Moon, H.-Y.; Yoon, J.H.; Ghim, J.; Hyun, H.; Kim, D.; Koh, A.; Berggren, P.-O.; et al. Apolipoprotein a1 increases mitochondrial biogenesis through AMP-activated protein kinase. *Cell. Signal.* **2015**, *27*, 1873–1881. [CrossRef]
47. Zhang, W.; Patil, S.; Chauhan, B.; Guo, S.; Powell, D.R.; Le, J.; Klotsas, A.; Matika, R.; Xiao, X.; Franks, R.; et al. FoxO1 Regulates Multiple Metabolic Pathways in the Liver: Effects on gluconeogenis, glycolytic, and lipogenic gene expression. *J. Biol. Chem.* **2006**, *281*, 10105–10117. [CrossRef]
48. Honma, M.; Sawada, S.; Ueno, Y.; Murakami, K.; Yamada, T.; Gao, J.; Kodama, S.; Izumi, T.; Takahashi, K.; Tsukita, S.; et al. Selective insulin resistance with differential expressions of IRS-1 and IRS-2 in human NAFLD livers. *Int. J. Obes.* **2018**, *42*, 1544–1555. [CrossRef] [PubMed]
49. Homedan, C.; Laafi, J.; Schmitt, C.; Gueguen, N.; Lefebvre, T.; Karim, Z.; Desquiret-Dumas, V.; Wetterwald, C.; Deybach, J.C.; Gouya, L.; et al. Acute intermittent porphyria causes hepatic mitochondrial energetic failure in a mouse model. *Int. J. Biochem. Cell Biol.* **2014**, *51*, 93–101. [CrossRef]
50. Yasuda, M.; Gan, L.; Chen, B.; Kadirvel, S.; Yu, C.; Phillips, J.D.; New, M.I.; Liebow, A.; Fitzgerald, K.; Querbes, W.; et al. RNAi-mediated silencing of hepatic Alas1 effectively prevents and treats the induced acute attacks in acute intermittent porphyria mice. *Proc. Natl. Acad. Sci. USA* **2014**, *111*, 7777–7782. [CrossRef]
51. Unzu, C.; Sampedro, A.; Mauleón, I.; Alegre, M.; Beattie, S.G.; De Salamanca, R.E.; Snapper, J.; Twisk, J.; Petry, H.; González-Aseguinolaza, G.; et al. Sustained enzymatic correction by rAAV-mediated liver gene therapy protects against induced motor neuropathy in acute porphyria mice. *Mol. Ther.* **2011**, *19*, 243–250. [CrossRef]
52. Ruan, X.; Li, Z.; Zhang, Y.; Yang, L.; Pan, Y.; Wang, Z.; Feng, G.-S.; Chen, Y. Apolipoprotein A-I possesses an anti-obesity effect associated with increase of energy expenditure and up-regulation of UCP1 in brown fat. *J. Cell. Mol. Med.* **2011**, *15*, 763–772. [CrossRef]

biomedicines

MDPI

Article

Hypertrophy and Insulin Resistance of Epicardial Adipose Tissue Adipocytes: Association with the Coronary Artery Disease Severity

Natalia V. Naryzhnaya *, Olga A. Koshelskaya, Irina V. Kologrivova, Olga A. Kharitonova, Vladimir V. Evtushenko and Alla A. Boshchenko

Cardiology Research Institute, Tomsk National Research Medical Center, Russian Academy of Science, 634050 Tomsk, Russia; oshel@live.ru (O.A.K.); ikologrivova@gmail.com (I.V.K.); hoa@cardio-tomsk.ru (O.A.K.); evtushenko.vladimir@gmail.com (V.V.E.); bosh@cardio-tomsk.ru (A.A.B.)
* Correspondence: natalynar@yandex.ru; Tel.: +7-3822262174

Abstract: Changes in the structural and functional characteristics of the epicardial adipose tissue (EAT) are recognized as one of the factors in the development of cardiometabolic diseases. However, the generally accepted quantitative assessment of the accumulation of EAT does not reflect the size of adipocyte and presence of adipocyte hypertrophy in this fat depot. Overall contribution of adipocyte hypertrophy to the development and progression of coronary atherosclerosis remains unexplored. Objective: To compare the morphological characteristics of EAT adipocyte and its sensitivity to insulin with the CAD severity, as well as to identify potential factors involved in the realization of this relationship. The present study involved 24 patients (m/f 16/8) aged 53–72 years with stable CAD, who underwent coronary artery bypass graft surgery. Adipocytes were isolated enzymatically from EAT explants obtained during the operation. The severity of CAD was assessed by calculating the Gensini score according to selective coronary angiography. Insulin resistance of EAT adipocytes was evaluated by reactivity to insulin. In patients with an average size of EAT adipocytes equal to or exceeding the median (87 μm) the percentage of hypertrophic adipocytes was twice as high as in patients in whom the average size of adipocytes was less than 87 μm. This group of patients was also characterized by the higher rate of the Gensini score, lower adiponectin levels, and more severe violation of carbohydrate metabolism. We have revealed direct nonparametric correlation between the size of EAT adipocytes and the Gensini score ($r_s = 0.56$, $p = 0.00047$). The number of hypertrophic EAT adipocytes showed a direct nonparametric correlation with the Gensini score ($r_s = 0.6$, $p = 0.002$). Inverse nonparametric correlations were found between the serum adiponectin level and size ($r_s = -0.60$, $p = 0.001$), hypertrophy of adipocytes ($r_s = -0.67$, $p = 0.00$), and Gensini score ($r_s = -0.81$, $p = 0.00007$). An inverse nonparametric correlation was found between the Gensini score and sensitivity of EAT adipocytes to insulin, estimated by the intracellular redox response ($r_s = -0.90$, $p = 0.037$) and decrease in lipolysis rate upon insulin addition ($r_s = -0.40$, $p = 0.05$). The intracellular redox response of adipocytes to insulin was directly correlated with fasting insulin and inversely with postprandial insulin. Our data indicate that the size and degree of hypertrophy of the epicardial adipocytes are related to the CAD severity. According to our results, insulin resistance of adipocytes may be considered as one of the factors mediating this relationship.

Keywords: epicardial adipose tissue; hypertrophy of adipocytes; CAD severity; adipokines; insulin resistance

Citation: Naryzhnaya, N.V.; Koshelskaya, O.A.; Kologrivova, I.V.; Kharitonova, O.A.; Evtushenko, V.V.; Boshchenko, A.A. Hypertrophy and Insulin Resistance of Epicardial Adipose Tissue Adipocytes: Association with the Coronary Artery Disease Severity. *Biomedicines* **2021**, *9*, 64. https://doi.org/10.3390/biomedicines9010064

Received: 24 December 2020
Accepted: 8 January 2021
Published: 11 January 2021

1. Introduction

Epicardial adipose tissue (EAT) has been widely proven to be an important cardiovascular risk factor due to its pronounced metabolic and humoral activity adversely affecting structural and functional state of the coronary arteries. There has been an association established of EAT with cardiovascular disease [1–3], hypertension [4], diabetic status [5], and insulin resistance [6].

However, in all the above-mentioned studies, authors evaluated only quantitative characteristics of EAT (its thickness or volume) in respect to the development of the cardiometabolic diseases, while the potential interconnection between morphological parameters of EAT adipocytes and coronary atherosclerosis severity was not studied. In a limited number of studies, it was shown that neither the body mass index nor the EAT depot volume are related to the size of epicardial adipocytes [7]. Differences in the size of EAT adipocytes in individuals with and without CAD (coronary artery disease) are also reported [8]. However, there is no information about possible association of hypertrophy of EAT adipocytes with the severity of CAD. It remains unclear whether the atherogenic effect of EAT is the result of its paracrine or systemic effects.

According to the results of the clinical trials, mechanisms mediating the relationship between the accumulation of EAT and the presence and severity of CAD include impaired balance of the adipokines expressed by this fat depot [9], activation of the local and systemic inflammation [10], and the development of adipose tissue fibrosis [3]. Meanwhile, the nature of factors that realize the atherogenic effects of the hypertrophied adipocytes of EAT was investigated in just a very few studies [7,8].

Insulin resistance is a well-known pathogenic factor in the development of metabolic syndrome and cardiovascular disease [11]. Excessive accumulation of EAT in insulin-resistant patients has been demonstrated [5,6]. In epicardial adipocytes, even under physiological conditions, insulin-dependent glucose uptake and anti-lipolytic function of insulin are reduced compared to adipocytes of the subcutaneous fat depot, which is associated with the need to maintain high lipolysis activity [12]. Some publications report an even more pronounced decrease in insulin sensitivity of the epicardial adipocyte pool in patients with type 2 diabetes [13,14] and low glucose transporter GLUT4 expression on epicardial adipocytes from CAD patients [15], but there is no information on the possible connection of this process with the severity of coronary atherosclerosis.

The purpose of this study: To compare the morphological characteristics of EAT adipocyte and its sensitivity to insulin with the severity of CAD in patients undergoing coronary artery bypass graft surgery, and to identify potential factors mediating this interconnection.

Hypothesis: Morphological and functional characteristics of the epicardial adipose tissue adipocytes are interrelated with the severity of coronary artery disease (CAD). We assumed that adipocyte's hypertrophy and its level of insulin resistance are independently related to the severity of coronary atherosclerosis.

2. Experimental Section

The present pilot study was performed at Cardiology Research Institute, Tomsk National Research Medical Center, Russian Academy of Science, Tomsk, Russian Federation. The study's protocol was approved by the local ethics committee, protocol nr. 146 from 16 July 2016.

2.1. Study Participants and Clinical Characteristics of Patients

Clinical Characteristics of Patients

Twenty-four patients with stable CAD, who underwent coronary artery bypass graft surgery, comprising 16 men and 8 women aged 53–72 years, were included in the study. All subjects gave their written informed consent before being enrolled in the study.

The exclusion criteria were: Age above 75 years; presence of acute atherosclerotic complications over the past 6 months; presence of concomitant diseases, including cancer, infections, chronic obstructive pulmonary disease, mental disorders, connective tissue diseases, renal insufficiency, and liver dysfunction. Hypertension was defined as a systolic blood pressure of ≥140 mmHg and/or diastolic blood pressure of ≥90 mmHg, or as the current use of antihypertensive medication. Diabetes was defined as HbA1c concentration ≥6.5% or fasting plasma glucose level >7 mM, or the current use of antidiabetic medication.

All patients received optimal therapy. The proportion of smokers and patients with metabolic syndrome was high. The clinical characteristics of patients are presented in Table 1.

Table 1. Clinical characteristics of patients depending on the hypertrophy of adipocytes of epicardial adipose tissue.

Parameters	Total Sample (*n* = 24)	Patients with Non-Hypertrophied EAT Adipocyte, (*n* = 12)	Patients with Hypertrophied EAT Adipocyte, (*n* = 12)	*p*
Gender (m/f)	16/8	7/5	9/3	0.1
Age, years	62 (53–72)	62 (53–71)	59 (55–72)	0.64
History of myocardial infarction, *n* (%)	9 (37.5%)	5 (42%)	4 (33%)	0.1
Hypertension, *n* (%)	24 (100%)	12 (100%)	12 (100%)	1
Diabetes mellitus, *n* (%)	7 (29.2%)	3 (25%)	4 (33%)	0.9
Duration of hypertension, years	15 (10; 21)	15 (10; 20)	20 (12; 23)	0.3
Gensini score, points *	70 (28; 99)	32 (25.75; 78)	82 (52.5; 140.75)	0.024
Duration of CAD, years	2 (1; 7)	2 (1; 11)	5 (2; 6)	0.29
Systolic blood pressure, mmHg	136 (127; 142)	130 (123; 141)	140(135; 144)	0.38
Diastolic blood pressure, mmHg	80 (73; 85)	77.5 (70; 83)	81 (74; 86)	0.49
Smoking, *n* (%)	11 (46%)	6 (50%)	5 (42%)	0.1
Obesity, *n* (%)	12 (50%)	4 (33%)	8 (66%)	0.1
BMI, kg/m^2	30 (27; 31)	28.1 (25.5; 30.3)	31.2 (29.8; 35.4)	0.028
Waist circumference, cm	104 (98; 110)	100 (99; 105)	109 (103; 117)	0.0083
Waist-to-hip ratio	1 (0.93; 1.04)	1 (0.9; 1.02)	1 (0.93; 1.09)	0.21
Fat mass, kg	30.55 (26.4; 37.4)	27.7 (24.4; 36.7)	30.7 (27.5; 38.1)	0.53
Fat free mass, kg	57.6 (47.1; 61.7)	57.5 (47.1; 58.3)	57.7 (47.0; 62.3)	0.65
Skeletal muscle mass, kg	26.3 (19.7; 28.7)	24.3 (22.4; 27.2)	26.8 (18.0; 29.2)	0.82
EAT thickness, mm	4.5 (4.1; 5.4)	4.85(4.36; 5.6)	4.35 (3.88; 4.9)	0.22
EAT adipocytes size, µm	86.9 (80.97; 89.31)	80.96 (78.8; 85.75)	89.31 (88.06; 90.39	0.000037
% EAT adipocytes >100 µm	14 (9.32; 18.65)	9.32 (5.91; 11.87)	18.65 (16.08; 26.88)	0.000014
Fasting glucose, mM	5.7 (5.13; 6.13)	5.25 (5.1; 5.65)	5.98 (5.75; 7.3)	0.046
Fasting insulin, µIU/mL	5.7 (5.13; 6.13)	8.22 (6.07; 9.15)	4.95 (3.5; 5.38)	0.0039
Postprandial glucose, mM	7.1(5.7; 7.8)	7.025 (5.7; 7.7)	7.1 (5.7; 7.89)	0.96
Postprandial insulin, µIU/mL	15.45 (11.4; 21.17)	14.13 (13.34; 18.1)	16.76 (11.4; 21.17)	0.87
Glycated hemoglobin, %	6.35 (5.54; 6.92)	6.7 (5.65; 7.32)	5.91 (5.46; 6.5)	0.10
HOMA-IR	1.6 (1.15; 2.01)	1.84 (1.43; 2.29)	1.38 (0.76; 1.76)	0.088
Total cholesterol, mM	3.88 (3.25; 4.58)	3.53 (3.1; 4.28)	4.21 (3.71; 4.79)	0.23
TG, mM	1.35 (1.12; 1.58)	1.24 (0.94; 1.41)	1.44 (1.23; 1.85)	0.078
HDL, mM	1.04 (0.92; 1.21)	1.0 (0.83; 1.18)	1.06 (0.99; 1.36)	0.20
LDL, mM	2.11 (1.66; 2.55)	2.0 (1.57; 2.63)	2.25 (1.76; 2.55)	0.62
Atherogenic index	2.14 (1.5; 2.6)	2.22 (1.36; 2.9)	2.05 (1.52; 2.37)	0.58

Note: data are presented as median (Me) and interquartile range ($Q_{25\%}$; $Q_{75\%}$); *p*—significance level of differences between genders (Mann–Whitney U-test). CAD, coronary artery disease, EAT, epicardial adipose tissue; BMI, body mass index; HOMA-IR, homeostatic model assessment of insulin resistance; TG, triacylglycerols; LDL, low density lipoprotein; HDL, high density lipoprotein.

All patients underwent selective coronary angiography on a Artis one angiographic complex and Digitron-3NAC computer system (Siemens Shenzhen Magnetic Resonance Ltd., Shenzhen, China). The severity of CAD was assessed by the value of the Gensini score [16]. In the blood serum, the content of leptin (Mediagnost, Reutlingen, Germany), adiponectin (Assaypro, St. Charles, MO, USA) and insulin (AccuBind kits, Diagnostic System Laboratories, Lake Forest, California, USA) were determined by enzyme-linked immunosorbent assay (ELISA). The level of glucose was detected by hexokinase assay (EKF diagnostic, Leipzig, Germany). Enzyme colorimetric method was used to estimate serum concentration of total cholesterol, triacylglycerol, high-density lipoprotein (HDL) cholesterol (Diakon, Pushchino, Russia). Concentration of low-density lipoprotein (LDL) cholesterol was calculated using formula [LDL] = [Total cholesterol] − [Triacylglycerol (TG)] − [HDL]. Atherogenic index was calculated using formula [AI] = ([Total cholesterol] − [HDL])/[HDL].

Anthropometric measurements were performed to assess total obesity according to the level of body mass index (BMI) and abdominal obesity according to the size of the waist circumference, hip circumference, and of the waist-to-hip ratio (WHR). In 16 randomly selected patients, body composition was assessed by Bioelectrical Impedance Analysis.

EAT thickness was measured on the free wall of the right ventricle in a still image at the end diastole on the parasternal long-axis view in 3 cardiac cycles. EAT thickness was measured at the point of perpendicular orientation of the ultrasound beam on the free wall of the right ventricle, using the aortic annulus as an anatomic landmark [17,18]. The thickest point of EAT was measured in each cycle. The EAT thickness was calculated as an average value from echocardiographic views in 3 cardiac cycles.

2.2. Adipose Tissue Explants

The material for the study were the explants of the epicardial (EAT) adipose tissue weighing 0.2–1 g obtained during the CABG surgery. Epicardial fat tissue explants were taken from the tissue surrounding the proximal parts of the right coronary artery. In all cases, electrocoagulation and other types of thermal and wave effects on tissues were not used for biopsy. Samples were placed in M199 medium and delivered to the laboratory within 15 min.

Adipose tissue cells were isolated enzymatically, in sterile conditions (laminar box BAVp-01- "Laminar-s" −1.5, ZAO "Laminar systems", Miass, Russia) [19]. The tissue was minced, incubated for 35–40 min at 37 °C, and underwent constant gentle stirring (10 rpm) in 5 mL of type I collagenase sterile solution (PanEco, Moscow, Russia) 1 mg/mL in Krebs-Ringer buffer (2 mM D-glucose, 135 mM NaCl, 2.2 mM $CaCl_2 \cdot 2H_2O$, 1.25 mM $MgSO_4 \cdot 7H_2O$, 0.45 mM KH_2PO_4, 2.17 mM Na_2HPO_4, 25 mM HEPES, 3.5% BSA, 0.2 mM adenosine). Five milliliters of Krebs-Ringer buffer (37 °C) were added to the digested tissue to neutralize collagenase. The cell suspension was filtered through a nylon filter (Falcon™ Cell strainer, pore diameter 100 μm), washed three times with 10 mL of warm Krebs-Ringer buffer (37 °C). After each was, cells were allowed to float, and wash solution was discarded. The number and size of the adipocytes obtained were counted using light microscopy (Axio Observer.Z1, Carl Zeiss Surgical GmbH, Oberkochen, Germany). Cells were stained with Hoechst 33,342 (5 μg/mL, stains nucleus of viable cells) and propidium iodide (10 μg/mL, Sigma-Aldrich, St. Louis, MO, USA, stains nucleus of dead cells) to distinguish viable cells from dead cells (Figure 1) [20]. Samples with viability lower than 95% were excluded from the study. The remaining cells' samples did not differ significantly in the percentage of viable cells.

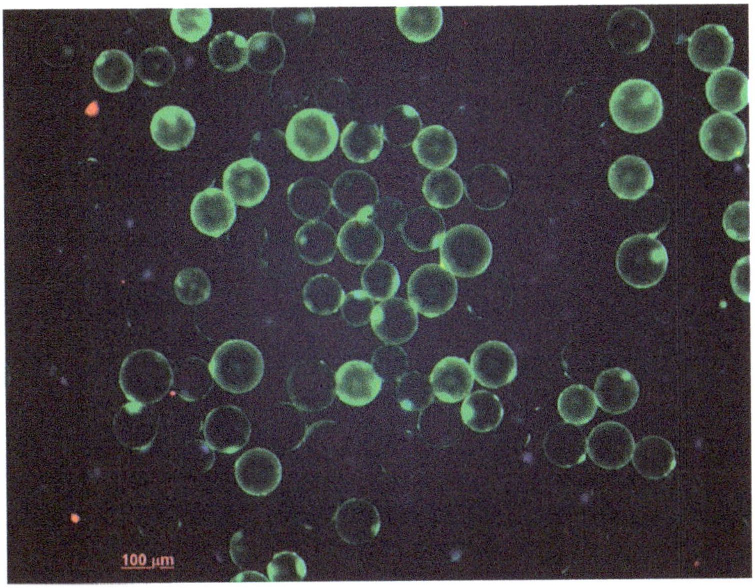

(**A**)

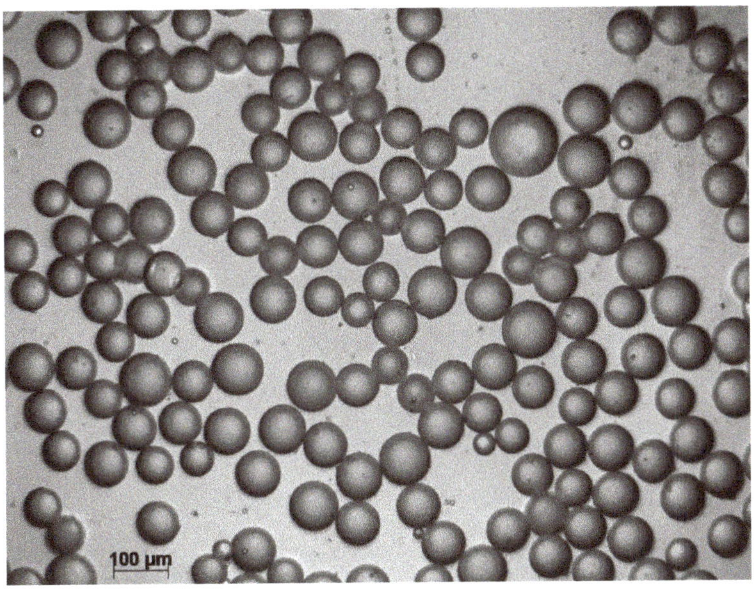

(**B**)

Figure 1. Accumulation of reactive oxygen species (ROS) and viability of adipocytes in epicardial adipose tissue (EAT) culture. (**A**) Fluorescence staining. Dyes: Green—2,3-dihydrodichlorofluorescein (ROS), red—Propidium iodid (dead cells), blue—Hoechst 33,342 (viable cells). (**B**) Corresponding light microscopy of EAT adipocytes. Magnification × 200. EAT, epicardial adipose tissue.

The median values of the size of EAT adipocytes in CAD patients were 86.9 (80.97; 89.31) µm. Adipocytes with a diameter of more than 100 µm were classified as hypertrophic. The proportion of adipocytes with size exceeding 100 µm was used as a value of adipocytes' hypertrophy. The median of this parameter constituted 14 (9.3; 18.7) %. The whole sample was divided into two groups according to the median of EAT adipocyte size and hypertrophy: If the average adipocyte size did not exceed 87 μm and percentage of hypertrophied adipocytes was less than 14%, the patient was assigned to the group of patients with non-hypertrophic adipocytes; if the average adipocyte size exceeded 87 μm and percentage of hypertrophied adipocytes was more than 14, the patient was assigned to the group of patients with hypertrophic adipocytes. All clinical parameters were calculated to these two groups (Table 1).

The sensitivity of adipocytes to insulin was estimated by increase of the production of reactive oxygen species (insulin-dependent ROS generation) [21] and by inhibition of lipolysis in response to the insulin addition to the incubation medium [13,22]. In the first case, adipocytes in 200 μL Krebs-Ringer buffer (1.25×10^6 cells/mL) were added to the two wells of a 96-well plate (500,000 cells per well) and were incubated for 30 min in the presence of 125 μM 2,3-dihydrodichlorofluorescein diacetate (DCF-DA) in a microplate reader (INFINITE 200M; Tecan, Grödig, Austria) at 37 °C for the intracellular uptake and deesterification of DCF-DA to DCF in viable adipocytes. The initial fluorescence of DCF was measured at a wavelength of λex = 500, λem = 530; 20 nM of insulin was added into one of the two wells, adipocytes were incubated for 120 min inside a microplate reader at 37 °C, and the fluorescence was measured as described above. The increase in fluorescence relative to the initial values and the increase in gain under the influence of insulin were evaluated. The study was carried out in duplicates. The cell medium after 120 min of incubation of adipocytes in the previous method was harvested and used to study the inhibition of lipolysis by insulin. The medium was degreased by the Folch reaction and glycerol was determined with an EGLY-200 kit (Gentaur).

Statistical analysis was performed using the software package "Statistica" 13.0 (StatSoft Inc., Tulsa, OK, USA). The normality of the distribution of sample data was checked by the Shapiro–Wilk test. The median and interquartile ranges of the 25th and 75th percentiles were used to describe data when data distribution differed from normal. The significance of differences between quantitative indicators in the absence of normality of data distribution was verified with the Mann–Whitney test to pair comparisons; one-way ANOVA followed by Duncan's post-hoc test were used to multiple comparisons in independent groups. When the sample data did not have normal distribution, Spearman's rank correlation coefficient (r_s) was adopted. Logistic regression was used to estimate the association between presence of EAT adipocytes hypertrophy, carbohydrate metabolism, adiponectin level, and atherosclerosis severity. All statistical hypotheses were accepted according to the achieved significance level $p < 0.05$.

To correct the possible modulation of the Gensini score by the gender factor, we adjusted the data for this indicator. In addition, the adipokine content was adjusted for sex and BMI.

3. Results

All parameters were studied depending on gender (Table 2) and ranges of BMI values (Table 3). We revealed significant differences in terms of leptin levels and adiponectin/leptin ratio between men and women (Table 2). In addition, significant differences were found in leptin levels in patients with high degrees of obesity (Table 3). To correct these effects, the adipokine content was adjusted by sex and by BMI. There were no significant differences in the Gensini score, adipocyte size, and degree of their hypertrophy depending on gender and BMI.

Table 2. Clinical characteristics of coronary artery disease (CAD) patients depending on gender.

Parameters	Total Sample (n = 24)	Men (n = 16)	Women (n = 8)	p
Age, years	62 (53–72)	59 (53–71)	63 (56–72)	0,21
BMI, kg/m^2	30 (27; 31)	29 (26; 31)	31 (30; 33)	0.11
Waist circumference, cm	104 (98; 110)	107 (99; 112)	101 (95; 107)	0.35
Waist-to-hip ratio	1 (0.93; 1.04)	1.02 (0.96; 1.05)	0.93 (0.91; 1.03)	0.14
Fat mass, kg	30.55 (26.4; 37.4)	30.3 (26.4; 36.5)	37.4 (28.5; 44.5)	0.33
Fat free mass, kg	57.6 (47.1; 61.7)	58.95 (52.6; 62.05	44.2 (44.0; 51.0)	0.025
Skeletal muscle mass, kg	26.3 (19.7; 28.7)	27.0 (23.35; 29.25	16.2 (15.10; 22.45)	0.95
EAT adipocytes size, μm	86.9 (80.97; 89.31)	87.3 (82.55; 88.87)	85.75 (80.10; 91.49)	0.92
% EAT adipocytes >100 μm	14 (9.32; 18.65)	13.86 (9.84; 18.43)	15.54 (8.16; 30.13)	0.6
% EAT adipocytes <50 μm	2.08 (0.84; 3.86)	1.84 (0.99; 3.1)	3.67 (0.43; 5.09)	0.49
Gensini score, points	70 (28; 99)	71 (32; 123)	35 (28; 82)	0.31
EAT thickness, mm	4.5 (4.1; 5.4)	4.36 (4.0; 5.0)	4.98 (4.5; 5.6)	0.19
Fasting glucose, mM	5.7 (5.1; 6.1)	5.7 (5.1; 6.3)	5.8 (5.4; 6.1)	0.61
Fasting insulin, μIU/mL	5.7 (5.13; 6.13)	5.49 (3.56; 7.62)	7.03 (5.01; 11.12)	0.14
Postprandial glucose, mM	7.1(5.7; 7.8)	6.7 (5.7; 7.7)	7.5 (6.8; 7.9)	0.38
Postprandial insulin, μIU/mL	15.45 (11.4; 21.17)	14.13 (11.87; 21.17)	16.76 (11.40; 17.12)	0.95
Glycated hemoglobin, %	6.35 (5.54; 6.92)	5.8 (5.46; 6.8)	6.69 (6.11; 9.09)	0.17
HOMA-IR	1.6 (1.15; 2.01)	1.6 (0.76; 1.77)	1.7 (1.27; 3.79)	0.23
Total cholesterol, mM	3.88 (3.25; 4.58)	3.87 (3.12; 4.38)	3.94 (3.58; 4.82)	0.54
TG, mM	1.35 (1.12; 1.58)	1.32(1.14; 1.44)	1.48 (1.01; 2.18)	0.43
HDL, mM	1.04 (0.92; 1.21)	1.05 (0.96; 1.18)	1.00 (0.87; 1.31)	0.83
LDL, mM	2.11 (1.66; 2.55)	1.98 (1.62; 2.57)	2.25 (1.93; 2.54)	0.56
Atherogenic index	2.14 (1.5; 2.6)	2.03 (1.36; 2.8)	2.24 (1.67; 2.39)	0.6
Adiponectin, μg/mL	7.25 (4.85; 9.91)	7.64 (4.77; 10.07)	6.70 (4.85; 9.92)	0.73
Leptin, ng/mL	17.49 (8.63; 27.01)	11.33 (7.36; 17.49)	40.25 (23.68; 67.58)	0.0017
Adiponectin/leptin	0.46 (0.17; 0.80)	0.8 (0.37; 1.08)	0.13 (0.08; 0.20)	0.0077

Note: data are presented as median (Me) and interquartile range ($Q_{25\%}$; $Q_{75\%}$); p—significance level of differences between genders (Mann–Whitney U-test). CAD, coronary artery disease, EAT, epicardial adipose tissue; BMI, body mass index; HOMA-IR, homeostatic model assessment of insulin resistance; TG, triacylglycerols; LDL, low density lipoprotein; HDL, high density lipoprotein.

Table 3. Clinical characteristics of CAD patients depending on BMI.

Parameters	Total Sample (n = 24)	BMI < 25 (n = 3)	25 < BMI < 30 (n = 9)	30 < BMI < 35 (n = 8)	35 < BMI < 40 (n = 4)
Age, years	62 (53–72)	59 (55–71)	61 (53–67)	63 (55–71)	63 (57–72)
BMI, kg/m^2	30 (27; 31)	24.6 (20.3; 24.7)	28.1 (27.4; 29.1) *,&,‡	31.2 (30.8; 31.4) *,#,‡	37.6 (36.1; 39.5) *,#,&
Waist circumference, cm	104 (98; 110)	98 (76; 100)	102 (97; 106)	109 (104; 110)	119 (105; 120)*
Waist-to-hip ratio	1 (0.93; 1.04)	1.02 (0.84; 1.02)	0.94 (0.9; 0.98)	1.04 (1.01; 1.08)	0.95 (0.92; 1.02)
EAT adipocytes size, μm	87 (81; 89)	87 (79; 87)	85 (79; 88)	87 (83; 89)	90 (89; 92)
% EAT adipocytes >100 μm	14 (9.3; 18.7)	12.3 (5.7; 18.2)	11.4 (5.9; 15.2)	13.9 (9.4; 17.9)	26.9 (23.5; 30.1) *,#
Gensini score, points	70 (28; 99)	110 (44; 156)	28 (24.5; 121)	69.5 (32; 73.5)	62 (34.5; 146)
EAT thickness, mm	4.5 (4.1; 5.4)	4.1 (3.5; 4.9)	4.35 (4.0; 4.4)	5.35 (4.60; 5.94)	4.65 (4.18; 4.9)
Fasting glucose, mM	5.7 (5.1; 6.1)	5.1 (4.3; 5.15)	5.3 (5.1; 6.0)	5.85 (5.55; 7.11)	6.24 (5.79; 7.6)

Table 3. Cont.

Parameters	**Total Sample (n = 24)**	**BMI < 25 (n = 3)**	**25 < BMI < 30 (n = 9)**	**30 < BMI < 35 (n = 8)**	**35 < BMI < 40 (n = 4)**
Fasting insulin, μIU/mL	5.7 (5.13; 6.13)	2.47 (1.85; 5.62)	7.48 (4.98; 8.6)	5.93 (4.97; 11.1)	5.71 (4.97; 6.9)
Postprandial glucose, mM	7.1(5.7; 7.8)	5.7 (4.9; 7.7)	6.9 (5.7; 7.9)	7.24 (5.58; 7.90)	7.34 (7.3; 7.37)
Postprandial insulin, μIU/mL	15.45 (11.4; 21.17)	18.1 (10.08; 20.1)	13.59 (11.06; 32.31)	15.62 (11.40; 41.67)	14.32 (11.87; 16.76)
Glycated hemoglobin, %	6.35 (5.54; 6.92)	5.65 (5.27; 7.32)	6.07 (5.59; 6.43)	6.8 (5.57; 9.09)	6.5 (5.5; 6.8)
HOMA-IR	1.6 (1.15; 2.01)	0.47 (0.42; 1.27)	1.7 (1.15; 2.03)	1.68 (1.38; 3.63)	1.59 (1.28; 2.37)
Total cholesterol, mM	3.88 (3.25; 4.58)	3.28 (3.22; 5.12)	3.76 (3.58; 4.31)	4.21 (2.68; 4.89)	3.9 (3.03; 4.42)
TG, mM	1.35 (1.12; 1.58)	1.37 (0.77; 1.45)	1.27 (1.14; 1.50)	1.43 (1.12; 2.01)	1.49 (1.19; 1.72)
HDL, mM	1.04 (0.92; 1.21)	1.09 (0.64; 1.18)	1.04 (1.01; 1.36)	0.9 (0.81; 1.26)	1.06 (1.01; 1.15)
LDL, mM	2.11 (1.66; 2.55)	1,95 (1.75; 3.36)	2.04 (1.68; 2.42)	2.13 (1.39; 2.56)	2.27 (1.92; 2.48)
Atherogenic index	2.14 (1.5; 2.6)	3,05 (1.48; 3.09)	2,12 (1.24; 2.45)	1.87 (1.37; 2.58)	2.17 (1.84; 2.25)
Adiponectin, μg/mL	7.25 (4.85; 9.91)	8.58 (6.26; 24.38)	5.73 (4.6; 11.94)	7.46 (4.85; 9.92)	6.89 (6.14; 7.64)
Leptin, ng/mL	17.49 (8.63; 27.01)	5.96 (1.52; 6.40)	11.64 (8.63; 23.65)	27.83 (16.74; 48.52)	48.23 (20.72; 75.74) *,#
Adiponectin/leptin	0.46 (0.17; 0.80)	1.34 (1.05; 1.04)	0.56 (0.25; 0.8)	0.18 (0.11; 0.46)	0.23 (0.08; 0.37)

Note: data are presented as median (Me) and interquartile range ($Q_{25\%}$; $Q_{75\%}$); *—significance level of differences to BMI < 25; #—significance level of differences to 25 < BMI < 30; &—significance level of differences to 30 < BMI < 35; ‡—significance level of differences to 35 < BMI < 40; $p < 0.05$, one-way ANOVA followed by Duncan's post-hoc test. CAD, coronary artery disease, EAT, epicardial adipose tissue; BMI, body mass index; HOMA-IR, homeostatic model assessment of insulin resistance; TG, triacylglycerols; LDL, low density lipoprotein; HDL, high density lipoprotein.

The median values of the size of EAT adipocytes in CAD patients were 86.9 (80.97; 89.31) μm. The proportion of adipocytes with size exceeding 100 μm was used as a value of adipocytes' hypertrophy. The median of this parameter constituted 14 (9.3; 18.7)% (Table 1). Thus, the medians of the EAT adipocyte size in both groups differed by 8.35 μm (p = 0.000037), and the proportion of hypertrophic adipocytes in group of the patients with hypertrophied EAT adipocytes was twice the value of this parameter in group of patients with non-hypertrophied EAT adipocytes (Figure 2, Table 1).

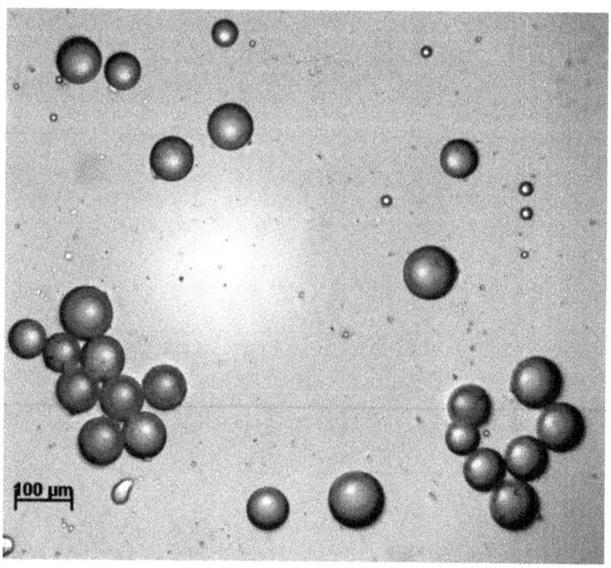

(**A**)

Figure 2. *Cont.*

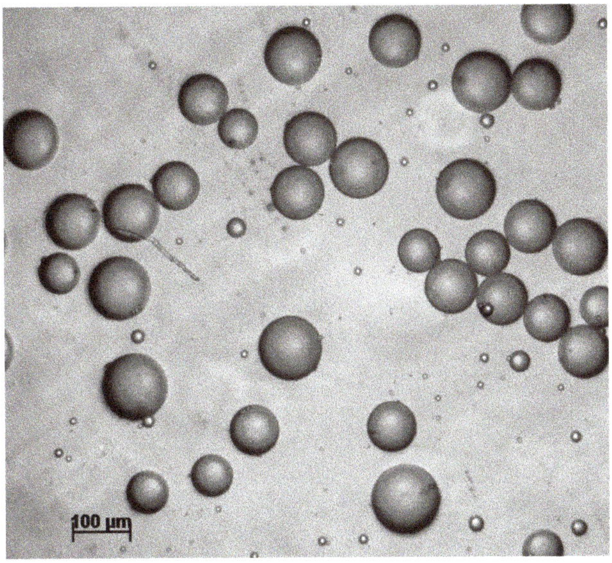

(B)

Figure 2. The representative snapshots of EAT adipocytes of patients from groups of patients with non-hypertrophied EAT adipocytes and with hypertrophied EAT adipocytes. (**A**) Patient from group of patients with non-hypertrophied EAT adipocytes, median of adipocytes diameter is 80.27 µm; (**B**) patient from group of patients with hypertrophied EAT adipocytes, median of adipocytes diameter is 89.88 µm. Light microscopy, magnification × 200. EAT, epicardial adipose tissue.

The level of fasting glucose in patients from the second group exceeded the values in group of patients with non-hypertrophied EAT adipocytes by 14% (p = 0.46, Table 1). However, the level of fasting insulin was 1.7 times lower in group of patients with hypertrophied EAT adipocytes compared to group of patients with non-hypertrophied EAT adipocytes (p = 0.0039, Table 1). The parameters of postprandial glucose, postprandial insulin, glycated hemoglobin. or HOMA-IR did not differ between groups of patients with hypertrophied and non-hypertrophied EAT adipocytes. We did not reveal any differences in parameters of lipid profile, namely triacylglycerols, total cholesterol and its fractions, as well as atherogenic index (Table 1).

Patients from group with hypertrophied EAT adipocytes also had higher values of body mass index, waist circumference (Table 1) and Gensini score (Table 4) values than those in group of patients with non- hypertrophied EAT adipocytes.

Table 4. Coronary artery disease severity and adipokine's levels adjusted to gender and BMI depending on the hypertrophy of adipocytes of epicardial adipose tissue.

Parameters	Total Sample (n = 24)	Patients with Non-Hypertrophied EAT Adipocyte (1) (n = 12)	Patients with Hypertrophied EAT Adipocyte (2) (n = 12)	p
* Gensini score, points	69.5 (46.3; 138.3)	55.28 (35.50; 78.58)	121.52 (67.18; 162.69)	0.05
# Serum adiponectin, µg/mL	7.31 (5.73; 9.59)	8.29 (7.31; 14.81)	6.15 (5.40; 7.46)	0.039
# Serum leptin, ng/mL	17.47 (10.73; 22.35)	17.47 (12.71; 20.78)	18.11 (9.25; 40.38)	0.69
# Serum adiponectin/leptin	0.48 (0.25; 0.752)	0.57 (0.48; 0.75)	0.24 (0.14; 0.65)	0.06
Fasting glucose, mM	5.7 (5.13; 6.13)	5.25 (5.1; 5.65)	5.98 (5.75; 7.3)	0.046
Fasting insulin, µIU/mL	5.51 (4.78; 8.22)	8.22 (6.07; 9.15)	4.95 (3.5; 5.38)	0.0039

Note: data are presented as median Me and interquartile range ($Q_{25\%}$; $Q_{75\%}$); *—data were adjusted to gender; #—data were adjusted to gender and BMI; p—significance level of differences between the patients with non-hypertrophied EAT adipocyte and patients with hypertrophied EAT adipocyte (Mann–Whitney U-test). BMI, body mass index, EAT, epicardial adipose tissue.

The results allow us to conclude that the average EAT adipocyte size equal to or exceeding 87 µm and the proportion of hypertrophied adipocytes of more than 14% are associated with a significantly greater severity of CAD. The results of the study showed that in the group of patients with hypertrophied adipocytes, the EAT serum adiponectin adjusted to gender and BMI was 1.34 times lower than in patients with non-hypertrophic adipocytes. Based on these results, we hypothesized that adiponectin, as well as indicators of carbohydrate metabolism, may be signs associated with adipocyte hypertrophy.

Using logistic regression, a mathematical model was built that made it possible to identify significant factors influencing the likelihood of adipocyte hypertrophy EAT (Table 5):

$$P = 1/(1 + e^{-z}),$$

$$z = -2.57 + 11\ \text{FGluc} - 6.7\ \text{FIns} - 4.05\ \text{Adip} + 0.0334\ \text{GS}$$

where P—the probability of belonging to the second group; FGluc—fasting glucose, mM; FIns—fasting insulin, µIU/mL; Adip—serum adiponectin, µg/mL, adjusted to gender and BMI; GS—Gensini score, points, adjusted to gender.

Table 5. Logistic regression results.

Parameters	Regression Coefficient	p
Fasting glucose, mM	11	<0.0001
Fasting insulin, µIU/mL	−6.7	<0.0001
Serum adiponectin, µg/mL	−4.05	<0.0001
Gensini score, points	0.0334	0.048

Note: Serum adiponectin level adjusted to gender and BMI; Gensini score adjusted to gender.

As can be seen, the greatest association of EAT adipocyte hypertrophy is observed with an increase in fasting glucose, a decrease in fasting insulin, and a decrease in adiponectin serum. The Gensini score is less associated with EAT hypertrophy (Table 5).

The results of the study showed that the size and degree of the EAT adipocytes hypertrophy positively correlated with BMI, waist, and hip circumferences (Table 6). Interestingly, the size of EAT adipocyte and the severity of its hypertrophy were characterized by positive correlations with the Gensini score (r_s = 0.52, p = 0.009 and r_s = 0.41, p = 0.044, respectively) (Table 6, Figure 3), while there were no correlations between the size or hypertrophy of epicardial adipocyte and the thickness of the EAT. None of the anthropometric measurements of obesity, excluding EAT adipocyte size, correlated with

Gensini score. Direct correlations were found between the fasting glucose level and the size of EAT adipocytes as well as with the degree of adipocyte's hypertrophy (Table 6).

Table 6. Correlation between the size of EAT adipocyte and the degree of its hypertrophy with the measurements of obesity, carbohydrates metabolism, and Gensini score.

	EAT Adipocytes Size		EAT Adipocytes Hypertrophy	
Parameters	r_s	p	r_s	p
BMI, kg/m^2	0.45	0.028	0.59	0.002
Waist circumference, cm	0.38	0.063	0.34	0.11
Hip circumference, cm	0.41	0.049	0.51	0.01
Waist-to-hip ratio	0.16	0.46	0.011	0.96
EAT thickness, mm	−0.33	0.11	−0.30	0.16
* Gensini score, points	0.56	0.00047	0.6	0.002
CAD duration, years	0.29	0.16	0.25	0.23
Fasting glucose, mM	0.43	0.034	0.43	0.037
Fasting insulin, μIU/mL	−0.37	0.076	−0.23	0.28
Postprandial glucose, mM	0.10	0.67	0.08	0.73
Postprandial insulin, μIU/mL	−0.25	0.32	−0.12	0.63
Glycated hemoglobin, %	−0.13	0.59	−0.07	0.78
HOMA-IR	−0.21	0.38	−0.07	0.79
Total cholesterol, mM	0.17	0.42	0.17	0.43
TG, mM	0.34	0.11	0.3	0.15
HDL, mM	0.20	0.35	0.18	0.40
LDL, mM	0.11	0.61	0.12	0.57
Atherogenic index	−0.05	0.83	−0.02	0.93

*—data were adjusted to gender; r_s, Spearman rank correlation coefficient; CAD, coronary artery disease; EAT, epicardial adipose tissue; BMI, body mass index; HOMA-IR, homeostatic model assessment of insulin resistance; TG, triacylglycerols; LDL, low density lipoprotein; HDL, high density lipoprotein.

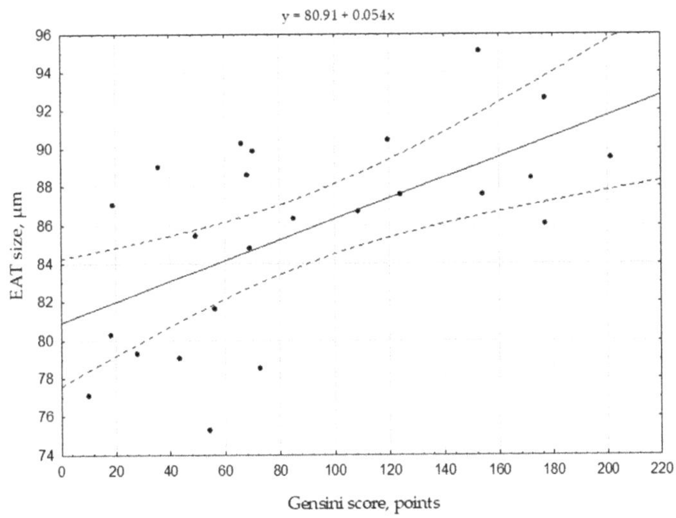

(A)

Figure 3. *Cont.*

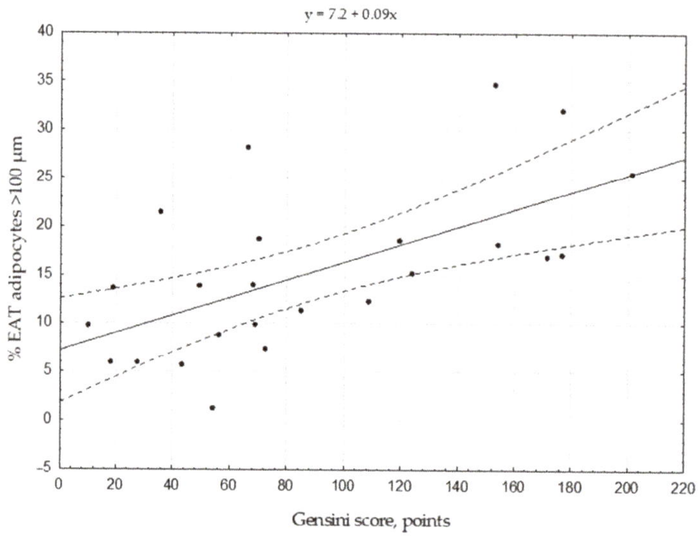

(**B**)

Figure 3. Scatter plots displaying the size of EAT adipocyte (**A**) and the proportion of hypertrophied adipocytes (**B**) related to the Gensini score. Note: EAT, epicardial adipose tissue. Gensini score was adjusted to gender.

Further, we attempted to detect factors that are potentially involved in the realization of the relationships between the severity of CAD, the size of EAT adipocyte, and the degree of adipocyte hypertrophy.

In patients with an average adipocyte size less than 87 µm, the serum adiponectin levels were significantly higher than in patients with average EAT adipocyte size exceeding 87 µm (Table 4).

The severity of CAD, estimated by the Gensini score, inversely correlated with adiponectin concentrations in the blood serum (Table 7). In addition, inverse correlations were found between the serum adiponectin levels and the size and degree of hypertrophy of EAT adipocyte: $r_s = -0.53$ and $r_s = -0.59$, respectively (Table 7), whereas correlations between the adipocyte's size and EAT adipocyte hypertrophy with serum leptin were absent.

Table 7. Correlation between serum adiponectin, leptin, EAT adipocyte size, EAT adipocyte hypertrophy, and Gensini score.

	EAT Adipocyte Size		EAT Adipocyte Hypertrophy		Gensini Score	
Parameters	r_s	p	r_s	p	r_s	p
Serum adiponectin	−0.60	0.01	−0.67	0.0029	−0.81	0.00007
Serum leptin	0.08	0.75	0.027	0.91	0.30	0.23
Serum adiponectin/leptin	−0.50	0.04	−0.48	0.047	−0.65	0.0044

Note: data are presented as median (Me) and interquartile range ($Q_{25\%}$; $Q_{75\%}$); r_s—Spearman rank correlation coefficient; adipokines' levels were adjusted to gender and body mass index; Gensini score was adjusted to gender; EAT, epicardial adipose tissue.

To assess the potential correlations between the CAD severity the functional characteristics of EAT adipocytes, we evaluated the degree of insulin-dependent ROS generation and insulin-dependent inhibition of lipolysis, as parameters reflecting the sensitivity of adipocytes to insulin.

ROS generation increased 1.11 (1.00; 1.33) times in response to insulin addition to EAT adipocytes' suspension from CAD patients compared to the control sample incubated without addition of insulin (Figure 4). Moreover, we have observed inhibition of lipolysis in response to insulin (a decrease in glycerol production compared to control values) by 1.42 (1.33; 1.78) times. Inverse correlations were found between the sensitivity of the EAT adipocytes to insulin (insulin-dependent ROS generation and insulin-dependent inhibition of lipolysis) and the values of the Gensini score (Table 6).

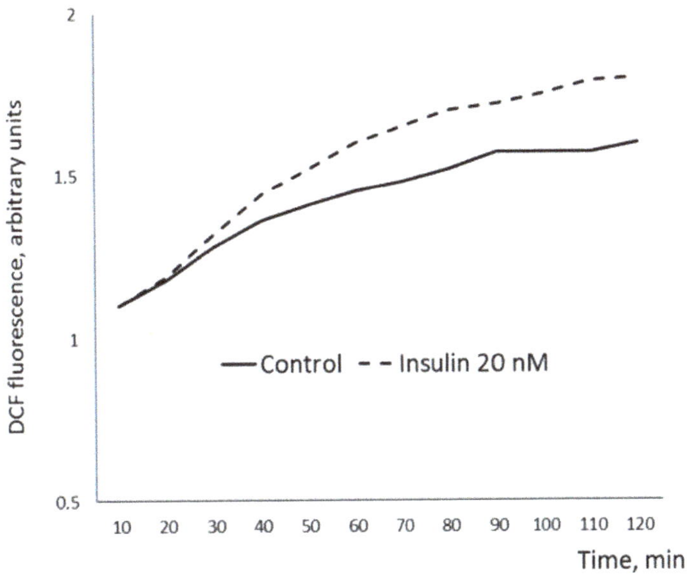

Figure 4. Representative dynamics of insulin-dependent ROS generation in EAT adipocytes. X axis—time, min; Y axis—2,3-dihydrodichlorofluorescein (DCF) fluorescence, arbitrary units, reference to blank. EAT, epicardial adipose tissue; ROS, reactive oxygen spices.

Direct correlation was found between insulin-dependent ROS generation and fasting insulin levels in blood serum inverse correlation was found between insulin-dependent ROS generation and postprandial insulin (Table 8).

Table 8. The correlation between insulin sensitivity and insulin resistance of EAT adipocyte with Gensini score or postprandial insulin.

Parameter	r_s	p
Insulin-dependent ROS generation/Gensini score	−0.90	0.037
Insulin-dependent inhibition of lipolysis/Gensini score	−0.40	0.051
Insulin-dependent ROS generation/Fasting insulin	0.90	0.037
Insulin-dependent ROS generation/Postprandial insulin	−0.90	0.037

Note: ROS, reactive oxygen spices; r_s Spearman rank correlation coefficient; Gensini score was adjusted to gender. EAT, epicardial adipose tissue; ROS, reactive oxygen spices.

Due to the small volume of the explant and the impossibility of carry out this study in all patients, it was impossible to assess the insulin-dependent ROS production and inhibition of lipolysis by groups depending on the size of the adipocyte.

4. Discussion

Our results showed that the size and hypertrophy of EAT adipocytes reflects the degree of systemic obesity and has a direct relationship with the severity of CAD. We have found that the adipocyte's size and its degree of hypertrophy is associated with the level of fasting glucose. Since the fasting level of glucose was not significantly elevated, the development of EAT adipocyte's hypertrophy might be assumed to precede the formation of systemic metabolic impairments. Fang L. et al. have shown that manifestation of diabetes mellitus type 2, on the contrary, is associated with an increase of the % small cells in subcutaneous and omental depots of adipose tissue, but EAT adipocytes have not been studied in this work [23].

According to our data, no relationships were found between the thickness of EAT and an average adipocyte size of the EAT. This result is in accordance with the recent studies showing that the thickness of EAT is determined not only by adipocytes' hypertrophy, but is also dependent upon an increase of the stromal component of this tissue [7]. Adipocytes hyperplasia has been also shown to play an important role in increase of the volume of this fat depot [24]. It can be assumed that absence of direct relationships between the adipocytes' size and EAT thickness is a consequence of the different severity of hyperplasia in patients. Differences in the degree of adipocyte hyperplasia in diabetic and non-diabetic patients were also found for another visceral fat depot—omentum [23]. However, the role of the intensity of preadipocytes differentiation (hyperplasia) of EAT in the development of CAD has not yet been studied. For the first time, we showed a significant relationship between the severity of CAD, estimated by the value of the Gensini score, the average size of EAT adipocytes, and the degree of their hypertrophy. This observation suggests that not only the quantitative determination of EAT (thickness or its volume) in itself, but also the morphological characteristics of epicardial adipocytes, reflecting the severity of their hypertrophy, namely, the average adipocyte size and proportion of cells with a size exceeding 100 μm, can be of high importance in assessing the pathological association of the epicardial fat depot with the development and progression of coronary atherosclerosis, as it allows to exclude influence of accumulation of the fibrotic component of EAT.

Adipokine imbalance, adiponectin content decrease in particular, is considered as one of the factors in the formation of coronary atherosclerosis [8,25,26]. However, the literature on the relationship of adiponectin with the formation of atherosclerosis is contradictory. Thus, the established point of view on the beneficial effects of adiponectin is based on the following data: Decrease of this parameter in patients with coronary artery disease [25]; decreased expression of adiponectin gene in epicardial adipose tissue in patients with cardiovascular disease [8]; inverse correlation of adiponectin levels and atherogenic blood plasma index [26]. Polymorphism of the gene encoding adiponectin is directly associated with the development of coronary atherosclerosis [25]. However, a meta-analysis showed that in the group of patients with elevated adiponectin levels, there was no change in the frequency of adverse cardiovascular events [27]. Moreover, distinct studies have not confirmed the difference in the severity of atherosclerosis in patients with high and low adiponectin levels [28]. One of the meta-analysis did not show a connection between transcription of adipokine genes by adipocytes of EAT and the presence of cardiovascular diseases [29]. These data indicate that information about the relationship of the adipokine profile with the degree of cardiovascular risk is incomplete.

In our study, we documented the inverse relationship of the serum adiponectin levels and the severity of CAD, which supports the point of view about the negative effect of low adiponectin on the formation of coronary atherosclerosis. It was shown that the EAT thickness is inversely related to the serum adiponectin levels regardless of the presence of CAD, as it is detected both in patients with documented CAD [30], and in patients without CAD [31]. The same pattern was found in patients with metabolic syndrome [32]. Nevertheless, there also exists evidence of the absence of such a relationship [33]. To date, comprehensive information on the ratio of the size of the adipocyte and serum adiponectin was not available in the literature. In our study, inverse correlations were established between the serum adiponectin levels, the size of the EAT adipocyte, and the degree of its hypertrophy. In addition, data were obtained on the existence of an inverse relationship between the size of EAT adipocyte, concentration of serum adiponectin, and the severity of CAD.

The mechanism of anti-atherogenic effect of adiponectin, according to the literature, includes the improvement of endothelial function and monocytes-endothelium interactions; inhibition of smooth muscle cell proliferation; reduced macrophage uptake of cholesterol and suppression of foam cells' formation [9]. In addition to the abovementioned effects of adiponectin, it can be assumed that its anti-atherogenic effect can be achieved by improving the metabolic function of adipocytes. It was found that a decrease in the production of adiponectin in EAT is associated with a violation of oxidative phosphorylation of epicardial adipocytes and is associated with the severity of coronary atherosclerosis [34]. In addition, adiponectin was found to stimulate glucose utilization and fatty acid oxidation via adenosine monophosphate activated protein kinase (AMPK) [35]. A number of experimental and clinical studies have shown an increase in insulin-dependent uptake of glucose by adipocytes under the influence of adiponectin [36,37] and the association between a decrease in adiponectin and insulin resistance [38], of which importance in the formation of cardiovascular pathology is now generally appreciated [11]. However, the question on the role of insulin resistance of epicardial fat adipocytes in the development of atherosclerosis remains poorly understood.

This study found a direct correlation between insulin resistance of EAT adipocytes and the severity of coronary atherosclerosis which is consistent with previous studies. Even more so, we demonstrated that insulin sensitivity of EAT adipocyte directly correlates with decreased fasting insulin concentration and inversely—with its postprandial production. Thus, we are the first to show that the degree of EAT adipocytes' insulin resistance is associated with the severity of atherosclerosis and systemic imbalance of insulin production independently from the size and hypertrophy of adipocytes.

EAT is known to be characterized by the certain basic level of insulin resistance. Even under the normal conditions, the expression of GLUT4 is reduced in EAT compared to subcutaneous adipose tissue. Thus, insulin-dependent glucose uptake and anti-lipolytic function of insulin in this fat depot are reduced. Adipocytes of CAD patients are characterized by an even more pronounced decrease in GLUT4 and an increased content of retinol-binding protein-4 (RBP4), associated with the development of insulin resistance [13,39]. Epicardial adipocytes are characterized by reduced expression of genes regulating lipid metabolism, in particular lipoprotein lipase being dependent on the size of the adipocyte, which leads to stimulation of lipolysis in hypertrophic adipocytes [40]. Violation of insulin sensitivity of EAT adipocytes can lead to even greater decrease in the ability of the epicardial fat depot to accumulate fatty acids and, thereby, regulate fatty acids' flow to the myocardium. This mechanism may cause, in particular, the absence of excessive accumulation of EAT in a certain cohort of CAD patients. It should be noted, however, that this assumption needs further proof. Indirect evidence of the important role of EAT adipocytes' insulin resistance in the formation of cardiovascular disease is provided by the fact that pharmacological increase in the sensitivity of adipocytes to glucose, induced by the sodium glucose cotransporter-2 inhibitor (SGLT2) dapagliflozin [41], reduces the risk of adverse cardiovascular events in patients with type 2 diabetes mellitus [42].

Our study is one of the few in which the cellular mechanisms of the relationships between EAT and the severity of coronary atherosclerosis are addressed in a clinical setting. Further studies involving a comparison group and prospective observation are required for substantiating the expansion of the cluster of the metabolic syndrome and improving the stratification of cardiovascular risk in patients with visceral obesity.

Limitations of this study are its cross-sectional nature and small sample of patients, which did not allow to study potential gender differences in relationships between morphological and functional characteristics of the epicardial adipose tissue adipocytes and CAD severity. The available amount of EAT collected from several patients was insufficient to perform all experiments within a portion of the same sample. All the recruited patients continuously received statins, which are known to affect adipocyte size. We presume that this might have affected absence of associations between the parameters of lipid profile, adipokines' concentrations and severity of atherosclerosis.

5. Conclusions

Morphological and functional characteristics of adipocytes from EAT are interrelated with the severity of atherosclerotic lesions of the coronary arteries. Violation of insulin sensitivity of EAT adipocytes and an imbalance in the secretion of adipokines are among the most plausible factors determining this relationship. Adipocyte's hypertrophy and its level of insulin resistance are independently related to the severity of coronary atherosclerosis. Hypertrophic adipocytes size are associated with insulin resistance, low plasma circulating adiponectin levels and CAD severity.

Author Contributions: Conceptualization, O.A.K. (Olga A. Koshelskaya), N.V.N., and A.A.B.; methodology, N.V.N., O.A.K. (Olga A. Koshelskaya), I.V.K., and V.V.E.; formal analysis, I.V.K. and O.A.K. (Olga A. Kharitonova); investigation, I.V.K., O.A.K. (Olga A. Kharitonova), and V.V.E.; data curation, O.A.K. (Olga A. Kharitonova); writing—original draft preparation, N.V.N. and O.A.K. (Olga A Koshelskaya); writing—review and editing, N.V.N., O.A.K. (Olga A. Koshelskaya), and I.V.K.; visualization, N.V.N.; supervision, project administration, A.A.B. All authors have read and agreed to the published version of the manuscript.

Funding: The study was carried out in the framework of the fundamental research No. AAAA-A15-115123110026-3.

Institutional Review Board Statement: The study was conducted according to the guidelines of the Declaration of Helsinki, and approved by the Institutional Ethics Committee of Cardiology Research Institute, Tomsk National Research Medical Center, Russian Academy of Science, Tomsk, Russian Federation, protocol nr. 146 from 16 July 2016.

Informed Consent Statement: Informed consent was obtained from all subjects involved in the study.

Data Availability Statement: The data presented in this study are available on request from the corresponding author.

Acknowledgments: The authors express their gratitude to Suslova Tatyana, Evtushenko Alexey, Andreev Sergey and Margolis Natalia for research assistance.

Conflicts of Interest: The authors declare no conflict of interest.

Abbreviations

BMI, body mass index; CAD, coronary artery disease; EAT, epicardial adipose tissue; HDL, high density lipoprotein; HOMA-IR, homeostatic model assessment of insulin resistance; LDL, low density lipoprotein; ROS, reactive oxygen spices; TG, triacylglycerol; WHR, waist-to-hip ratio.

References

1. Erkan, A.F.; Tanindi, A.; Kocaman, S.A.; Ugurlu, M.; Tore, H.F. Epicardial Adipose Tissue Thickness Is an Independent Predictor of Critical and Complex Coronary Artery Disease by Gensini and Syntax Scores Epicardial. *Texas Heart Inst. J.* **2016**, *43*, 29–37. [CrossRef] [PubMed]
2. Mancio, J.; Azevedo, D.; Saraiva, F.; Azevedo, A.I.; Pires-Morais, G.; Leite-Moreira, A.; Falcao-Pires, I.; Lunet, N.; Bettencourt, N. Epicardial Adipose Tissue Volume Assessed by Computed Tomography and Coronary Artery Disease: A Systematic Review and Meta-Analysis. *Eur. Heart J. Cardiovasc. Imaging* **2018**, *19*, 490–497. [CrossRef] [PubMed]
3. Villasante Fricke, A.C.; Iacobellis, G. Epicardial Adipose Tissue: Clinical Biomarker of Cardio-Metabolic Risk. *Int. J. Mol. Sci.* **2019**, *20*, 5989. [CrossRef] [PubMed]
4. Sharma, A.M. Mediastinal Fat, Insulin Resistance, and Hypertension. *Hypertension* **2004**, *44*, 117–118. [CrossRef]
5. Li, Y.; Liu, B.; Li, Y.; Jing, X.; Deng, S.; Yan, Y.; She, Q. Epicardial Fat Tissue in Patients with Diabetes Mellitus: A Systematic Review and Meta-Analysis. *Cardiovasc. Diabetol.* **2019**, *18*, 3. [CrossRef]
6. Iacobellis, G.; Leonetti, F. Epicardial Adipose Tissue and Insulin Resistance in Obese Subjects. *J. Clin. Endocrinol. Metab.* **2005**, *90*, 6300–6302. [CrossRef]
7. Aitken-Buck, H.M.; Moharram, M.; Babakr, A.A.; Reijers, R.; Van Hout, I.; Fomison-Nurse, I.C.; Sugunesegran, R.; Bhagwat, K.; Davis, P.J.; Bunton, R.W.; et al. Relationship between Epicardial Adipose Tissue Thickness and Epicardial Adipocyte Size with Increasing Body Mass Index. *Adipocyte* **2019**, *8*, 412–420. [CrossRef]

8. Vianello, E.; Dozio, E.; Arnaboldi, F.; Marazzi, M.G.; Martinelli, C.; Lamont, J.; Tacchini, L.; Sigrüner, A.; Schmitz, G.; Corsi Romanelli, M.M. Epicardial Adipocyte Hypertrophy: Association with M1-Polarization and Toll-like Receptor Pathways in Coronary Artery Disease Patients. *Nutr. Metab. Cardiovasc. Dis.* **2016**, *26*, 246–253. [CrossRef]
9. Yanai, H.; Yoshida, H. Beneficial Effects of Adiponectin on Glucose and Lipid Metabolism and Atherosclerotic Progression: Mechanisms and Perspectives. *Int. J. Mol. Sci.* **2019**, *20*, 1190. [CrossRef]
10. Koshelskaya, O.A.; Suslova, T.E.; Kologrivova, I.V.; Margolis, N.Y.; Zhuravleva, O.A.; Kharitonova, O.A.; Kravchenko, E.S.; Vinnitskaya, I.V.; Karpov, R.S. Epicardial Fat Thickness and Biomarkers of Inflammation in Patients with Stable Coronary Artery Disease: Correlation with the Severity of Coronary Atherosclerosis. *Russ. J. Cardiol.* **2019**, *4*, 20–26. [CrossRef]
11. Reaven, G.M. Insulin Resistance: The Link Between Obesity and Cardiovascular Disease. *Endocrinol. Metab. Clin. N. Am.* **2008**, *37*, 581–601. [CrossRef] [PubMed]
12. Patel, V.B.; Shah, S.; Verma, S.; Oudit, G.Y. Epicardial Adipose Tissue as a Metabolic Transducer: Role in Heart Failure and Coronary Artery Disease. *Heart Fail. Rev.* **2017**, *22*, 889–902. [CrossRef] [PubMed]
13. Burgeiro, A.; Fuhrmann, A.; Cherian, S.; Espinoza, D.; Jarak, I.; Carvalho, R.A.; Loureiro, M.; Patrício, M.; Antunes, M.; Carvalho, E. Glucose Uptake and Lipid Metabolism Are Impaired in Epicardial Adipose Tissue from Heart Failure Patients with or without Diabetes. *Am. J. Physiol. Endocrinol. Metab.* **2016**, *310*, E550–E564. [CrossRef] [PubMed]
14. Iacobellis, G.; Barbaro, G. Epicardial Adipose Tissue Feeding and Overfeeding the Heart. *Nutrition* **2019**, *59*, 1–6. [CrossRef]
15. Salgado-Somoza, A.; Teijeira-Fernández, E.; Rubio, J.; Couso, E.; González-Juanatey, J.R.; Eiras, S. Coronary Artery Disease Is Associated with Higher Epicardial Retinol-Binding Protein 4 (RBP4) and Lower Glucose Transporter (GLUT) 4 Levels in Epicardial and Subcutaneous Adipose Tissue. *Clin. Endocrinol.* **2012**, *76*, 51–58. [CrossRef]
16. Gensini, G.G. A More Meaningful Scoring System for Determining the Severity of Coronary Heart Disease. *Am. J. Cardiol.* **1983**, *51*, 606. [CrossRef]
17. Iacobellis, G.; Assael, F.; Ribaudo, M.C.; Zappaterreno, A.; Alessi, G.; Di Mario, U.; Leonetti, F. Epicardial Fat from Echocardiography: A New Method for Visceral Adipose Tissue Prediction. *Obes. Res.* **2003**, *11*, 304–310. [CrossRef]
18. Hwang, J.-W.; Choi, U.-J.; Ahn, S.-G.; Lim, H.-S.; Kang, S.-J.; Choi, B.-J.; Choi, S.-Y.; Yoon, M.-H.; Hwang, G.-S.; Tahk, S.-J.; et al. Echocardiographic Plains Reflecting Total Amount of Epicardial Adipose Tissue as Risk Factor of Coronary Artery Disease. *J. Cardiovasc. Ultrasound* **2008**, *16*, 17. [CrossRef]
19. Thalmann, S.; Juge-Aubry, C.E.; Meier, C.A. Explant Cultures of White Adipose Tissue. In *Adipose Tissue Protocols*; Yang, K., Ed.; Humana Press: Totowa, NJ, USA, 2008.
20. Suga, H.; Matsumoto, D.; Inoue, K.; Shigeura, T.; Eto, H.; Aoi, N.; Kato, H.; Abe, H.; Yoshimura, K. Numerical Measurement of Viable and Nonviable Adipocytes and Other Cellular Components in Aspirated Fat Tissue. *Plast. Reconstr. Surg.* **2008**, *122*, 103–114. [CrossRef]
21. Steinhorn, B.; Sartoretto, J.L.; Sorrentino, A.; Romero, N.; Kalwa, H.; Dale Abel, E.; Michel, T. Insulin-Dependent Metabolic and Inotropic Responses in the Heart Are Modulated by Hydrogen Peroxide from NADPH-Oxidase Isoforms NOX2 and NOX. *Free Radic Biol. Med.* **2017**, *113*, 16–25. [CrossRef]
22. Viswanadha, S.; Londos, C. *Determination of Lipolysis in Isolated Primary Adipocytes BT—Adipose Tissue Protocols*; Yang, K., Ed.; Humana Press: Totowa, NJ, USA, 2008; pp. 299–306.
23. Fang, L.; Guo, F.; Zhou, L.; Stahl, R.; Grams, J. The Cell Size and Distribution of Adipocytes from Subcutaneous and Visceral Fat Is Associated with Type 2 Diabetes Mellitus in Humans. *Adipocyte* **2015**, *4*, 273–279. [CrossRef] [PubMed]
24. Haczeyni, F.; Bell-Anderson, K.S.; Farrell, G.C. Causes and Mechanisms of Adipocyte Enlargement and Adipose Expansion. *Obes. Rev.* **2018**, *19*, 406–420. [CrossRef] [PubMed]
25. Tong, G.; Wang, N.; Leng, J.; Tong, X.; Shen, Y.; Yang, J.; Ye, X.; Zhou, L.; Zhou, Y. Common Variants in Adiponectin Gene Are Associated with Coronary Artery Disease and Angiographical Severity of Coronary Atherosclerosis in Type 2 Diabetes. *Cardiovasc. Diabetol.* **2013**, *12*, 1–10. [CrossRef] [PubMed]
26. Kou, H.; Deng, J.; Gao, D.; Song, A.; Han, Z.; Wei, J.; Jin, X.; Ma, R.; Zheng, Q. Relationship among Adiponectin, Insulin Resistance and Atherosclerosis in Non-Diabetic Hypertensive Patients and Healthy Adults. *Clin. Exp. Hypertens.* **2018**, *40*, 656–663. [CrossRef] [PubMed]
27. Yang, L.; Li, B.; Zhao, Y.; Zhang, Z. Prognostic Value of Adiponectin Level in Patients with Coronary Artery Disease: A Systematic Review and Meta-Analysis. *Lipids Health Dis.* **2019**, *18*, 227. [CrossRef]
28. Amirzadegan, A.; Shakarami, A.; Borumand, M.A.; Davoodi, G.; Ghaffari-Marandi, N.; Jalali, A.; Mazurek, T.; Zhang, L.F.; Zalewski, A.; Mannion, J.D.; et al. Human Epicardial Adipose Tissue Is a Source of Inflammatory Mediators. *Circulation* **2003**, *108*, 2460–2466. [CrossRef]
29. Maghbooli, Z.; Hossein-Nezhad, A. Transcriptome and Molecular Endocrinology Aspects of Epicardial Adipose Tissue in Cardiovascular Diseases: A Systematic Review and Meta-Analysis of Observational Studies. *BioMed Res. Int.* **2015**. [CrossRef]
30. Yun, K.H.; Rhee, S.J.; Yoo, N.J.; Oh, S.K.; Kim, N.-H.; Jeong, J.-W.; Park, D.-S.; Park, H.-Y. Relationship between the Echocardiographic Epicardial Adipose Tissue Thickness and Serum Adiponectin in Patients with Angina. *J. Cardiovasc. Ultrasound* **2009**, *17*, 121. [CrossRef]
31. Vrselja, Z.; Šram, M.; Andrijevic, D.; Takač, B.; Lekšan, I.; Radić, R.; Curic, G. Transcardial Gradient of Adiponectin, Interleukin-6 and Tumor Necrosis Factor-α in Overweight Coronary Artery Disease Patients. *Cytokine* **2015**, *76*, 321–327. [CrossRef]

32. Lima-Martínez, M.M.; López-Mendez, G.; Odreman, R.; Donis, J.H.; Paoli, M. Epicardial Adipose Tissue Thickness and Its Association with Adiponectin in Metabolic Syndrome Patients from Mérida, Venezuela. *Arq. Bras. Endocrinol. Metabol.* **2014**, *58*, 352–361. [CrossRef]
33. Yañez-Rivera, T.G.; Baños-Gonzalez, M.A.; Ble-Castillo, J.L.; Torres-Hernandez, M.E.; Torres-Lopez, J.E.; Borrayo-Sanchez, G. Relationship between Epicardial Adipose Tissue, Coronary Artery Disease and Adiponectin in a Mexican Population. *Cardiovasc. Ultrasound* **2014**, *12*, 35. [CrossRef] [PubMed]
34. Nakajima, T.; Yokota, T.; Shingu, Y.; Yamada, A.; Iba, Y.; Ujihira, K.; Wakasa, S.; Ooka, T.; Takada, S.; Shirakawa, R.; et al. Impaired Mitochondrial Oxidative Phosphorylation Capacity in Epicardial Adipose Tissue Is Associated with Decreased Concentration of Adiponectin and Severity of Coronary Atherosclerosis. *Sci. Rep.* **2019**, *9*, 1–10. [CrossRef] [PubMed]
35. Yamauchi, T.; Kamon, J.; Minokoshi, Y.; Ito, Y.; Waki, H.; Uchida, S.; Yamashita, S.; Noda, M.; Kita, S.; Ueki, K.; et al. Adiponectin Stimulates Glucose Utilization and Fatty-Acid Oxidation by Activating AMP-Activated Protein Kinase. *Nat. Med.* **2002**, *8*, 1288–1295. [CrossRef] [PubMed]
36. Wang, C.; Mao, X.; Wang, L.; Liu, M.; Wetzel, M.D.; Guan, K.-L.; Dong, L.Q.; Liu, F. Adiponectin Sensitizes Insulin Signaling by Reducing P70 S6 Kinase-Mediated Serine Phosphorylation of IRS-1. *J. Biol. Chem.* **2007**, *282*, 7991–7996. [CrossRef] [PubMed]
37. Danielsson, A.; Öst, A.; Lystedt, E.; Kjolhede, P.; Gustavsson, J.; Nystrom, F.H.; Strålfors, P. Insulin Resistance in Human Adipocytes Occurs Downstream of IRS1 after Surgical Cell Isolation but at the Level of Phosphorylation of IRS1 in Type 2 Diabetes. *FEBS J.* **2005**, *272*, 141–151. [CrossRef] [PubMed]
38. Yadav, A.; Kataria, M.A.; Saini, V.; Yadav, A. Role of Leptin and Adiponectin in Insulin Resistance. *Clin. Chim. Acta* **2013**, *417*, 80–84. [CrossRef] [PubMed]
39. Salgado-Somoza, A.; Teijeira-Fernández, E.; Fernández, Á.L.; González-Juanatey, J.R.; Eiras, S. Changes in Lipid Transport-Involved Proteins of Epicardial Adipose Tissue Associated with Coronary Artery Disease. *Atherosclerosis* **2012**, *224*, 492–499. [CrossRef]
40. Barber, M.C.; Ward, R.J.; Richards, S.E.; Salter, A.M.; Buttery, P.J.; Vernon, R.G.; Travers, M.T. Ovine Adipose Tissue Monounsaturated Fat Content Is Correlated to Depot-Specific Expression of the Stearoyl-CoA Desaturase Gene. *J. Anim. Sci.* **2000**, *78*, 62–68. [CrossRef]
41. Díaz-Rodríguez, E.; Agra, R.M.; Fernández, Á.L.; Adrio, B.; García-Caballero, T.; González-Juanatey, J.R.; Eiras, S. Effects of Dapagliflozin on Human Epicardial Adipose Tissue: Modulation of Insulin Resistance, Inflammatory Chemokine Production, and Differentiation Ability. *Cardiovasc. Res.* **2018**, *114*, 336–346. [CrossRef]
42. Nyström, T.; Bodegard, J.; Nathanson, D.; Thuresson, M.; Norhammar, A.; Eriksson, J.W. Novel Oral Glucose-Lowering Drugs Are Associated with Lower Risk of All-Cause Mortality, Cardiovascular Events and Severe Hypoglycaemia Compared with Insulin in Patients with Type 2 Diabetes. *Diabetes Obes. Metab.* **2017**, *19*, 831–841. [CrossRef]

biomedicines

MDPI

Review

Rethinking Fragility Fractures in Type 2 Diabetes: The Link between Hyperinsulinaemia and Osteofragilitas

Isabella D. Cooper [1,*], Kenneth H. Brookler [2] and Catherine A. P. Crofts [3]

1 Translational Physiology Research Group, School of Life Sciences, University of Westminster, 115 New Cavendish Street, London W1W 6UW, UK
2 Research Collaborator, Aerospace Medicine and Vestibular Research Laboratory, Mayo Clinic, Scottsdale, AZ 85259, USA; brookler.kenneth@mayo.edu
3 School of Public Health and Interdisciplinary Studies, Faculty of Health and Environmental Sciences, Auckland University of Technology, Auckland 0627, New Zealand; catherine.crofts@aut.ac.nz
* Correspondence: bellamitochondria@gmail.com

Abstract: Patients with type 2 diabetes mellitus (T2DM) and/or cardiovascular disease (CVD), conditions of hyperinsulinaemia, have lower levels of osteocalcin and bone remodelling, and increased rates of fragility fractures. Unlike osteoporosis with lower bone mineral density (BMD), T2DM bone fragility "hyperinsulinaemia-osteofragilitas" phenotype presents with normal to increased BMD. Hyperinsulinaemia and insulin resistance positively associate with increased BMD and fragility fractures. Hyperinsulinaemia enforces glucose fuelling, which decreases NAD+-dependent antioxidant activity. This increases reactive oxygen species and mitochondrial fission, and decreases oxidative phosphorylation high-energy production capacity, required for osteoblasto/cytogenesis. Osteocytes directly mineralise and resorb bone, and inhibit mineralisation of their lacunocanalicular space via pyrophosphate. Hyperinsulinaemia decreases vitamin D availability via adipocyte sequestration, reducing dendrite connectivity, and compromising osteocyte viability. Decreased bone remodelling and micropetrosis ensues. Trapped/entombed magnesium within micropetrosis fossilisation spaces propagates magnesium deficiency (MgD), potentiating hyperinsulinaemia and decreases vitamin D transport. Vitamin D deficiency reduces osteocalcin synthesis and favours osteocyte apoptosis. Carbohydrate restriction/fasting/ketosis increases beta-oxidation, ketolysis, NAD+-dependent antioxidant activity, osteocyte viability and osteocalcin, and decreases excess insulin exposure. Osteocalcin is required for hydroxyapatite alignment, conferring bone structural integrity, decreasing fracture risk and improving metabolic/endocrine homeodynamics. Patients presenting with fracture and normal BMD should be investigated for T2DM and hyperinsulinaemia.

Keywords: hyperinsulinaemia; beta hydroxybutyrate; osteoporosis; type 2 diabetes; fragility fractures; bone mineral density; osteocalin; vitamin D; collagen; hydroxyapatite

Citation: Cooper, I.D.; Brookler, K.H.; Crofts, C.A.P. Rethinking Fragility Fractures in Type 2 Diabetes: The Link between Hyperinsulinaemia and Osteofragilitas. *Biomedicines* **2021**, *9*, 1165. https://doi.org/10.3390/biomedicines9091165

Academic Editor: Susan J. Burke

Received: 22 July 2021
Accepted: 2 September 2021
Published: 6 September 2021

1. Introduction

In 2010, 3.5 million fragility fractures were sustained in 27 European Union countries, with an estimated economic burden cost of €37 billion. It is predicted that by 2025, the economic burden will continue to increase by 25% [1]. Further, in the largest five European Union countries plus Sweden (EU6), fragility fractures are predicted to increase from 2.7 million in 2017 to 3.3 million by 2030, with a €37.5 billion annual fracture-related cost in 2017, expected to increase by 27% in 2030 [2].

Increased fragility fractures are well documented in patients with type 2 diabetes mellitus (T2DM), a condition of chronic hyperinsulinaemia [3–7]. Decreased skeletal bone mineral density (L-BMD) is the phenotype of "classical" osteoporosis [8,9]. A higher BMD is considered to confer greater bone strength (fracture resistance). However, the T2DM bone fragility phenotype more often presents with normal to increased BMD (H-BMD) [7,9–14], and positively tracks with increased fracture risk [4,5,7,12,15]. In a cross-sectional non-intervention study of 146 Caucasian non-diabetic postmenopausal women with a mean age of 60 ± 2.7 years, HOMA-IR was found to be positively associated with volumetric bone mineral density (vBMD) [11]. In addition, increased insulin resistance (IR) and treatment with exogenous insulin therapy is positively associated with higher BMD and increased fragility fractures. A study following 5994 T2DM males ≥ 65 years of age, found a higher non-vertebral fracture risk in those using insulin (HR 1.74, 95% CI 1.13, 2.69) who also had a higher BMD [15]. Furthermore, in a population-based matched cohort study investigating primary care records of 2979 insulin users and 14,895 non-users, T2DM patients exposed to insulin therapy to manage glycaemia were found to have a 38% excess fracture risk [12]. In both H-BMD-associated T2DM "osteofragilitas" (bone fragility) and L-BMD osteoporosis phenotypes, there is an increase in bone fragility and a loss in tensile and/or torsion strength, and bone ductility, resulting in higher rates of fractures.

Hyperinsulinaemia drives the pathogenesis of T2DM, which may precede hyperglycaemia by up to 24 years [6,16]. Hyperinsulinaemia decreases osteoblastogenesis and propagates poorer-quality collagen production, a problem further compounded by hyperglycaemia increasing glycation damage on new or existing bone collagen. Hyperinsulinaemia drives chronic osteocyte distress via excess ceramide synthesis, which increases cellular reactive oxygen species (ROS) [17–20], leading to a unique type of mineralisation within their lacunae, coined by Bell, Kayser and Jones as "living fossilisation" [21–25]. Concomitantly, hyperinsulinaemia decreases osteoclastogenesis, thus impeding the bone resorption needed to enable homeodynamic bone remodelling—a marker of good health [7,26] —and this results in a form of hyperinsulinaemia-hyperglycaemia pseudo-osteopetrosis. Combined, these hyperinsulinaemia-driven effects result in the increased bone mineral density seen in people with T2DM.

Osteocytes are the backbone of bone health, and consequently major players in whole-body metabolism [27]. Osteocytes are able to directly mineralise and resorb bone and are central mediators in the regulatory control of osteoblasts and osteoclasts [28–36]. Chronic hyperinsulinaemia diminishes the replenishment of osteocytes and drives the living fossilisation of the ones in existence, leading to the loss of the osteocyte's dynamic orchestration of bone remodelling [21–25,37]. We propose chronic hyperinsulinaemia provides a plausible explanation, a unifying theory of the mechanisms of action, for the increased BMD and bone fragility "osteofragilitas", that leads to the increased fracture rates seen in the T2DM bone phenotype.

2. Osteocytes: Mediators of Bone Remodelling and Metabolic Heath

Within the adult skeleton, osteocytes comprise 90% to 95% of the total bone cells [37]. Osteocytes are terminally differentiated cells from the osteoblast lineage. They are the heavyweight lifters in the living dynamic bone tissue that not only provides a physical scaffold for the body, but is also fundamental in endocrine regulation and whole-body metabolism [14,38,39]. Osteocytes embed within the bone after collagen formation by osteoblasts, of which some of these osteoblasts are fated to differentiate into the embedding osteocytes [37]. The osteocytes, along with their neighbouring osteoblasts, continue to form mineralised bone onto the collagen scaffold, in the process forming the hydroxyapatite lacuna chamber around the embedding osteocytes [27]. Continual morphological changes take place in the process of osteocytogenesis, leading to cells that bear little resemblance in structure to their predecessors, appearing visually more like neurons. Osteocytes have on average 50, and up to 100 dendrites, extending through highly connected intricate tunnels, called the canaliculi. The canaliculi are formed and maintained by the occupying osteocytes [40], and enable physical connection to other osteocytes and to the outside surface of bone, to osteoblasts, osteoclasts and the vasculature [41].

The lacunocanalicular system is a fluid-filled space, separated from the mineralised component of bone and maintained by the resident osteocytes [22]. The osteocytes are the workhorse physical mechano-sensors and mediators of bone remodelling, sensing and responding to mechanical stress via their strategic distribution and network of a vast number of connected dendritic processes that enable intercellular communication. Their network forms the essential antenna to detect fluid shear stress, which enables the translation of mechanical stimuli into biochemical messages. Osteocytes, in turn, feed these signals forward via autocrine and paracrine mechanisms, to elicit homeostatic adaptive responses, from regulating bone remodelling in order to provide sensitive and continual changes to whole-body mineral needs, to effecting distant organ responses via endocrine signalling [22,41].

3. Dendritic Connectivity Is Essential for Function and Viability

Osteocytes connect to one another via their dendritic processes which form the osteocyte-lacunocanalicular network, and their health and viability are dictated by their dendritic connectivity [41,42]. A compromise in osteocyte health diminishes their ability to actively inhibit mineralisation of their pericellular space [25,28], consequently reducing their connectivity. This further compromises the health of deeper osteocytes that become cut off from receiving nutrients which are no longer able to be delivered through fluid movement via the canalicular tunnels. Furthermore, osteocyte connectivity is required for the transduction of load-induced fluid flow that enables their mechanical sensory system to decrease apoptosis and increase osteocytogenesis [43]. Maintenance of dendritic connectivity allows osteocyte-directed regulation of osteoblasts and osteoclasts, in addition to their own capacity to directly resorb bone [29,30,34–36].

It is highly likely that osteocytes rather than osteoclasts are responsible for bone resorption in the basal condition, as a function of whole-body mineral homeostasis, and that osteoclast bone resorption serves to function in the acute action/need/stress response [34]. Osteocytes from lactating mice have been shown to express markers thought to be specific to osteoclasts. Their osteocyte lacunae were found to be enlarged, suggesting localised bone resorption, providing evidence of osteocytic-osteolysis [44]. Patients with T2DM have decreased levels of carboxy-terminal collagen crosslinks (CTX) bone resorption marker, indicating a lower bone turnover [13]. Osteocytes regulate both osteoblast and osteoclast differentiation and function, and are thus master regulators of dynamic bone remodelling, a function of healthy physiology (Figure 1) [45].

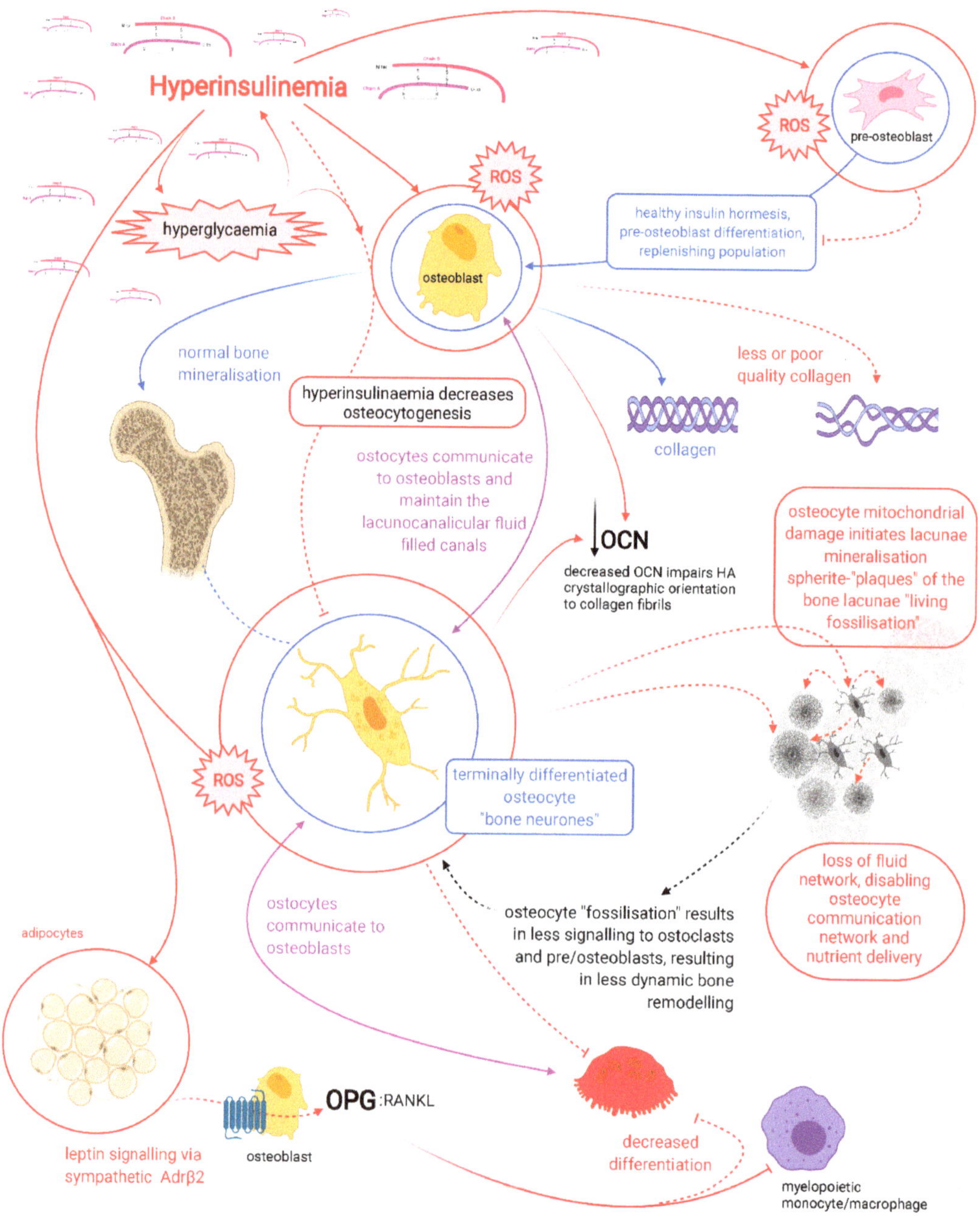

Figure 1. Schematic representation showing the dynamic role of osteocytes in the regulation of healthy and dysregulated bone. Beta-adrenergic receptor (Adrβ2), hydroxyapatite (HA), osteocalcin (OCN), osteoprotegerin (OPG), receptor activator of nuclear factor kappa-β ligand (RANKL), and reactive oxygen species (ROS). Red lines indicate hyperinsulinaemia-driven pathology pathways. Healthy physiology indicated with blue and purple lines.

4. Hyperglycaemia Increases Advanced Glycation End-Product Formation in Bone Collagen

The anabolic hormone insulin is required at the basal level for healthy bone formation [3]. However, chronic hyperinsulinaemia surpasses this threshold in a dose and duration manner. When coupled with hyperglycaemia, it results in the production of poorer-quality, more rigid collagen and glycation damage of existing collagen [46,47]. Together, they drive increasing BMD by promoting skeletal mineral acquisition that is fragile in structure [22,48]. Hyperglycaemia and hyperinsulinaemia increase advanced glycation end-product (AGE) formation. Increased glycation on fibrillar collagen negatively affects bone quality [14]. In human tissue, the most abundant AGE is glucosepane, a lysine–arginine cross-linking, that forms the major AGE in bone type 1 collagen. Hyperglycaemia is one of the leading causes of AGE formation, affecting the structural and biochemical properties of protein binding sites, rendering them unrecognisable to other proteins and enzymes [47].

Hyperglycaemia-driven AGE formation of glucosepane in bone collagen causes a decrease in viscoelasticity and increases the production of a stiffer collagen, resulting in negative effects on the mechanical properties of load-bearing collagen in bone. This causes bone toughness to decrease, while a greater accumulation of AGE in bone results in increased fracture risk [47]. Furthermore, an increase in stiffer/rigid collagen production occurs due to increased glycation effects on the vasculature, leading to increased hypoxia in the microenvironment [14]. Hypoxia then compromises the osteoblasts' capacity to generate sufficient ATP for collagen synthesis and for differentiation into osteocytes (Figure 1). Both of which are very energy intensive processes, requiring an efficient mitochondrial capacity to generate ATP via oxidative phosphorylation (OxPhos) [49–52]. In addition, hyperinsulinaemia and hyperglycaemia inhibit beta-oxidation, whilst increasing mitochondrial (mt) reactive oxygen species (ROS) formation [53]. This leads to increased H_2O_2 production, causing damage to intracellular protein synthesis machinery and consequently synthesis of poorer quality collagen. The typical methodology of assessing bone quality is via dual x-ray absorptiometry, however, this method is unable to detect the collagen aspect of bone quality [14]. As a result, there is an increased frequency in missing early detection of hyperinsulinaemia-hyperglycaemia osteofragilitas fracture risk, which typically does not present with L-BMD. This suggests that BMD alone is a poor proxy/diagnostic marker for fracture risk in hyperinsulinaemic individuals.

Hyperinsulinaemia "enforces" cellular glucose substrate fuelling [54], and downregulates beta-oxidation by increasing intracellular ceramide production [20]. Excess ceramide production compromises mtOxPhos capacity, by increasing dynamin-related protein 1 (Drp1) synthesis. Drp1 functions to increase mitochondrial fission, in addition to increasing the production of mtROS such as: superoxide (O_2^-), hydroxyl radical (^-OH) and hydrogen peroxide (H_2O_2) [17,20,55]. Concomitantly, ATP production from glucose oxidation reduces the intracellular pool of nicotinamide adenine dinucleotide (NAD+), consuming four NAD+ in the production of two acetyl moieties, in comparison to beta-oxidation, ketolysis or oxidation of acetoacetate, which consume two, one and zero respectively (Figure 2) [56]. Thus, ATP synthesis that is increasingly reliant on glucose oxidation, negatively impacts the availability of NAD+.

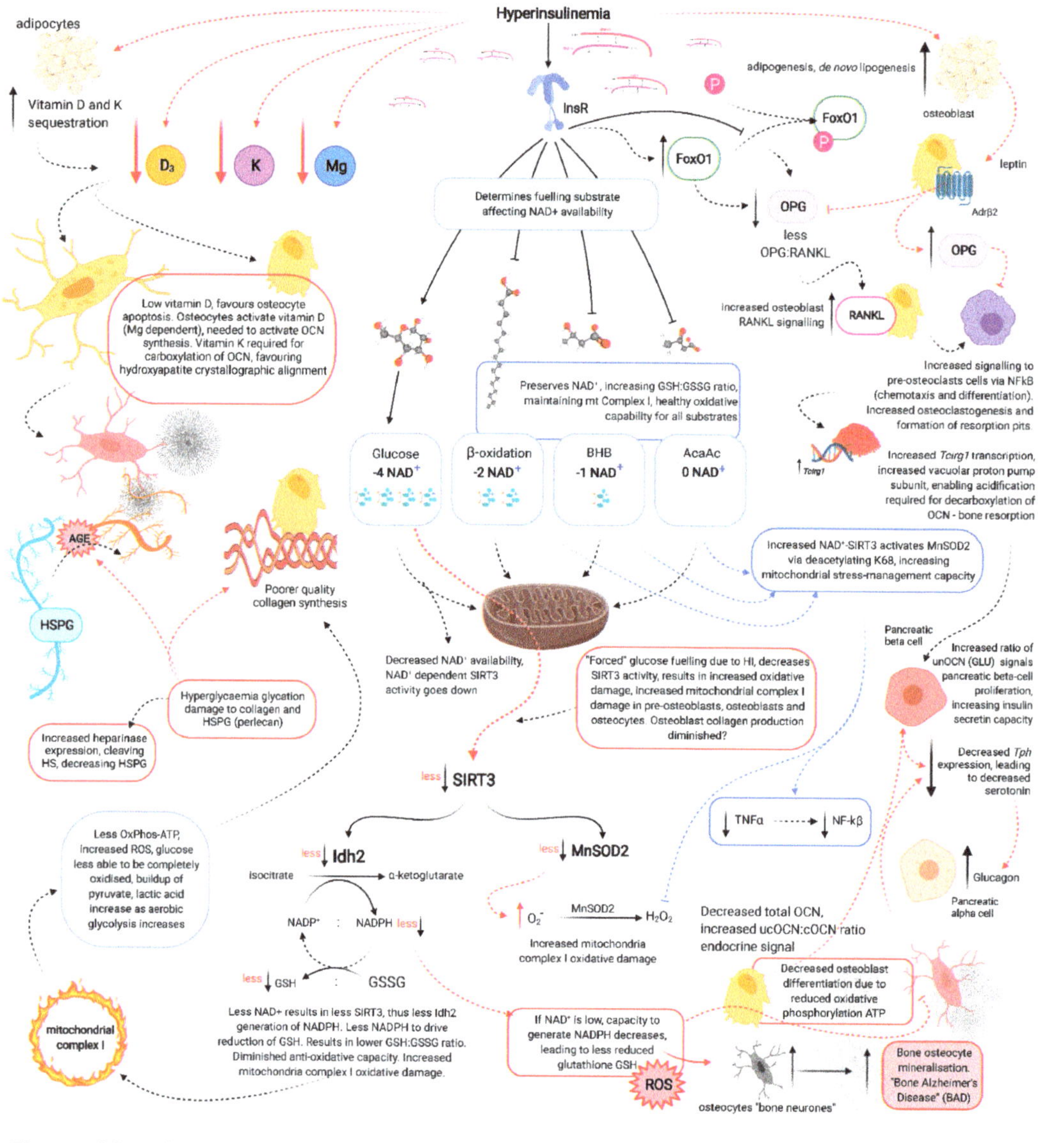

Figure 2. Schematic representation of hyperinsulinaemia effects on cellular oxidative state, and bone homeodynamics. Acetoacetate (AcAc), adenosine triphosphate (ATP), advanced glycation end-products (AGE), beta-adrenergic receptor 2 (AdrB2), beta-hydroxybutyrate (BHB), heparan sulphate (HS), heparan sulphate proteoglycan (HSPG), hydrogen peroxide (H_2O_2), hyperinsulinaemia (HI), insulin receptor (InsR), isocitrate dehydrogenase 2 (Idh2), forkhead box O1 (FoxO1), glutathione oxidised form (GSSG), glutathione reduced form (GSH), lysine 68 (K68), magnesium (Mg), manganese superoxide dismutase 2 (MnSOD2), nicotinamide adenine dinucleotide (NAD+), nicotinamide adenine dinucleotide phosphate (NADP+), nuclear factor-kB (NF-kB), osteocalcin carboxylated (Gla-OCN or cOCN), osteocalcin un(der)carboxylated (Glu-OCN or ucOCN), osteoprotegerin (OPG), oxidative phosphorylation (OxPhos), receptor activator of nuclear factor-kB ligand (RANKL), reactive oxygen species (ROS), sirtuin 3 (SIRT3), superoxide (O_2^-), tumour necrosis factor α (TNFα), T cell immune regulator 1 (*Tcirg1*), tryptophan hydroxylase (*Tph*), vitamin D (D_3), and vitamin K (K).

Sirtuin-3 (SIRT3) regulates the synthesis of endogenous antioxidant enzymes such as mitochondrial manganese superoxide dismutase (MnSOD2) and NADPH-dependent production of reduced glutathione (GSH) [56,57]. An increased reliance on glucose fuelling and its effect on NAD+ availability, leads to a reduction in SIRT3 activity, since SIRT3 is NAD+ dependent. This in turn decreases signals for the transcription and synthesis of MnSOD2 and GSH [56,57]. Furthermore, hyperinsulinaemia, diminishes the oxidative buffering capacity of the cellular redox antioxidant GSH [58]. Beta-oxidation activity dramatically increases in osteoblasts as they mature. If beta-oxidation is diminished and/or inhibited, metabolic demand cannot be met, resulting in a decrease in precursor cellular differentiation capability, as osteoblastogenesis and osteocytogenesis are energy intensive processes [49,51,52]. Bone neither stores nor synthesises any significant amount of fat. Thus, fatty acids delivered via chylomicron remnants (CR) and low-density lipoproteins (LDL) to bone are more likely to be used for ATP synthesis via beta-oxidation [51].

5. Hyperinsulinaemia Increases Osteocyte Mitochondrial Fission and Disassociation from the Endoplasmic Reticulum

Healthy osteocytes are able to transfer their functional mitochondria to neighbouring distressed osteocytes via their physically connected dendrites [42]. Mitochondrial transfer between osteocytes declines with age and with decreased dendritic connections [22,28,42,59]. Osteocytic rescue of their distressed neighbours via mitochondrial transfer is dependent on their mitochondria associating with the endoplasmic reticulum (ER) and on their dendrite connectivity, which requires the maintenance of their lacunocanalicular tunnels via inhibiting excessive mineralisation [21,29,31,42]. Hyperinsulinaemia increases mtROS production via ceramide synthesis which increases mitochondrial fission and ER stress [20,60]. This results in less mitochondrial fusion, which is necessary for beta-oxidation, and also leads to decreasing the osteocytes' capability to transfer mitochondria to adjacent distressed osteocytes. This osteocytic mitochondria transfer process is dependent on mitochondrial association with the ER (ER-mito), driven by the protein guanosine triphosphatase mitofusin 2 (Mfn2), which localises to the outer mitochondrial membrane (OMM), enabling the tethering of the mitochondrion to the ER [42]. Reductions in Mfn2 expression leads to reductions in mitochondrial fusion and distribution, which further decreases the ability of healthier osteocytes to rescue their distressed neighbours [22]. Hyperinsulinaemia increases the Drp1:Mfn2 ratio. As a result this favours mitochondrial fission and ER-mito disassociation, leading to decreased OxPhos ATP synthesis and increased mtROS production [20,61]. Interestingly, osteocytes are under a form of social control, as they must receive signals from other cells for their survival, which is reliant on their connectivity [48]. A fall in the number of viable osteocytes or dendritic connections, below an essential minimum number (a threshold), or excessive lacunocanalicular mineralisation, may result in a severely compromised osteocyte-network signal transduction capacity, which in turn would impair dynamic bone remodelling [48]. Osteocytes actively maintain the lacunae-canalicular space, enabling fluid flow dynamics that ensures the provision of nutrients and signalling molecules such as growth factors and cytokines, in addition to facilitating the removal of waste. This is essential for the deeper embedded osteocytes and osteoblasts, to remain viable and function healthily [29]. Chronic hyperinsulinaemia may contribute to significant detrimental remodelling of dendrites and the lacunae-canalicular space, where the number (threshold) of osteocytes affected may result in a profoundly negative outcome.

6. MnSOD2 and SIRT3 Required for Osteoblastogenesis and Osteocytogenesis

Chronic hyperinsulinaemia and hyperglycaemia drives the pathogenesis of T2DM, chronic kidney disease (CKD) and atherosclerosis, conditions that are associated with increased rates of fractures independent of L-BMD. Hyperinsulinaemia with hyperglycaemia likely causes a decrease in osteocyte population and/or functional capacity leading to hyperinsulinaemia-*osteofragilitas* [4,38,39,62,63]. Glucose restriction increases osteoblast-to-osteocyte specification, and conversely hyperglycaemia reduces osteoblast gene expression of *Osx*, *Bglap* or *Dkk1*, leading to a reduction in osteocytogenesis [37]. Furthermore, hyperglycaemia decreases osteocyte connectivity and population. Viable osteocytes are needed to signal pre-osteoblastogenesis and osteocytogenesis. Therefore, hyperglycaemia's negative impact on osteocyte numbers and dendritic connectivity, may result in a detrimental impact on the maintenance of the osteocyte-lacunocanalicular network. As a result, this would impair the osteocytes' role in dynamic bone remodelling in the basal state. This is likely a contributing factor to the lowered state of bone remodelling seen in hyperinsulinaemic T2DM and CVD patients [45].

Hyperinsulinaemia/glycaemia mediated decreases in NAD+ availability, consequently decreases antioxidants MnSOD2 and SIRT3 activity, this leads to a negative impact on osteoblast differentiation [64]. Both MnSOD2 and SIRT3 are required to regulate mitochondrial stress, to enable cellular differentiation and bone formation [64]. MnSOD2 dismutes mitochondrial superoxide, thus protecting complexes I and II of the mitochondrial electron transport chain (ETC), where OxPhos occurs. Glucose restriction increases ketolysis, which increases the redox span between complex I and III [54,65]. This decreases electron leakage at these sites, which reduces the formation of superoxide. Glucose restriction concomitantly leads to an increase in NAD+ dependent SIRT3 activity, which increases deacetylation of MnSOD2 lysine residues: 53, 68 and 89 (K53, K68, K89), consuming NAD+ in the process [56,57,66–69]. This consequently upregulates MnSOD2 antioxidant activity, resulting in improved ROS management [70]. With improved ROS management and mt-OxPhos capacity, due to glucose restriction and ketolysis, this maintains osteocytogenesis, viability and connectivity. Maintenance of osteocyte viability knock-on effect results in sustained dynamic bone remodelling, as well as regulation of osteoblast and osteoclast differentiation and activity.

7. Glucose Restriction Increases Glutathione Activity and Improved Cell Viability

Glucose restriction decreases insulin signalling, thereby enabling increased beta-oxidation and ketolysis, preserving NAD+ availability to enable upregulated SIRT3 activity which increases NADPH production [71]. This drives increased glutathione reductase activity, which increases the ratio of reduced to oxidised glutathione (GSH:GSSG). Thus enhancing the intracellular antioxidant capacity to combat ROS damage (Figure 2) [72].

Calorie restriction decreases glucose and insulin levels, whilst increasing fatty acid oxidation and ketogenesis. This is the metabolic phenotype of fasting. Mice under lower levels of glucose and insulin show a marked increase in SIRT3 expression and activity, consequently preventing age-related hearing loss, a condition associated with ageing, increased hyperinsulinaemia, and osteofragilitas conditions [73]. In a mouse model, elevated insulin and high glucose compromised mitochondria in osteocytes, increasing both cytoplasmic and mitochondrial ROS, driving an inverse correlation between glutathione and mitochondrial ROS. Additionally, an increasing compromise in osteocyte mitochondrial-function strongly correlates with increased impairments in skeletal health [74].

8. Glucose Restriction Enables Osteocytogenesis; Hyperglycaemia Inhibits It

Hyperglycaemia inhibits osteoblast differentiation into osteocytes, and reduces the osteoblasts bone mineralisation capacity. In an in vitro model, IDG-SW3 cells differentiated into osteoblasts were cultured in 1, 5 or 25 mmol/L of glucose, using an alizarin red assay, it was shown that the osteoblasts cultured in the higher glucose concentration had a decreased ability to form calcium deposits [37]. Furthermore, osteocytes cultured in 25 mmol/L of glucose, showed a decrease in osteoblast-to-osteocyte transition, assessed using osteocytic *Dmp*-GFP gene expression reporter. A negative correlation was found between glucose availability and osteocytogenesis, where osteoblast-to-osteocyte transition increased as glucose concentration decreased, assessed via fluorescence microscopy/cell sorting of *Dmp*-GFP expression. In short, glucose restriction increased osteocyte gene expression [37]. These results indicate hyperglycaemia may disturb, if not diminish osteoblast and osteocyte replenishment potential. Hyperinsulinaemia and hyperglycaemia impair osteoblastogenesis, osteocytogenesis and osteoclastogenesis, resulting in decreased dynamic bone remodelling. Patients with T2DM and CVD show marked decrease in bone remodelling [7,51].

In a situation of hyperinsulinaemia which potentiates and consolidates hyperglycaemia, osteoblast bone formation may become compromised over time, while osteoblast-to-osteocyte transition is also diminished. In this scenario, one would expect less evidence of bone mineralisation. However, if in the same hyperinsulinaemic and hyperglycaemic state, the existing osteocytes are also producing elevated levels of mtROS, osteocyte death would potentially incur vacancies in their lacunocanalicular space, which would result in a state of heightened structural fragility. This may phenotypically appear as L-BMD osteoporosis. However, hypothetically in a hyperinsulinaemic state, the "bone neurone" sensory osteocyte makes a "margin-call" [25,75,76], to fossilize instead of leaving an empty and thus structurally fragile space, so as to ensure structural rigidity at the very least.

Alternatively, the rationale may be simpler, where the osteocyte fossilisation occurs in the hyperinsulinaemic state, so as to aid in mineral accretion during times recognised as "feasting and energy abundance". A plausible explanation would be hyperinsulinaemia occurring in summer and autumn, when there is greater availability of high carbohydrate foods, leading to increased levels and duration of insulin signals to the cells that there is a food abundance period. Insulin mediates de novo adipogenesis, chronic insulin signals to adipocytes to increase storage of fuel in the form of lipids (de novo lipogenesis) and to bone cells to save minerals "for a rainy day", in expectation of typically less food abundance in winter. A similar physiological adaptation example for increasing storage of fuel, minerals and vitamins, is during pregnancy, which has a natural state of mild hyperinsulinaemia, in-order to provide for post-partum lactation [35,77].

Glucose restriction would historically normally occur during wintertime, with concomitant lowered insulin levels, as a result of decreased food abundance, especially the farinaceous kind. A low glucose and low insulin environment would facilitate bone remodelling and improved resorption, as beta-hydroxybutyrate (BHB) inhibits mineralisation, while acetoacetate activates mineralisation. Bone resorption increases either via increased osteoclast activity, or higher levels of BHB which decreases the localised pH [78]. This releases undercarboxylated osteocalcin (OCN), which would negatively regulate BHB synthesis through its action on regulating basal insulin secretion [79,80]. This keeps ketones in the physiologically healthy range [81]. The low glucose and insulin state that facilitates dynamic bone remodelling, produces opened-up spaces for osteoblasts to then begin matrix synthesis and renewed bone mineralisation activity. This is followed by transition into a new generation of fresh osteocytes to maintain the newer sections of the lacuna-canalicular network. However, if a glucose restriction phase does not occur, over time the pre-osteoblast population may become compromised, whilst over-fossilisation simultaneously occurs. Subsequent chronic hyperinsulinaemia induces excess osteocytic ROS production to a degree that osteocytes are no longer able to actively inhibit mineralisation of their lacunocanalicular space, which is required to enable their survival, as well as those they are further connected with. Combined, these effects may drive increasing bone density, while making the bones both brittle and fragile.

9. Pyrophosphate and Sclerostin

Osteocytes produce pyrophosphate (ePPi), an inhibitor of mineralisation and carbonate solubiliser of bone mineral and matrix, thus regulating mineralisation and the circulatory systemic levels of calcium and phosphate [82]. Hyperinsulinaemia increases mtROS and cellular pathology, compromising osteocyte health, leading to mitochondrial mediated living fossilisation of osteocytes within their lacunae. Magnesium-dependent spherite-mediated osteocyte mineralisation provides evidence to support this, where higher levels of magnesium are found in these micropetrosis fossilisation spaces formerly inhabited by osteocytes [23]. This potentially traps/entombs magnesium within the living fossilised bone; subsequently contributing to magnesium deficiency (MgD). Furthermore,

MgD has its own effects on potentiating hyperinsulinaemia [83–85]. Culminating in a vicious feedforward cycle.

While bisphosphonates are known to work at inhibiting osteoclast bone resorption, they likely also work as an ePPi analogue. By assisting in inhibiting excessive mineralisation of the osteocytes lacunae, helping to maintain fluid flow and osteocyte connectivity where struggling osteocytes may fail to do so, thus saving the day in "assisting" the osteocytes in their job to be able to regulate dynamic bone turnover, a marker of metabolic health [45,86].

A density threshold level of osteocytes is required to maintain homeostatic bone remodelling, from sensing damage in-order to initiate repair processes [48], to mineral storage during pregnancy or in preparation for a winter with decreased food availability. Greater lacunae mineralisation has been detected with advanced age and/or untreated osteoporosis. Increased lacunae calcium content coupled with poorer quality matrix, contributes to increased bone brittleness and subsequent fragility [22,46].

Osteocytes synthesise sclerostin, a 22.5 kDa protein, which reduces osteoblast differentiation, consequently downregulating bone metabolism. Although there is a clear relationship between hyperinsulinaemia and sclerostin levels, further research is required. Theoretically, a decreased osteocyte mass should predict lower sclerostin levels. However, elevated sclerostin levels have been found in patients with CVD and chronic kidney disease (CKD), two groups who typically have osteofragilitas. Furthermore, osteogenic differentiation of vascular smooth muscle cells (VSMC) in CVD and CKD patients have been detected, where there is an increase in expression of sclerostin in aortic valve tissue [87–90]. This means caution is required when assessing bone and metabolic health via sclerostin levels.

10. Osteocytes Produce Alkaline Phosphatase

Alkaline phosphatase (ALP) is an extracellular membrane bound ectometalloenzyme that catalyses the hydrolysis of inorganic pyrophosphate (PPi) to phosphate (Pi) at an alkaline pH. Phosphate forms part of calcium hydroxyapatite crystals, and an increase in phosphate promotes mineralisation. Gene mutation of tissue-nonspecific alkaline phosphatase (TNAP) results in hypophosphatasia under-mineralisation, demonstrating the integral role of ALP in bone mineralisation [91].

Nucleotide pyrophosphatase phosphodiesterase (NPP1) inhibits the action of ALP by increasing the concentration of calcification inhibitor pyrophosphate (ePPi) [70]. Insulin and the fed-state reduces NPP1 gene (*Enpp1*) expression. Interestingly, fasting has been shown to increase its expression. In short, fasting increases the concentration of ePPi via NPP1, leading to inhibition of excessive mineralisation of the osteocyte lacunocanalicular space, which results in maintaining osteocyte viability, dendritic connectivity and consequent dynamic bone remodelling capacity. Insulin action leads to decreased ePPi concentration, subsequently decreasing the osteocytes ability to inhibit bone mineralisation, thus increased mineralisation of the osteocytes lacunocanalicular space occurs. The fasted state provokes the opposite effect, inhibition of mineralisation, and potentially enhances physiological levels of bone resorption via increased beta-hydroxybutyrate (BHB) [92].

If autumn were a time for humans to accumulate stored energy in preparation for a fasting period during winter, where foods available during autumn increase insulin secretion, activating energy storage mechanisms, logically mineral storage would also be required and may also be facilitated via insulin's action on inhibiting NPP1 production of ePPi. This may provide an evolutionary explanation, where seasonal hyperinsulinaemia propagates increased osteocyte-lacunae mineralisation during autumn, as an adaptive survival mechanism. This would then be followed by a winter of fasting, which would subsequently lead to increased NPP1 activity, relinquishing back into the system the stored minerals. Hyperinsulinaemia T2DM could be described as a metabolic phenotype reflecting a constantly "fed-state", the ever-lasting autumn. Thus, providing an explanation of the normal to increased BMD observed in people with T2DM.

11. Osteocalcin

Patients with T2DM and insulin resistance have significantly lower levels of circulating osteocalcin (OCN) than healthy controls [71,93–96]. OCN is a non-collagenous protein, largely synthesized by osteoblasts and osteocytes that retain their expression of OCN [97]. Levels serve as a marker of osteoblast and osteocyte health. Serum OCN levels positively correlates with: dynamic bone remodelling, decreased insulin resistance (IR), and reduced T2DM and CVD risk [96,98–100]. Much published research show OCN increases insulin secretion, with conclusions stating decreased OCN levels would result in decreased insulin synthesis and secretion, and result in impaired glucose homeostasis. However, when assessed in humans, low levels of OCN tracks with hyperinsulinaemia [96,101]. Ergo providing evidence that production of high levels of insulin does not require high levels of OCN.

12. Osteocalcin Endocrine Effects

Interestingly, OCN significantly increases insulin-independent glucose uptake, and even more so in the presence of insulin. This leads to increased insulin sensitivity, through reducing the amount of insulin required to facilitate glucose uptake [98]. Additionally, OCN production increases the expression of mitochondrial uncoupling protein 1 (UCP1) in adipocytes, leading to increased thermogenesis and mitochondrial biogenesis, thus increasing glucose and fatty acid oxidation capacity. Furthermore, OCN increases adipocyte production of adiponectin. However, caution is required when interpreting experiments involving administering exogenous OCN in animal studies, in which these animals are fed diets that do not induce hyperinsulinaemia (obesogenic for that species), that would mimic the main human causal factor for T2DM. In the hyperinsulinaemic state, adiponectin receptors Adr1/2 are downregulated [102]. The signalling dynamics and results elucidated from animal studies in which exogenous OCN is provided to metabolically normal, or genetically induced OCN deficient and/or OCN receptor KO mice, would likely not be the same as what would occur in hyperinsulinaemic humans with low OCN. For example, healthy insulin levels and insulin sensitivity plus exogenous OCn = X, whilst hyperinsulinaemia and insulin resistance plus exogenous OCn = Y.

When wild-type mice (healthy) were administered exogenous OCN, their adipose tissues were observed to have upregulated peroxisome proliferator-activated receptor gamma coactivator 1-alpha (PGC1α) and adiponectin. Both adiponectin and PGC1α expression and activation, lead to increased beta-oxidation and correspond to improved metabolism, insulin sensitivity and glucose homeostasis [103]. Effectively, metabolically healthy mice respond in a physiologically normal way to the OCN. The elevated OCN in this context, is thus able to further effect (positive feedforward) mechanisms that consolidate better glucose/fatty acid oxidation and ROS management. The marker then becomes a maker of good health. OCN levels and the carboxylated-to-undercarboxylated ratio (cOCN:ucOCN or Gla:Glu) act as surrogate markers of osteoblast and osteocyte health, and consequently bone quality, which are then able to actively function in endocrine homeostasis and metabolic health.

The effect of OCN on adipocytes include: improved insulin independent glucose uptake, increased adiponectin synthesis and "energy wastage" through thermogenesis, and decreased inflammatory cytokine production, leading to increased skeletal and muscular insulin sensitivity [98]. Greater insulin sensitivity decreases the amount of insulin required to achieve glucose homeostasis. Individuals who maintain normo-insulin levels via restricting carbohydrate intake, would likely maintain healthier osteoblasts and osteocytes. This enables healthy endogenous OCN synthesis, as is seen in healthy controls. In the context of normo-insulin and insulin sensitivity plus OCN, the result is X. Which is OCN increasing adiponectin synthesis and signalling in a none hyperinsulinaemia and normoglycaemia contextual environment. Furthermore, a reduced requirement and secretion of insulin, decreases excessive mitochondrial ROS production and subsequent downstream cellular pathophysiological adaptations.

13. Carboxylation of Osteocalcin

Post translational modification alters OCN into one of 3 isoforms, which affects bioavailability and activity of the bone-derived hormone. γ-carboxylation of OCN occurs before secretion from osteoblasts and osteocytes. γ-glutamyl carboxylase (GGCX) enzyme activity is dependent on availability of its cofactor vitamin K (in the reduced state), in order to carboxylate OCN on glutamic acid residues: Glu17, 21 and 24. The carboxylated (cOCN or Gla-OCN) form, is the most abundant form in bone extracellular matrix as carboxylation increases OCN affinity for the mineral component of bone hydroxyapatite [79,98]. The undercarboxylated and uncarboxylated forms, (ucOCN or Glu-OCN) are considered the biologically active hormone isoforms. All three isoforms are present in the blood [39]. OCN concentration in human blood ranges between 10 to 40 ng/mL [39]. Interestingly, in vitro experiments by Hill et al. show OCN in both carboxylated and uncarboxylated forms are biologically active, both are able to modulate glucose uptake and increase insulin sensitivity. However, the ucOCN form was more effective [98].

14. Osteocalcin and Insulin

Much research has shown that insulin signalling induces osteoblasts and osteocytes to produce OCN and in vivo studies indicate that plasma OCN stimulates an increase in pancreatic beta cell differentiation/proliferation via increasing cyclin D1, D2 and Cdk4 gene expression, proteins involved in cell division [103]. Additionally, OCN directly and indirectly via glucagon-like peptide-1 (GLP-1), increases insulin production capacity and secretion upon glucose stimulus [103]. While this is shown in in vitro and animal studies, T1DM patients who are given exogenous insulin should technically gain in increased endogenous OCN synthesis, this would then be expected to stimulate pancreatic beta-cell proliferation and subsequent endogenous insulin secretion capabilities. However, T1DM patients do not appear to gain in the upregulation of endogenous insulin production. This may be due to a lack of pancreatic precursor beta-cells, although it has been shown that both T1DM and late-stage T2DM patients do have some functioning precursor pancreatic beta-cells [100,104].

Wei et al. investigated the potential role of OCN as a means to stimulate pancreatic beta-cell proliferation, given T1DM patients retain a small residual population of functional beta-cells [104–106], the logic of their hypothesis seems plausible. T2DM patients have low OCN levels too. However, this is with high insulin levels in non-insulin dependent diabetes mellites (NIDDM). The question remains then, would these patients benefit from exogenous OCN therapy? Or similar to conditions of T2DM, patients given exogenous insulin, serves only to mask the downstream problem (hyperglycaemia), while increasing hyperinsulinaemia and insulin resistance, driving the disease further [93,107]. If OCN increases insulin secretion, then would we not see higher levels of OCN in hyperinsulinaemia pathologies such as T2DM, CVD and MetS? On the contrary, those with normal insulin levels have significantly higher OCN levels [93,107]. This may indicate OCN levels and the carboxylation ratio, are firstly markers and then contributory makers of bone fracture resistance, and metabolic and endocrine health.

Using a mouse model, Ferron et al. showed that intermittent injections of OCN improved glucose metabolism, and increased skeletal mitochondrial content which led to improved glucose and fatty acid oxidation capacity. Elevated OCN levels in the absence of hyperinsulinaemia also increased energy expenditure, corroborating other research showing OCN signalling via increasing adiponectin production, leads to increases in brown fat UCP1, resulting in increased thermogenesis. As fat and glucose is oxidised more efficiently, independent of insulin mediated glucose uptake, metabolic markers consequently improve [108].

Evidence suggests the metabolic phenotype of low insulin and glucose levels facilitates maintaining healthy osteoblasts and osteocytes, dendritic connectivity and maintenance of the lacunae-canalicular network. This leads to the metabolic healthy phenotype of higher levels of OCN synthesis and carboxylation capacity, resulting in fracture-resistant bone, with osteocytes maintaining basal bone remodelling. Furthermore, the resultant ability to endogenously produce OCN, enables bone to participate in its endocrine-action on other tissues and organs, further contributing to improved glucose homeostasis and insulin sensitivity [109]. Restriction of dietary carbohydrate intake simultaneously maintains low glucose levels and minimises additional exogenous stimulus on insulin secretion, thus maintaining both markers in the low healthy physiological range.

15. cOCN Levels Determine Hydroxyapatite Alignment Formation

Moriishi et al. in *PLOS Genetics*, demonstrated using mouse model OCN knockouts of the two mouse genes for OCN: *Bglap* and *Bglap2*, unlike humans who poses only one OCN gene, that bone apatite crystallite alignment is dependent on carboxylated OCN [110]. In the OCN-deficient mice, bone strength decreased, supporting the classical biology phrase—structure dictates function. In this case, bone strength is determined by its structural quality, which includes bone mass and quality of crystallite alignment with collagen fibres. These structural features, combined, determine bone fracture resistance or osteofragilitas [111]. It is important to note that Moriishi was not investigating OCN deficiency in the context of hyperinsulinaemia, which likely would further contribute other detrimental factors, such as poorer collagen synthesis and increased collagen glycation damage.

Hyperinsulinaemia and hyperglycaemia propagation of impaired osteoblastogenesis and osteocytogenesis results in decreased OCN production capacity, which impairs hydroxyapatite crystallographic orientation to collagen fibrils. Combined with poorer quality collagen synthesis and increased glycated collagen, the summative results may lead to compounding effects on bone fragility via compromised structural quality, that is independent of BMD. The sum of all of these dysregulated/impaired conditions, likely leads to the increased fracture rates seen in patients with the T2DM hyperinsulinaemia-osteofragilitas phenotype.

16. The Acute Stress Response

Responses to acute stress increases the ucOCN:cOCN (Glu:Gla) OCN ratio, this decreases OCN bone affinity, which reduces the well-formed structural alignment of hydroxyapatite that is needed for bone fracture resistance [110]. Elevated glucocorticoids increase the synthesis of inflammatory cytokines and tumour necrosis factor α (TNFα). Acute stress signals induce glutamate release from neurites, which competitively inhibits GGCX, resulting in decreased post-translational modification-carboxylation of OCN in osteoblastic lineage cells before cellular release [112,113]. The glutamate concentrations that were shown to achieve this, were on par to that found in glutaminergic synapses [113,114], which were able to cause osteoblasts to increases their release of ucOCN (Glu). This downregulates parasympathetic tone and consequently allows sympathetic signals to be propagated in the absence of suppression. Berger et al. demonstrated that intravenously injected Glu-OCN into WT mice, caused an immediate and significant downregulation of parasympathetic nervous system activity. This causes decreased contraction of the trachea rings via GPRC6A receptor and acetylcholine signalling, with decreased gastrin levels and increased heart rate [113].

Increased ucOCN (Glu) released due to the acute stress response (ASR), leads to inhibition of parasympathetic tone via reducing acetylcholine (ACH) synthesis, release and recycling [113]. Viewed from an evolutionary standpoint, in the event of potential physical trauma, a high stress situation, which may result in a wound, in the insulin sensitive state, the increased ASR inducing ucOCN (Glu) release may be evolutionarily advantageous. ASR ucOCN (Glu) release may induce acute hyperinsulinaemia in-order to effect multiple actions such as facilitating rapid skeletal muscle glucose uptake and aid to inhibit anti-coagulation activity via elevating plasminogen activator inhibitor type 1 (PAI-1) thus disturbing fibrinolysis [53,115,116].

Acute hyperglycaemia provides extra glucose for skeletal muscle, and also increases hepatic clotting factors and clotting activation [115], thus aiding in prevention of excessive blood-loss/haemorrhage from any potential wound and concomitant rapid hypotension due to hypovolemia. Furthermore, elevated blood glucose due to glucocorticoid induced hepatic glycogenolysis and gluconeogenesis, independently stimulates insulin secretion. The ASR induces: acute hyperglycaemia that increases coagulation capacity, bone release of ucOCN (Glu) from osteoblast lineage cells and osteoclastic resorption, together with hyperglycaemia and ucOCN (Glu) induced acute hyperinsulinaemia, which facilitates inhibition of the breakdown of clots. The plausible evolutionary purpose for this, would be to help enable the stabilisation of wound clotting and thus prevent potentially fatal haemorrhage.

17. Osteocalcin Regulation of Ketosis

Fasting increases ketogenesis and plasma BHB which increases bone resorption ability via lowering the localised pH level as well as inhibiting/regulating osteoblast mineralisation activity [78]. This would increase plasma OCN levels that would go on to stimulate pancreatic beta cell proliferation and insulin production capacity. Natural diurnal cortisol signalling leads to hepatic glycogenolysis and release of glucose to the system, in turn stimulating insulin secretion. Together, through BHB effect on OCN release, they provide the stimulus for insulin secretion to act as a feedback mechanism to regulate ketogenesis.

OCN is considered to protect against obesity, improve glucose uptake and enhance insulin sensitivity, either directly or via OCN stimulated adiponectin secretion [97,100,103,117–119]. Osteocytes comprise the largest population of bone cells, making them likely the biggest producer of OCN than is currently understood. Logically, a substantial loss of osteocytes would result in decreased OCN production and lead to subsequent increases in adiposity. This phenotype is seen in HI/T2DM/CVD patients, who have lower plasma OCN and adiponectin, with increased or normal BMD. Bone mineralisation is possible without OCN and may be enhanced in the hyperinsulinaemic state [110]. Micropetrosis/living-fossilisation of the osteocytic lacuna-canalicular has been demonstrated to be increased in hyperinsulinaemia pathologies [22,23,25,28]. BHB mediated inhibition of mineralisation may be analogous to ePPi and bisphosphonates that help osteocytes maintain their lacunocanalicular network. Inhibition of lacunocanalicular mineralisation concurrent with maintaining osteocyte viability, results in correctly-formed fracture resistant bone that is also dynamically remodelled, a function of healthy bone metabolism leading to wider effects on whole body metabolism.

18. Osteocalcin and Insulin Resistance/Hyperinsulinaemia

In vitro and in vivo KO studies, and studies administering exogenous OCN either orally or intravenously to mice or rats, present results that suggest unOCN (Glu) is necessary for pancreatic beta-cell proliferation. OCN is recognised and activates the GPRC6A receptor on pancreatic beta cells [93,120]. This receptor has other ligands; therefore, caution should be applied in interpreting results from genetic GPRC6A receptor KO studies. For example, with *Gprc6a*$^{-/-}$ mouse pancreatic beta-cells, knocking out the receptor may have potentially also ablated the role of other ligands in activating this promiscuous receptor and its subsequent downstream intracellular signal cascades. Consequently, there is a possibility of producing different combinatory cell signal results based on ligand type [121]. Furthermore, OCN crosses the blood-brain barrier, and studies have shown a brain target receptor of OCN is *Gpr158*. However, it is likely that this is not the only brain OCN receptor, as studies have shown areas of brain that do not express *Gpr158* whilst still having OCN activity, indicating other receptors at play [100].

In a mouse study, KO of beta-cell *Gprc6a*$^{-/-}$ gene expression which encodes the GPRC6A protein receptor, rendered the beta-cells unable to produce sufficient insulin. As a result, these mice had metabolic abnormalities with a similar phenotype to *Ocn*$^{-/-}$ KO mice [98,104,120]. However, as the GPRC6A receptor is able to be activated by other ligands, absence of the receptor does not necessarily mean that a lack of ucOCN (Glu) signalling induces the metabolic abnormality. It may be due to the KO inadvertently rendering other ligands to be unable to signal the beta-cells via the "universal multiligand" GPRC6A receptor [121]. An example of the conflicting information from these in vitro and in vivo model studies is found in humans with hyperinsulinaemia pathologies. In these conditions, studies have shown there is a significant inverse correlation between serum OCN levels and: fasting insulin and glucose, BMI, HOMA-IR, leptin and triglycerides ($p < 0.001$), while higher OCN levels positively correlates with adiponectin levels ($p < 0.001$) [93,98]. These real-world investigations appear to contradict the animal model studies, which expect lower OCN levels to predict lower insulin secretion.

Osteocalcin exerts a large amount of its effects via adiponectin [98]. However, under high insulin conditions, adiponectin receptors are downregulated via insulin activating the PI3K/FOXO1 signal transduction pathway. This diminishes the osteocalcin/adiponectin induction of 5′ adenosine monophosphate-activated protein kinase (AMPK), PGC-1α, mitochondrial biogenesis and increased thermogenesis, that all act to facilitate glucose uptake and oxidation independent of insulin [102].

19. Vitamin D and Magnesium

Vitamin D is required for osteocyte viability and dendrite connectivity [44]. Hyperinsulinaemia reduces vitamin D availability by increasing sequestration of the lipophilic hormone into adipocytes [122]. Patients with pathologies of hyperinsulinaemia, including T2DM, CVD, obesity, MetS and metabolic cancers, are associated with having a lower vitamin D status [95,123]. In addition, hyperinsulinaemia promotes magnesium deficiency (MgD) which decreases vitamin D transport in the blood [84]. Hyperinsulinaemia decreases hydroxylation of inactive 25-OHD$_3$-cacidiol to active 1,25(OH)$_2$D$_3$-calcitriol by lowering 1-alpha hydroxylase (CYP27B1) activity. Hyperinsulinaemia increases mitochondrial (mt) ROS generation and NAD+ depletion, both decrease NADPH availability [20,56,57]. CYP27B1 activity is NADPH and Mg dependent [84,85,124]. Thus, hyperinsulinaemia decreases cellular capacity to activate vitamin D via lowering CYP27B1 hydroxylase activity.

Chronic hyperinsulinaemia dysregulated vitamin D metabolism negatively affects osteocyte health and their subsequent ability to perform dynamic perilacunar remodelling [44]. A study investigating 783 young northern European males ages 20 to 29 years, found a significant association in vitamin D deficiency with peak bone mass. In participants that had inadequate vitamin D status, parathyroid hormone (PTH) levels and bone-specific alkaline phosphatase (BAP) were higher [125]. Interestingly, a reduction in pre-osteoblast beta-oxidation capacity leads to decreased parathyroid hormone (PTH) sensitivity [126], a potential contributor to the elevated levels of PTH seen in HI/T2DM/CVD patients [127,128].

With the increasing prevalence of pre/diabetes, overweight and obesity in children and adolescents, there are serious implications in long-term health. The negative effect of low vitamin D levels on osteocyte viability consequently negatively affects the OCN-producing cell population; this results in wider implications in bone fracture resistance and metabolic/endocrine health. Vitamin D insufficiency is common in the paediatric population, and is shown to be both a marker and/or maker in increasing the risk of developing pre-diabetes (phenotype 3, stage 1/2, [71]). Early lifestyle interventions that improve vitamin D status, starting from childhood, may contribute to reducing rates of low-energy fractures, and to improving longer-term metabolic health [129].

Insulin is required for healthy bone mineralisation, as seen in insulin insufficient T1DM [130]. Puberty and pregnancy are two stages of development where a natural state of hyperinsulinaemia occurs [77,131,132]. Hyperinsulinaemia enables increased growth in bones during puberty, and mineral accretion during pregnancy, which may be in the form of micropetrosis (osteocyte-lacunae mineralisation), to ensure adequate provision of minerals for nursing offspring during lactation. These phases of life lend a physiological adaptive explanation as to why we would see an increase in BMD in T2DM hyperinsulinaemia [30,32,133]. However, puberty and pregnancy are for a limited duration of time and come with either growth in bones (puberty), or lactation and subsequent resorption of bone minerals. Whereas T2DM hyperinsulinaemia may go undetected for many years, a pernicious chronic elevation, leading to excessive micropetrosis and living fossilisation. The resultant disconnecting of osteocytes from one another, impairs their ability to: sense and transmit information, modulate one another, and directly/indirectly modulate bone turnover [31,62,134]. This hypothesis provides a plausible explanation as to why we would see normal to H-BMD in T2DM not conferring fracture resistance.

Rolvien et al., found a 43% greater decrease in empty lacunae fraction with a decrease in the number of viable osteocytes in vitamin D deficient versus replete human iliac crest biopsies ($p < 0.001$) [44]. Given that hyperinsulinaemia decreases vitamin D availability, activation and transport, this is likely to profoundly influence osteocyte directed bone remodelling, as osteocyte viability and thus population is vitamin D dependent. Interestingly, osteocytes express the CYP27B1 enzyme, and are able to directly activate 25(OH)D_3-calcidiol to 1,25$(OH)_2D_3$-calcitriol, providing a localised refined control in provision of active vitamin D availability. One of calcitriol's many roles is to regulate/suppress nuclear factor-kB (NFkB) pathway signalling in the adaptive immune system [135,136]. Theoretically, given osteocytes modulate osteoclastogenesis from the myelopoietic monocyte/macrophage cell lineage, it stands to reason that their ability to directly activate vitamin D serves to locally modulate/fine tune NFkB signalling, which would function in the regulation of osteoclastogenesis. However, this needs to be investigated further.

20. Hyperinsulinaemia Decreases Vitamin D Availability, Decreasing Osteocalcin Synthesis

Bioactive 1,25$(OH)_2D_3$-calcitriol stimulates OCN transcription. Vitamin D response elements are found within the osteoblast/osteocyte 600-nucleotide OCN gene regulatory-sequence transcription start site [137]. Hyperinsulinaemia increases de novo lipogenesis, and drives lipophilic vitamin D to be accreted into adipocytes [122], decreasing availability. Vitamin D deficiency favours osteocyte apoptosis and reduced osteocyte connectivity, decreasing their viability further. Osteocyte connectivity is essential for their function in healthy bone remodelling, and to effect regulation of OCN production [44,135,138,139].

The osteocytes expression of CYP27B1, enabling localised direct activation of calcidiol to calcitriol [41,136], arguably facilitates osteocyte regulation of osteoblastic OCN production. Increased presence of differentiated osteocytes positively correlates with gene expression of OCN in osteoblasts, indicating osteocyte control of osteoblast OCN synthesis [140]. If osteocyte population/health is compromised, this may in turn compromise osteoblast OCN production. A cross-sectional study of 191 non-osteoporotic postmenopausal women found a significant negative correlation between serum OCN and insulin resistance (IR) and HbA1c, $p = 0.001$ and $p = 0.048$ respectively [95]. Hyperinsulinaemia patients have lower serum OCN levels than healthy persons, supporting this notion. However further investigations are needed.

21. Vitamin K, Osteocalcin Carboxylation and Hydroxyapatite Crystallite Alignment

Hyperinsulinaemia mediates de novo lipogenesis and adipogenesis, inhibits lipolysis, and drives adipocyte vitamin K sequestration [95]. This leads to decreased vitamin K availability for OCN carboxylation. Vitamin K_1 sequestration into adipocytes is found in higher concentrations in adipose tissue. The reduced form of vitamin K_1 is an essential co-factor for the γ-glutamyl carboxylase enzyme, to carboxylate OCN [112]. Children who have suffered low energy fractures versus never-suffered-fractures controls, have significantly lower levels of serum carboxylated OCN to uncarboxylated (cOCN:ucOCN) [117]. This corroborates OCN requirement for hydroxyapatite crystallite alignment with collagen for bone quality, which includes the mineralisation structural formation as well as BMD, which translates into fracture resistance [99,110].

Lower levels of vitamin K has also been found to be negatively correlated with ucOCN (Glu) levels in healthy women, where elevated ucOCN (Glu) is a known risk factor for increased fracture risk [141]. In a double blinded, randomised controlled trial with T2D patients, $n = 40$, between 30 to 70 years of age, patients were assigned to one of three groups: (a) placebo-a, $n = 16$, 1000 IU vitamin D3 + calcinated magnesium, (b) placebo-b, $n = 12$,100 ug vitamin K2 + calcinated magnesium and (c) the intervention group, $n = 12$, 1000 IU vitamin D3 + 100 ug vitamin K2. The vitamin D3 group had a significant decrease in serum ucOCN (Glu) 3.3 ± 1.7 ng/dL to 2.5 ± 1.5 ng/dL ($p = 0.026$), ucOCN:cOCN (Glu:Gla) ratio 7.0 ± 7.0 to 3.1 ± 1.7 ($p = 0.039$), glucose ($p < 0.001$) and pancreatic beta cell percentage function ($p = 0.041$). The vitamin K2 group saw a significant decrease in glycaemia ($p = 0.002$), pancreatic beta cell functional percentage ($p = 0.039$), HOMA-IR ($p = 0.041$) and a significant increase in cOCN (Gla) from 0.818 ± 0.567 ng/dL to 1.2 ± 1.1 ng/dL ($p = 0.041$). The group receiving both vitamin D3 and K2, saw a significant decrease in glycaemia ($p = 0.002$) with a decrease in pancreatic beta cell function ($p = 0.004$), and a significant decrease in ucOCN:cOCN (Glu:Gla) from 6.4 ± 4.2 to 4.3 ± 2.9 ratio ($p = 0.023$) [101]. An interesting aspect of this is the decrease in glycaemia in the context of a decrease in functional pancreatic beta cells, which seems counter-intuitive. This indicates decreased reliance on insulin for glucose homeostasis, which supports other studies that show both cOCN and ucOCN are biologically active and have a functional role in insulin-independent glucose uptake. In addition to increasing adipocyte and myocyte mitochondrial biogenesis and uncoupling proteins, leading to increased thermogenesis, likely via stimulating adiponectin synthesis [98].

22. Glycation Damage Decreases Vitamin K-Dependent Carboxylation of Osteocalcin

As described earlier, cOCN is required for healthy bone matrix hydroxyapatite crystallite alignment, which confers structural integrity to bone that decreases fracture risk. Carboxylation of OCN is mediated by the vitamin K dependent γ-glutamyl carboxylase enzyme [112]. Lipophilic vitamin K is transported via chylomicron remnants (CR) in the plasma [142]. Osteoblasts and osteocytes express the membrane receptor proteins apolipoprotein E (apoE) and LDL receptor related protein 1 (LRP-1). A ligand for osteoblast LRP-1 is the CR apoE protein. Osteoblasts are thus able to take up CR containing vitamin K via LRP-1 recognition and binding with the CR structural apoE protein leading to receptor mediated endocytosis. LRP-1 recognition of apoE is via the heparan sulfate proteoglycan (HSPG) mediated pathway [143,144]. Uptake of CR cargo: vitamin K and dietary lipids, are required for osteoblasts' high metabolic demand. Hyperinsulinaemia negatively impacts HSPG function and availability, via impairment of vitamin D regulation. Vitamin D regulates sulfate synthesis, required for heparan sulphate [145].

Neimeier et al. demonstrated in a murine in vivo model that osteoblast internalisation of CR, requires osteoblast membrane expression of endogenous apoE in order to tether CR in the first steps of endocytosis uptake. This occurs in a similar secretion-recapture mechanism performed by hepatocytes [142]. ApoE is highly susceptible to sugar moieties irreversibly attaching to NH_2-protein groups. This glycation damage interferes with CR attraction, receptor recognition and lipid-binding ability [144,146], and ultimately prevents the uptake of CR carrying vitamin K and fatty acids to osteoblasts.

23. Hyperinsulinaemia Decreases Heparan Sulphate Proteoglycans

Heparan sulphate proteoglycans are robust anticoagulants, buffering glycation damage. An increase in heparanase and a decrease in heparin sulphate is implicated in endothelial cell dysfunction [147]. Heparanase is an endoglycosidase that enzymatically cleaves glycosaminoglycan heparan sulphate [148]. Hyperglycaemia oxidative damage, resulting in AGE, and receptor for AGE (RAGE) production, increases heparanase expression [147,149]. Heparan sulphate is required for adipocyte to macrophage mitochondrial transfer, which is decreased in hyperinsulinaemia/obesity [150]. Hyperinsulinaemia drives lowered vitamin D hydroxylation/activation/transport, increasing sulphate wastage and oxidation damage to heparin sulphate proteoglycans [145]. The multiple mechanisms by which hyperinsulinaemia and hyperglycaemia decrease extracellular-localised heparan sulphate, contribute to impairing osteocytes and osteoblast uptake of CR cargo, including essential fatty acids for fuelling and vitamin K. These further drive cellular reliance on glucose fuelling, in turn depleting NAD+ and increasing ROS. Perlecan is a form of heparan sulphate proteoglycan that is essential in positioning, anchoring and receiving mechano-stimuli for osteocytes in their lacunocanalicular space. Hyperinsulinaemia and hyperglycaemia driven breakdown of HS via increased heparanase enzymatic cleavage and increased sulphate wastage, results in decreasing osteocyte perlecan, further contributing to harming osteocyte viability and function [31,62].

Thus the "perfect storm" is established in hyperinsulinaemia osteofragilitas, with increased glycation damage to: collagen rendering it rigid and unrecognisable to digestion enzymes, the vasculature causing increased hypoxia, and to apoE proteins that are essential for receptor mediated uptake of vitamin K. Without vitamin K, osteocalcin carboxylation status decreases, resulting in disordered hydroxyapatite crystallite formation and consequent increased bone fragility irrespective of BMD. Furthermore, glycation damage to apoE and HSPG would decrease delivery of fatty acids for the high metabolic demands required for osteoblast collagen synthesis and osteoblasto/cytogenesis. Impaired fatty acid delivery, along with insulin-mediated inhibition of beta-oxidation and increased mtROS generation, consequently drives poorer collagen production and decreased OCN synthesis. With reduced ability to access fatty acids for fuel, concomitant to increased hypoxia, osteoblasts and osteocytes are forced to be more reliant on glucose oxidation, leading to increased mtROS production, which decreases osteocytogenesis and osteocyte dendritic connectivity [59]. Dynamic bone remodelling decreases and a vicious cycle ensues.

24. OCN and the Brain

OCN protects against neural apoptosis and enhances hippocampus neurogenesis. OCN has been shown to induce changes in GABA and neurotransmitter gene expression, decreasing GABA and increasing serotonin and dopamine, ameliorating anxiety and depression, and improving learning capacity [151]. Oury et al. showed in mice studies and in ex vivo and explant studies; OCN affects neurotransmitter gene expression in the brain. They cleverly demonstrated that OCN passes through the BBB. OCN mediates these neural effects via a different receptor to GPRC6A, as they conducted the same experiments in *Gprc6a*$^{-/-}$ knockout and *Ocn*$^{-/-}$ mice. This study also showed administering of OCN rescued the neurological effects in learning. Mice pups born to *Ocn*$^{-/-}$ mothers suffered significant increased anxiety, depression and loss of learning capability. Furthermore, neurogenesis was negatively affected. The ability to learn was assessed via the Morris Water Maze test (MWM), which is considered a test that measures the function of the hippocampus. Mice born from *Ocn*$^{-/-}$, WT, *Esp*$^{-/-}$ mothers were tested 4 times a day for 10 days. The *Ocn*$^{-/-}$ mice results showed that, over the 10 days of "exposure to the activity," they were almost completely unable to learn [151].

Khrimian et al. provides evidence that OCN positively regulates hippocampal-dependent memory by activating the inositol 1,4,5-triphosphate (IP3) intracellular signal transduction cascade via binding to neuronal Gpr158 in the CA3 region of the hippocampus [152]. This area of research is in its infancy and warrants further investigation, given the increase in cognitive decline, Alzheimer's disease (AD) and Parkinson's disease (PD) that are increasingly recognised as conditions of hyperinsulinaemia and mitochondrial distress [153–155]. Increasing evidence is demonstrating a role of OCN in neurological health and disease [109,156,157]. This leads to further concerns with the increasing earlier rates of obesity, T2D and hyperinsulinaemia in women. Evidence suggests that maternal OCN levels may impact embryonic neurogenesis and rescue from apoptosis, leading to long term effects on offspring such as anxiety, depression and learning capacity [151]. AD and PD are associated with significant increased rates of fragility fractures, furthermore, fragility fractures are also associated with an increased rate in the development of dementia [158–161].

25. Osteocalcin Affects Satiety Regulation and Hepatic Glucose Output

OCN increases serum glucagon-like peptide-1 (GLP-1) levels [162]. GLP-1 suppresses pancreatic α cell secretion of glucagon leading to glycaemia regulation (decreasing hepatic glucose output). GLP-1 also slows gastric emptying leading to increased nutrient absorption, and inhibits food intake [163]. The slowing of gastric emptying and inhibition of food intake by GLP-1 is mediated via vagal circuits, as this effect was shown to be abrogated after truncal vagotomy in normal weight none-diabetic men [164].

OCN further modulates pancreatic α cell secretory profile via regulating gene expression of the rate limiting enzyme tryptophan hydroxylase (*Tph*) for serotonin (5-hydroxytryptamine) synthesis from 5-hydroxytryptophan (5-HTP) [151,154,165–168]. Under high and low glucose settings, the pancreatic alpha cell glucagon secretion decreases when serotonin levels are higher [71,168]. OCN levels are higher in healthy individuals and lower in people with T2D. This is potentially a contributory factor in lost-inhibition on glucagon secretion, therefore leading to increased hepatic glucose output observed in T2 diabetics. The consequent higher set point of hepatic glucose output, feeds forward in generating higher rates of glycation damage and hyperinsulinaemia.

Interestingly, OCN survives the digestive tract intact and is biologically active in the gut. Mizokami et al. demonstrated in mice, that oral ingestion of OCN was more effective at maintaining serum GLP-1 levels than via intraperitoneal injection. The digestive form likely works without absorption via stimulating GPRC6A on the apical side of the intestinal enterocytes, to stimulate synthesis of GLP-1, without the need to enter into the blood stream [162]. GLP-1 is rapidly degraded by cell-surface aminopeptidase dipeptidyl peptidase IV (DPPIV) [163]. Around 4% of oral OCN is absorbed into the blood stream, and raises serum GLP-1 as much as intravenous injection [162]. It may be that OCN action within the gut, on the apical surface side of epithelial cells works in addition to basal side activation, effectively sustaining synthesis for a longer duration of time, independent of raising serum OCN levels. GLP-1 agonists are currently an exciting area of research in the management of T2DM, aiding in decreasing hepatic glucose output.

26. Laboratory Identification

Patients presenting with low energy fracture(s) and normal to higher BMD should be investigated for T2DM and hyperinsulinemia. Furthermore, identification of hyperinsulinaemia-osteofragilitas before fracture occurrence will enable earlier intervention, and provide better patient understanding and compliance with clinical nutritional management. Laboratory identification biomarkers to be evaluated include: fasting insulin, glucose, BHB, OCN (ideally the cOCN:ucOCN ratio to be included), glucagon, GLP-1 and serotonin. If BHB levels are below 0.1 mmol/L and OCN levels are on the lower end or below the reference range, a 5-h OGTT with insulin sensitivity assay in-order to determine Kraft pattern and metabolic phenotype is advised [16,71].

An effective method to decrease excess insulin exposure whilst simultaneously improving glucose homeostasis, for hyperinsulinaemia with or without hyperglycaemia, is through lifestyle management. Carbohydrate restriction with adequate individualised support, should be considered the first line of treatment [53,169,170], with further focus on a diet that maximises nutrient density, especially: vitamin D3, K and magnesium.

27. Conclusions

Micropetrosis/living fossilisation increases with age. Further investigations need to be conducted into the presence and degrees of micropetrosis in hyperinsulinaemia-osteofragilitas. Furthermore, hyperglycaemia decreases osteoclastogenesis [7], which would inhibit the bone resorption that is necessary to effect bone remodelling, which would include the need for removal of glycated collagen and micropetrotic bone. However glycated collagen is rendered unrecognisable by enzymes for breakdown [47]. In addition to increased glycation of HbA1c and collagen in hyperinsulinaemia T2DM, MetS and CVD patients [146], there is also increased glycation of their ApoE and ApoB lipoproteins [146].

This renders chylomicron remnants carrying vitamin K, unable to be recognised by LRP-1 receptors on osteoblast and osteocytes [142]. Carboxylation capacity will be decreased as a result of reduced vitamin K dependent γ-glutamyl carboxylase activity, along with increased bone fragility and fracture risk due to poorer structural alignment of bone hydroxyapatite, which is dependent on carboxylated OCN status [110]. Osteocytes synthesize OCN and directly control osteoblast OCN synthesis. Osteocyte dendritic connectivity and viability, and their capacity to maintain the lacunocanalicular fluid-filled network is evidently essential, in order for them to perform their role in orchestrating and regulating dynamic bone remodelling. This is not only a marker and maker of healthy bone, but is also increasingly appearing as a key endocrine player in whole body metabolism, as well as many homeostatic feedback loops.

Pregnancy and seasonal feasting (autumn where fruit would be more abundant) are two natural events are recognised to induce a period of physiological hyperinsulinaemia [77], where it is plausible to reason that the hyperinsulinaemia phenotype would not only enable the increase in adipogenesis for energy storage, but also facilitate vitamin and mineral storage. Hyperinsulinaemia drives increased adipocyte sequestration of vitamin D and K. Vitamin D is required for osteocyte viability and dendritic health, and to regulate OCN synthesis. Healthy osteocytes synthesise OCN and stimulate osteoblast OCN synthesis. The carboxylation status of OCN is dependent on vitamin K availability. Hyperinsulinaemia decreases vitamin K circulation availability, whilst the increased glycation to proteins occurring in hyperinsulinaemic and hyperglycaemic patients renders vitamin K delivery and uptake via CR apoE/LRP-1 receptor-mediated uptake seriously impaired. Consequently, less OCN is synthesised and less is carboxylated, resulting in poorly formed bone hydroxyapatite crystallite alignment, yet not necessarily compromising BMD, resulting in hyperinsulinaemia-osteofragilitas. Furthermore, glycation damage is incurred on existing bone type 1 collagen, rendering it stiffer and unrecognisable to digestive enzymes. Glycation-induced inhibition of CR fatty acid delivery to osteoblasts and osteocytes results in decreased beta-oxidation, which is needed for the high energy demand for collagen synthesis—at the same time, increasing mtROS production due to an increased reliance on glucose-derived ATP synthesis, which depletes intracellular NAD+. This decreases mitochondrial antioxidant enzyme synthesis, culminating in the production of poorer quality collagen.

Both seasonal diet-induced hyperinsulinaemia and the time limited duration of pregnancy hyperinsulinaemia increase osteocyte spherite-mediated living fossilisation, thus entombing the osteocytes along with magnesium and other minerals within bone, increasing BMD in the process. In a healthy natural cycle, winter follows autumn and lactation follows pregnancy, where minerals would need to be released back into the system. However, chronic hyperinsulinaemia is a situation where winter never comes. The continual stimulant of insulin secretion largely due to dietarily derived farinaceous carbohydrates, perpetuates the physiological condition into a pathological condition as threshold limits are passed, such as the degree in loss of osteocytes and degree of living fossilisation of their lacunocanalicular system, that may no longer be easily recoverable. The consequences are further reaching than simply bone fracture resistance, as we are increasingly discovering bone's role in cardiovascular, muscular, renal and potentially neurological health. Keeping bone healthy has much wider implications in chronic diseases and the ageing process. Evidence suggests that carbohydrate restriction that maintains a healthy low insulin and glucose level, also positively maintains the viability of osteocytes and osteoblasts, that orchestrate dynamic bone remodelling, bone strength and fracture resilience. In turn, maintaining healthy osteocytes, likely feeds forward in positively modulating neurological, endocrine and metabolic health.

Author Contributions: I.D.C. performed the literature search, wrote the original draft, designed the figures, and reviewed and edited the final manuscript; K.H.B. contributed to writing and reviewed and edited the final manuscript; C.A.P.C. contributed to writing and reviewed and edited the final manuscript. All authors have read and agreed to the published version of the manuscript.

Funding: This research received no external funding.

Institutional Review Board Statement: Not applicable.

Informed Consent Statement: Not applicable.

Acknowledgments: Figures created with BioRender.com (2021).

Conflicts of Interest: The authors declare no conflict of interest.

References

1. Hernlund, E.; Svedbom, A.; Ivergård, M.; Compston, J.; Cooper, C.; Stenmark, J.; McCloskey, E.V.; Jönsson, B.; Kanis, J.A. Osteoporosis in the European Union: Medical management, epidemiology and economic burden: A report prepared in collaboration with the International Osteoporosis Foundation (IOF) and the European Federation of Pharmaceutical Industry Associations (EFPIA). *Arch. Osteoporos.* **2013**, *8*, 1–115. [CrossRef] [PubMed]
2. Borgström, F.; Karlsson, L.; Ortsäter, G.; Norton, N.; Halbout, P.; Cooper, C.; Lorentzon, M.; McCloskey, E.V.; Harvey, N.C.; Javaid, M.K.; et al. Fragility fractures in Europe: Burden, management and opportunities. *Arch. Osteoporos.* **2020**, *15*, 1–21. [CrossRef]
3. Moseley, K.F. Type 2 diabetes and bone fractures. *Curr. Opin. Endocrinol. Diabetes Obes.* **2012**, *19*, 128–135. [CrossRef]
4. Napoli, N.; Chandran, M.; Pierroz, D.D.; Abrahamsen, B.; Schwartz, A.V.; Ferrari, S.L. Mechanisms of diabetes mellitus-induced bone fragility. *Nat. Rev. Endocrinol.* **2017**, *13*, 208–219. [CrossRef]
5. Srikanthan, P.; Crandall, C.J.; Miller-Martinez, D.; Seeman, T.E.; Greendale, G.A.; Binkley, N.; Karlamangla, A.S. Insulin resistance and bone strength: Findings from the study of midlife in the United States. *J. Bone Miner. Res.* **2014**, *29*, 796–803. [CrossRef] [PubMed]
6. Crofts, C.A.P.; Zinn, C.; Wheldon, M.; Schofield, M. Hyperinsulinemia: A unifying theory of chronic disease? *Diabesity* **2015**, *1*, 34. [CrossRef]
7. Hu, Z.; Ma, C.; Liang, Y.; Zou, S.; Liu, X. Osteoclasts in bone regeneration under type 2 diabetes mellitus. *Acta Biomater.* **2019**, *84*, 402–413. [CrossRef]
8. Sipos, W.; Pietschmann, P.; Rauner, M.; Kerschan-Schindl, K.; Patsch, J. Pathophysiology of osteoporosis. *Wien. Med. Wochenschr.* **2009**, *159*, 230–234. [CrossRef] [PubMed]
9. Compston, J. Type 2 diabetes mellitus and bone. *J. Intern. Med.* **2018**, *283*, 140–153. [CrossRef] [PubMed]
10. Luisa, M.I.; Ruano, B. Bone disease in diabetes. *Curr. Diabetes Rev.* **2010**, *6*, 144–155. [CrossRef]
11. Shanbhogue, V.V.; Finkelstein, J.S.; Bouxsein, M.L.; Yu, E.W. Association between insulin resistance and bone structure in nondiabetic postmenopausal women. *J. Clin. Endocrinol. Metab.* **2016**, *101*, 3114–3122. [CrossRef] [PubMed]
12. Losada-Grande, E.; Hawley, S.; Soldevila, B.; Martinez-Laguna, D.; Nogues, X.; DIez-Perez, A.; Puig-Domingo, M.; Mauricio, D.; Prieto-Alhambra, D. Insulin use and excess fracture risk in patients with type 2 diabetes: Apropensity-matched cohort analysis. *Sci. Rep.* **2017**, *7*, 1–9. [CrossRef]
13. Sassi, F.; Buondonno, I.; Luppi, C.; Spertino, E.; Stratta, E.; Di Stefano, M.; Ravazzoli, M.; Isaia, G.; Trento, M.; Passera, P.; et al. Type 2 diabetes affects bone cells precursors and bone turnover. *BMC Endocr. Disord.* **2018**, *18*, 55. [CrossRef]
14. Cipriani, C.; Colangelo, L.; Santori, R.; Renella, M.; Mastrantonio, M.; Minisola, S.; Pepe, J. The interplay between bone and glucose metabolism. *Front. Endocrinol.* **2020**, *11*, 122. [CrossRef]
15. Napoli, N.; Strotmeyer, E.S.; Ensrud, K.E.; Sellmeyer, D.E.; Bauer, D.C.; Hoffman, A.R.; Dam, T.T.L.; Barrett-Connor, E.; Palermo, L.; Orwoll, E.S.; et al. Fracture risk in diabetic elderly men: The MrOS study. *Diabetologia* **2014**, *57*, 2057–2065. [CrossRef]
16. Crofts, C.; Schofield, G.; Zinn, C.; Wheldon, M.; Kraft, J. Identifying hyperinsulinaemia in the absence of impaired glucose tolerance: An examination of the Kraft database. *Diabetes Res. Clin. Pract.* **2016**, *118*, 50–57. [CrossRef]
17. Hill, P.A.; Tumber, A. Ceramide-induced cell death/survival in murine osteoblasts. *J. Endocrinol.* **2010**, *206*, 225–233. [CrossRef]
18. Kim, B.J.; Lee, J.Y.; Park, S.J.; Lee, S.H.; Kim, S.J.; Yoo, H.J.; De Pena, S.I.R.; McGee-Lawrence, M.; Isales, C.M.; Koh, J.M.; et al. Elevated Ceramides 18:0 and 24:1 with Aging are Associated with Hip Fracture Risk Through Increased Bone Resorption. *Aging (Albany. N. Y.)* **2019**, *11*, 9388–9404. [CrossRef]
19. Smith, M.E.; Tippetts, T.S.; Brassfield, E.S.; Tucker, B.J.; Ockey, A.; Swensen, A.C.; Anthonymuthu, T.S.; Washburn, T.D.; Kane, D.A.; Prince, J.T.; et al. Mitochondrial fission mediates ceramide-induced metabolic disruption in skeletal muscle. *Biochem. J.* **2013**, *456*, 427–439. [CrossRef]
20. Hansen, M.E.; Tippetts, T.S.; Anderson, M.C.; Holub, Z.E.; Moulton, E.R.; Swensen, A.C.; Prince, J.T.; Bikman, B.T. Insulin increases ceramide synthesis in skeletal muscle. *J. Diabetes Res.* **2014**, *2014*, 765784. [CrossRef]
21. Bell, L.S.; Kayser, M.; Jones, C. The mineralized osteocyte: A living fossil. *Am. J. Phys. Anthropol.* **2008**, *137*, 449–456. [CrossRef]
22. Busse, B.; Djonic, D.; Milovanovic, P.; Hahn, M.; Püschel, K.; Ritchie, R.O.; Djuric, M.; Amling, M. Decrease in the osteocyte lacunar density accompanied by hypermineralized lacunar occlusion reveals failure and delay of remodeling in aged human bone. *Aging Cell* **2010**, *9*, 1065–1075. [CrossRef]
23. Milovanovic, P.; Zimmermann, E.A.; vom Scheidt, A.; Hoffmann, B.; Sarau, G.; Yorgan, T.; Schweizer, M.; Amling, M.; Christiansen, S.; Busse, B. The formation of calcified nanospherites during micropetrosis represents a unique mineralization mechanism in aged human bone. *Small* **2017**, *13*, 1602215. [CrossRef]

24. Rolvien, T.; Schmidt, F.N.; Milovanovic, P.; Jähn, K.; Riedel, C.; Butscheidt, S.; Püschel, K.; Jeschke, A.; Amling, M.; Busse, B. Early bone tissue aging in human auditory ossicles is accompanied by excessive hypermineralization, osteocyte death and micropetrosis. *Sci. Rep.* **2018**, *8*, 1–11. [CrossRef]
25. Milovanovic, P.; Busse, B. Phenomenon of osteocyte lacunar mineralization: Indicator of former osteocyte death and a novel marker of impaired bone quality? *Endocr. Connect.* **2020**, *9*, R70–R80. [CrossRef] [PubMed]
26. Huang, S.; Kaw, M.; Harris, M.T.; Ebraheim, N.; Mcinerney, M.F.; Najjar, S.M.; Lecka-Czernik, B. Decreased osteoclastogenesis and high bone mass in mice with impaired insulin clearance due to liver-specific inactivation to CEACAM1. *Bone* **2009**, *46*, 1138–1145. [CrossRef] [PubMed]
27. Robling, A.G.; Bonewald, L.F. The osteocyte: New insights. *Annu. Rev. Physiol.* **2020**, *82*, 485–506. [CrossRef]
28. Carpentier, V.T.; Wong, J.; Yeap, Y.; Gan, C.; Sutton-Smith, P.; Badiei, A.; Fazzalari, N.L.; Kuliwaba, J.S. Increased proportion of hypermineralized osteocyte lacunae in osteoporotic and osteoarthritic human trabecular bone: Implications for bone remodeling. *Bone* **2012**, *50*, 688–694. [CrossRef] [PubMed]
29. Qing, H.; Bonewald, L.F. Osteocyte remodeling of the perilacunar and pericanalicular matrix. *Int. J. Oral Sci.* **2009**, *1*, 59–65. [CrossRef]
30. Qing, H.; Ardeshirpour, L.; Divieti Pajevic, P.; Dusevich, V.; Jähn, K.; Kato, S.; Wysolmerski, J.; Bonewald, L.F. Demonstration of osteocytic perilacunar/canalicular remodeling in mice during lactation. *J. Bone Miner. Res.* **2012**, *27*, 1018–1029. [CrossRef]
31. Weinkamer, R.; Kollmannsberger, P.; Fratzl, P. Towards a connectomic description of the osteocyte lacunocanalicular network in bone. *Curr. Osteoporos. Rep.* **2019**, *17*, 186–194. [CrossRef] [PubMed]
32. Jähn, K.; Kelkar, S.; Zhao, H.; Xie, Y.; Tiede-Lewis, L.A.M.; Dusevich, V.; Dallas, S.L.; Bonewald, L.F. Osteocytes acidify their microenvironment in response to PTHrP in vitro and in lactating mice in vivo. *J. Bone Miner. Res.* **2017**, *32*, 1761–1772. [CrossRef] [PubMed]
33. Kitaura, H.; Marahleh, A.; Ohori, F.; Noguchi, T.; Shen, W.-R.; Qi, J.; Nara, Y.; Pramusita, A.; Kinjo, R.; Mizoguchi, I. Osteocyte-related cytokines regulate osteoclast formation and bone resorption. *Int. J. Mol. Sci.* **2020**, *21*, 5169. [CrossRef]
34. Nango, N.; Kubota, S.; Hasegawa, T.; Yashiro, W.; Momose, A.; Matsuo, K. Osteocyte-directed bone demineralization along canaliculi. *Bone* **2016**, *84*, 279–288. [CrossRef]
35. Wysolmerski, J.J. Osteocytes remove and replace perilacunar mineral during reproductive cycles. *Bone* **2013**, *54*, 230–236. [CrossRef]
36. Hao, Z.; Ma, Y.; Wu, J.; Li, X.; Chen, H.; Shen, J.; Wang, H. Osteocytes regulate osteoblast differentiation and osteoclast activity through Interleukin-6 under mechanical loading. *RSC Adv.* **2017**, *7*, 50200–50209. [CrossRef]
37. Sánchez-de-Diego, C.; Artigas, N.; Pimenta-Lopes, C.; Valer, J.A.; Torrejon, B.; Gama-Pérez, P.; Villena, J.A.; Garcia-Roves, P.M.; Rosa, J.L.; Ventura, F. Glucose restriction promotes osteocyte specification by activating a PGC-1α-dependent transcriptional program. *iScience* **2019**, *15*, 79–94. [CrossRef]
38. Dallas, S.L.; Prideaux, M.; Bonewald, L.F. The osteocyte: An endocrine cell and more. *Endocr. Rev.* **2013**, *34*, 658–690. [CrossRef]
39. Ferron, M. Endocrine functions of bone. In *Principles of Endocrinology and Hormone Action*; Springer: Cham, Switzerland, 2018; pp. 1–27. [CrossRef]
40. Bonewald, L.F. The amazing osteocyte. *J. Bone Miner. Res.* **2011**, *26*, 229–238. [CrossRef] [PubMed]
41. Lanske, B.; Densmore, M.J.; Erben, R.G. Vitamin D endocrine system and osteocytes. *Bonekey Rep.* **2014**, *3*, 494. [CrossRef]
42. Gao, J.; Qin, A.; Liu, D.; Ruan, R.; Wang, Q.; Yuan, J.; Cheng, T.S.; Filipovska, A.; Papadimitriou, J.M.; Dai, K.; et al. Endoplasmic reticulum mediates mitochondrial transfer within the osteocyte dendritic network. *Sci. Adv.* **2019**, *5*, eaaw7215. [CrossRef] [PubMed]
43. Noble, B.S.; Peet, N.; Stevens, H.Y.; Brabbs, A.; Mosley, J.R.; Reilly, G.C.; Reeve, J.; Skerry, T.M.; Lanyon, L.E. Mechanical loading: Biphasic osteocyte survival and targeting of osteoclasts for bone destruction in rat cortical bone. *Am. J. Physiol.-Cell Physiol.* **2003**, *284*, C934–C943. [CrossRef] [PubMed]
44. Rolvien, T.; Krause, M.; Jeschke, A.; Yorgan, T.; Püschel, K.; Schinke, T.; Busse, B.; Demay, M.B.; Amling, M. Vitamin D regulates osteocyte survival and perilacunar remodeling inhuman and murine bone. *Bone* **2017**, *103*, 78–87. [CrossRef] [PubMed]
45. Liu, T.T.; Liu, D.M.; Xuan, Y.; Zhao, L.; Sun, L.H.; Zhao, D.D.; Wang, X.F.; He, Y.; Guo, X.Z.; Du, R.; et al. The association between the baseline bone resorption marker CTX and incident dysglycemia after 4 years. *Bone Res.* **2017**, *5*, 1–7. [CrossRef]
46. Saito, M.; Marumo, K. Collagen cross-links as a determinant of bone quality: A possible explanation for bone fragility in aging, osteoporosis, and diabetes mellitus. *Osteoporos. Int.* **2010**, *21*, 195–214. [CrossRef]
47. Gautieri, A.; Passini, F.S.; Silván, U.; Guizar-Sicairos, M.; Carimati, G.; Volpi, P.; Moretti, M.; Schoenhuber, H.; Redaelli, A.; Berli, M.; et al. Advanced glycation end-products: Mechanics of aged collagen from molecule to tissue. *Matrix Biol.* **2017**, *59*, 95–108. [CrossRef]
48. Vashishth, D.; Verborgt, O.; Divine, G.; Schaffler, M.B.; Fyhrie, D.P. Decline in osteocyte lacunar density in human cortical bone is associated with accumulation of microcracks with age. *Bone* **2000**, *26*, 375–380. [CrossRef]
49. Frey, J.L.; Li, Z.; Ellis, J.M.; Zhang, Q.; Farber, C.R.; Aja, S.; Wolfgang, M.J.; Clemens, T.L.; Riddle, R.C. Wnt-Lrp5 signaling regulates fatty acid metabolism in the osteoblast. *Mol. Cell. Biol.* **2015**, *35*, 1979–1991. [CrossRef]
50. Qian, G.; Fan, W.; Ahlemeyer, B.; Karnati, S.; Baumgart-Vogt, E. Peroxisomes in different skeletal cell types during intramembranous and endochondral ossification and their regulation during osteoblast differentiation by distinct peroxisome proliferator-activated receptors. *PLoS ONE* **2015**, *10*, e0143439. [CrossRef]

51. Kim, S.P.; Li, Z.; Zoch, M.L.; Frey, J.L.; Bowman, C.E.; Kushwaha, P.; Ryan, K.A.; Goh, B.C.; Scafidi, S.; Pickett, J.E.; et al. Fatty acid oxidation by the osteoblast is required for normal bone acquisition in a sex- and diet-dependent manner. *JCI Insight* **2017**, *2*, e92704. [CrossRef]
52. Kushwaha, P.; Wolfgang, M.J.; Riddle, R.C. Fatty acid metabolism by the osteoblast. *Bone* **2018**, *115*, 8–14. [CrossRef]
53. Cooper, I.D.; Crofts, C.A.P.; DiNicolantonio, J.J.; Malhotra, A.; Elliott, B.; Kyriakidou, Y.; Brookler, K.H. Relationships between hyperinsulinaemia, magnesium, vitamin D, thrombosis and COVID-19: Rationale for clinical management. *Open Hear.* **2020**, *7*, e001356. [CrossRef]
54. Veech, R.L. The therapeutic implications of ketone bodies: The effects of ketone bodies in pathological conditions: Ketosis, ketogenic diet, redox states, insulin resistance, and mitochondrial metabolism. *Prostaglandins Leukot. Essent. Fat. Acids* **2004**, *70*, 309–319. [CrossRef] [PubMed]
55. Hamanaka, R.B.; Chandel, N.S. Mitochondrial reactive oxygen species regulate cellular signaling and dictate biological outcomes. *Trends Biochem. Sci.* **2010**, *35*, 505–513. [CrossRef] [PubMed]
56. Newman, J.C.; Verdin, E. Ketone bodies as signaling metabolites. *Trends Endocrinol. Metab.* **2014**, *25*, 42–52. [CrossRef] [PubMed]
57. Newman, J.C.; Verdin, E. β-hydroxybutyrate: A signaling metabolite. *Annu. Rev. Nutr.* **2017**, *37*, 51–76. [CrossRef]
58. Anderson, E.J.; Lustig, M.E.; Boyle, K.E.; Woodlief, T.L.; Kane, D.A.; Lin, C.T.; Price, J.W.; Kang, L.; Rabinovitch, P.S.; Szeto, H.H.; et al. Mitochondrial H2O2 emission and cellular redox state link excess fat intake to insulin resistance in both rodents and humans. *J. Clin. Investig.* **2009**, *119*, 573–581. [CrossRef] [PubMed]
59. Kobayashi, K.; Nojiri, H.; Saita, Y.; Morikawa, D.; Ozawa, Y.; Watanabe, K.; Koike, M.; Asou, Y.; Shirasawa, T.; Yokote, K.; et al. Mitochondrial superoxide in osteocytes perturbs canalicular networks in the setting of age-related osteoporosis. *Sci. Rep.* **2015**, *5*, 1–11. [CrossRef] [PubMed]
60. Yu, T.; Robotham, J.L.; Yoon, Y. Increased production of reactive oxygen species in hyperglycemic conditions requires dynamic change of mitochondrial morphology. *Proc. Natl. Acad. Sci. USA.* **2006**, *103*, 2653–2658. [CrossRef]
61. Giacomello, M.; Pyakurel, A.; Glytsou, C.; Scorrano, L. The cell biology of mitochondrial membrane dynamics. *Nat. Rev. Mol. Cell Biol.* **2020**, *21*, 204–224. [CrossRef]
62. Lai, X.; Price, C.; Modla, S.; Thompson, W.R.; Caplan, J.; Kirn-Safran, C.B.; Wang, L. The dependences of osteocyte network on bone compartment, age, and disease. *Bone Res.* **2015**, *3*, 1–11. [CrossRef]
63. Yeung, S.M.H.; Binnenmars, S.H.; Gant, C.M.; Navis, G.; Gansevoort, R.T.; Bakker, S.J.L.; De Borst, M.H.; Laverman, G.D. Fibroblast growth factor 23 and mortality in patients with type 2 diabetes and normal or mildly impaired kidney function. *Diabetes Care* **2019**, *42*, 2151–2153. [CrossRef]
64. Kim, H.; Lee, Y.D.; Kim, H.J.; Lee, Z.H.; Kim, H.-H. SOD2 and Sirt3 control osteoclastogenesis by regulating mitochondrial ROS. *J. Bone Miner. Res.* **2017**, *32*, 397–406. [CrossRef]
65. Kashiwaya, Y.; Satos, K.; Tsuchiya, N.; Thomas, S.; Fells, D.A.; Veechn, R.L.; Passonneau, J. V Control of glucose utilization in working perfused rat heart. *J. Biol. Chem.* **1994**, *269*, 25502–25514. [CrossRef]
66. van de Ven, R.A.H.; Santos, D.; Haigis, M.C. Mitochondrial sirtuins and molecular mechanisms of aging. *Trends Mol. Med.* **2017**, *23*, 320–331. [CrossRef]
67. Qiu, X.; Brown, K.; Hirschey, M.D.; Verdin, E.; Chen, D. Calorie restriction reduces oxidative stress by SIRT3-mediated SOD2 activation. *Cell Metab.* **2010**, *12*, 662–667. [CrossRef]
68. Chen, Y.; Zhang, J.; Lin, Y.; Lei, Q.; Guan, K.L.; Zhao, S.; Xiong, Y. Tumour suppressor SIRT3 deacetylates and activates manganese superoxide dismutase to scavenge ROS. *EMBO Rep.* **2011**, *12*, 534–541. [CrossRef]
69. Dikalova, A.E.; Itani, H.A.; Nazarewicz, R.R.; McMaster, W.G.; Flynn, C.R.; Uzhachenko, R.; Fessel, J.P.; Gamboa, J.L.; Harrison, D.G.; Dikalov, S.I. Sirt3 impairment and SOD2 hyperacetylation in vascular oxidative stress and hypertension. *Circ. Res.* **2017**, *121*, 564–574. [CrossRef] [PubMed]
70. Golub, E.E.; Boesze-Battaglia, K. The role of alkaline phosphatase in mineralization. *Curr. Opin. Orthop.* **2007**, *18*, 444–448. [CrossRef]
71. Cooper, I.D.; Brookler, K.H.; Kyriakidou, Y.; Elliott, B.T.; Crofts, C.A.P. Metabolic phenotypes and step by step evolution of type 2 diabetes: A new paradigm. *Biomed* **2021**, *9*, 800.
72. Xu, Y.; Liu, L.; Nakamura, A.; Someya, S.; Miyakawa, T.; Tanokura, M. Studies on the regulatory mechanism of isocitrate dehydrogenase 2 using acetylation mimics. *Sci. Rep.* **2017**, *7*, 1–10. [CrossRef]
73. Someya, S.; Yu, W.; Hallows, W.C.; Xu, J.; Vann, J.M.; Leeuwenburgh, C.; Tanokura, M.; Denu, J.M.; Prolla, T.A. Sirt3 mediates reduction of oxidative damage and prevention of age-related hearing loss under caloric restriction. *Cell* **2010**, *143*, 802–812. [CrossRef]
74. Liu, Z.; Solesio, M.E.; Schaffler, M.B.; Frikha-Benayed, D.; Rosen, C.J.; Werner, H.; Kopchick, J.J.; Pavlov, E.V.; Abramov, A.Y.; Yakar, S. Mitochondrial function is compromised in cortical bone osteocytes of long-lived growth hormone receptor null mice. *J. Bone Miner. Res.* **2019**, *34*, 106–122. [CrossRef]
75. Spencer, G.; Genever, P. Long-term potentiation in bone-A role for glutamate in strain-induced cellular memory? *BMC Cell Biol.* **2003**, *4*, 9. [CrossRef]
76. Spencer, G.; Hitchcock, I.; Genever, P.G. Emerging neuroskeletal signalling pathways: A review. *FEBS Lett.* **2004**, *559*, 6–12. [CrossRef]
77. Sonagra, A.D. Normal pregnancy-A state of insulin resistance. *J. Clin. Diagnostic Res.* **2014**, *8*, CC01. [CrossRef]

78. Saito, A.; Yoshimura, K.; Miyamoto, Y.; Kaneko, K.; Chikazu, D.; Yamamoto, M.; Kamijo, R. Enhanced and suppressed mineralization by acetoacetate and β-hydroxybutyrate in osteoblast cultures. *Biochem. Biophys. Res. Commun.* **2016**, *473*, 537–544. [CrossRef]
79. Ferron, M.; Lacombe, J. Regulation of energy metabolism by the skeleton: Osteocalcin and beyond. *Arch. Biochem. Biophys.* **2014**, *561*, 137–146. [CrossRef]
80. Li, J.; Zhang, H.; Yang, C.; Li, Y.; Dai, Z. An overview of osteocalcin progress. *J. Bone Miner. Metab.* **2016**, *34*, 367–379. [CrossRef]
81. Grabacka, M.; Pierzchalska, M.; Dean, M.; Reiss, K. Regulation of ketone body metabolism and the role of PPARα. *Int. J. Mol. Sci.* **2016**, *17*, 2093. [CrossRef]
82. Arnett, T.R. Osteocytes: Regulating the mineral reserves? *J. Bone Miner. Res.* **2013**, *28*, 2433–2435. [CrossRef]
83. Gröber, U.; Schmidt, J.; Kisters, K. Magnesium in Prevention and Therapy. *Nutrients* **2015**, *7*, 8199–8226. [CrossRef] [PubMed]
84. DiNicolantonio, J.J.; O'Keefe, J.H.; Wilson, W. Subclinical magnesium deficiency: A principal driver of cardiovascular disease and a public health crisis. *Open Hear.* **2018**, *5*, e000668. [CrossRef]
85. Kostov, K. Effects of magnesium deficiency on mechanisms of insulin resistance in type 2 diabetes: Focusing on the processes of insulin secretion and signaling. *Int. J. Mol. Sci.* **2019**, *20*, 1351. [CrossRef]
86. Drake, M.T.; Clarke, B.L.; Khosla, S. Bisphosphonates: Mechanism of action and role in clinical practice. *Mayo. Clin. Proc.* **2008**, *83*, 1032–1045. [CrossRef]
87. Zhu, D.; Mackenzie, N.C.W.; Millán, J.L.; Farquharson, C.; MacRae, V.E. The appearance and modulation of osteocyte marker expression during calcification of vascular smooth muscle cells. *PLoS ONE* **2011**, *6*, e19595. [CrossRef]
88. Brandenburg, V.M.; Kramann, R.; Koos, R.; Krüger, T.; Schurgers, L.; Mühlenbruch, G.; Hübner, S.; Gladziwa, U.; Drechsler, C.; Ketteler, M. Relationship between sclerostin and cardiovascular calcification in hemodialysis patients: A cross-sectional study. *BMC Nephrol.* **2013**, *14*, 1–10. [CrossRef]
89. De Maré, A.; D'haese, P.C.; Verhulst, A. The role of sclerostin in bone and ectopic calcification. *Int. J. Mol. Sci.* **2020**, *21*, 3199. [CrossRef]
90. Tyson, J.; Bundy, K.; Roach, C.; Douglas, H.; Ventura, V.; Segars, M.F.; Schwartz, O.; Simpson, C.L. Mechanisms of the osteogenic switch of smooth muscle cells in vascular calcification: WNT signaling, BMPs, mechanotransduction, and EndMT. *Bioengineering* **2020**, *7*, 88. [CrossRef]
91. Song, L. Calcium and bone metabolism indices. *Adv. Clin. Chem.* **2017**, *82*, 1–46.
92. Ma, H.; Wang, P.; Jin, D.; Jia, T.; Mao, H.; Zhang, J.; Zhao, S. The hepatic ectonucleotide pyrophosphatase/phosphodiesterase 1 gene mRNA abundance is reduced by insulin and induced by dexamethasone. *Brazilian J. Med. Biol. Res.* **2018**, *51*. [CrossRef]
93. Saleem, U.; Mosley, T.H.; Kullo, I.J. Serum osteocalcin is associated with measures of insulin resistance, adipokine levels, and the presence of metabolic syndrome. *Arterioscler. Thromb. Vasc. Biol.* **2010**, *30*, 1474–1478. [CrossRef] [PubMed]
94. Razny, U.; Fedak, D.; Kiec-Wilk, B.; Goralska, J.; Gruca, A.; Zdzienicka, A.; Kiec-Klimczak, M.; Solnica, B.; Hubalewska-Dydejczyk, A.; Malczewska-Malec, M. Carboxylated and undercarboxylated osteocalcin in metabolic complications of human obesity and prediabetes. *Diabetes. Metab. Res. Rev.* **2017**, *33*, e2862. [CrossRef] [PubMed]
95. Guney, G.; Sener-Simsek, B.; Tokmak, A.; Yucel, A.; Buyukkagnici, U.; Yilmaz, N.; Engin-Ustun, Y.; Ozgu-Erdinc, A.S. Assessment of the relationship between serum vitamin D and osteocalcin levels with metabolic syndrome in non-osteoporotic postmenopausal women. *Geburtshilfe Frauenheilkd.* **2019**, *79*, 293–299. [CrossRef]
96. Riquelme-Gallego, B.; García-Molina, L.; Cano-Ibáñez, N.; Sánchez-Delgado, G.; Andújar-Vera, F.; García-Fontana, C.; González-Salvatierra, S.; García-Recio, E.; Martínez-Ruiz, V.; Bueno-Cavanillas, A.; et al. Circulating undercarboxylated osteocalcin as estimator of cardiovascular and type 2 diabetes risk in metabolic syndrome patients. *Sci. Rep.* **2020**, *10*, 1–10.
97. Wei, J.; Karsenty, G. An overview of the metabolic functions of osteocalcin. *Rev. Endocr. Metab. Disord.* **2015**, *16*, 93–98. [CrossRef]
98. Hill, H.; Grams, J.; Walton, R.G.; Liu, J.; Moellering, D.R.; Garvey, W.T. Carboxylated and uncarboxylated forms of osteocalcin directly modulate the glucose transport system and inflammation in adipocytes. *Horm. Metab. Res.* **2014**, *46*, 341–347. [CrossRef]
99. Popko, J.; Karpiński, M.; Chojnowska, S.; Maresz, K.; Milewski, R.; Badmaev, V.; Schurgers, L.J. Decreased levels of circulating carboxylated osteocalcin in children with low energy fractures: A pilot study. *Nutrients* **2018**, *10*, 734. [CrossRef]
100. Moser, S.C.; van der Eerden, B.C.J. Osteocalcin—A versatile bone-derived hormone. *Front. Endocrinol.* **2019**, *9*, 794. [CrossRef]
101. Aguayo-Ruiz, J.I.; García-Cobián, T.A.; Pascoe-González, S.; Sánchez-Enríquez, S.; Llamas-Covarrubias, I.M.; García-Iglesias, T.; López-Quintero, A.; Llamas-Covarrubias, M.A.; Trujillo-Quiroz, J.; Rivera-Leon, E.A. Effect of supplementation with vitamins D3 and K2 on undercarboxylated osteocalcin and insulin serum levels in patients with type 2 diabetes mellitus: A randomized, double-blind, clinical trial. *Diabetol. Metab. Syndr.* **2020**, *12*, 73. [CrossRef]
102. Tsuchida, A.; Yamauchi, T.; Ito, Y.; Hada, Y.; Maki, T.; Takekawa, S.; Kamon, J.; Kobayashi, M.; Suzuki, R.; Hara, K.; et al. Insulin/Foxo1 pathway regulates expression levels of adiponectin receptors and adiponectin sensitivity. *J. Biol. Chem.* **2004**, *279*, 30817–30822. [CrossRef]
103. Ferron, M.; Hinoi, E.; Karsenty, G.; Ducy, P. Osteocalcin differentially regulates β cell and adipocyte gene expression and affects the development of metabolic diseases in wild-type mice. *Proc. Natl. Acad. Sci. USA.* **2008**, *105*, 5266–5270. [CrossRef] [PubMed]
104. Wei, J.; Hanna, T.; Suda, N.; Karsenty, G.; Ducy, P. Osteocalcin promotes β-cell proliferation during development and adulthood through Gprc6a. *Diabetes* **2014**, *63*, 1021–1031. [CrossRef]
105. Meier, J.J.; Bhushan, A.; Butler, A.E.; Rizza, R.A.; Butler, P.C. Sustained beta cell apoptosis in patients with long-standing type 1 diabetes: Indirect evidence for islet regeneration? *Diabetologia* **2005**, *48*, 2221–2228. [CrossRef]

106. Muoio, D.M.; Newgard, C.B. Mechanisms of disease: Molecular and metabolic mechanisms of insulin resistance and β-cell failure in type 2 diabetes. *Nat. Rev. Mol. Cell Biol.* **2008**, *9*, 193–205. [CrossRef]
107. ACCORD; Gerstein, H.C.; Miller, M.E.; Byington, R.P.; Goff, D.C.; Bigger, J.T.; Buse, J.B.; Cushman, W.C.; Genuth, S.; Ismail-Beigi, F.; et al. Effects of intensive glucose lowering in type 2 diabetes. *N. Engl. J. Med.* **2008**, *358*, 2545–2559. [PubMed]
108. Ferron, M.; McKee, M.D.; Levine, R.L.; Ducy, P.; Karsenty, G. Intermittent injections of osteocalcin improve glucose metabolism and prevent type 2 diabetes in mice. *Bone* **2012**, *50*, 568–575. [CrossRef]
109. Gerosa, L.; Lombardi, G. Bone-to-brain: A round trip in the adaptation to mechanical stimuli. *Front. Physiol.* **2021**, *12*, 565. [CrossRef]
110. Moriishi, T.; Ozasa, R.; Ishimoto, T.; Nakano, T.; Hasegawa, T.; Miyazaki, T.; Liu, W.; Fukuyama, R.; Wang, Y.; Komori, H.; et al. Osteocalcin is necessary for the alignment of apatite crystallites, but not glucose metabolism, testosterone synthesis, or muscle mass. *PLoS Genet.* **2020**, *16*, e1008586. [CrossRef] [PubMed]
111. Manolagas, S.C. Osteocalcin promotes bone mineralization but is not a hormone. *PLoS Genet.* **2020**, *16*, e1008714. [CrossRef] [PubMed]
112. Presnell, S.R.; Stafford, D.W. The vitamin K-dependent carboxylase*. *Thromb Haemost* **2002**, *87*, 937–946. [CrossRef]
113. Berger, J.M.; Singh, P.; Khrimian, L.; Morgan, D.A.; Chowdhury, S.; Arteaga-Solis, E.; Horvath, T.L.; Domingos, A.I.; Marsland, A.L.; Yadav, V.K.; et al. Mediation of the acute stress response by the skeleton. *Cell Metab.* **2019**, *30*, 890–902.e8. [CrossRef] [PubMed]
114. Clements, J.D.; Lester, R.A.J.; Tong, G.; Jahr, C.E.; Westbrook, G.L. The time course of glutamate in the synaptic cleft. *Science* **1992**, *258*, 1498–1501. [CrossRef]
115. Stegenga, M.E.; Van Der Crabben, S.N.; Levi, M.; De Vos, A.F.; Tanck, M.W.; Sauerwein, H.P.; Van Der Poll, T. Hyperglycemia stimulates coagulation, whereas hyperinsulinemia impairs fibrinolysis in healthy humans. *Diabetes* **2006**, *55*, 1807–1812. [CrossRef]
116. Perkins, J.M.; Joy, N.G.; Tate, D.B.; Davis, S.N. Acute effects of hyperinsulinemia and hyperglycemia on vascular inflammatory biomarkers and endothelial function in overweight and obese humans. *Am. J. Physiol. Metab.* **2015**, *309*, E168–E176. [CrossRef]
117. Shea, M.K.; Booth, S.L.; Gundberg, C.M.; Peterson, J.W.; Waddell, C.; Dawson-Hughes, B.; Saltzman, E. Adulthood obesity is positively associated with adipose tissue concentrations of vitamin K and inversely associated with circulating indicators of vitamin K status in men and women. *J. Nutr.* **2010**, *140*, 1029–1034. [CrossRef] [PubMed]
118. Kirk, B.; Feehan, J.; Lombardi, G.; Duque, G. Muscle, Bone, And fat crosstalk: The biological role of myokines, osteokines, and adipokines. *Curr. Osteoporos. Rep.* **2020**, *18*, 388–400. [CrossRef]
119. Bilotta, F.L.; Arcidiacono, B.; Messineo, S.; Greco, M.; Chiefari, E.; Britti, D.; Nakanishi, T.; Foti, D.P.; Brunetti, A. Insulin and osteocalcin: Further evidence for a mutual cross-talk. *Endocrine* **2018**, *59*, 622–632. [CrossRef]
120. Pi, M.; Kapoor, K.; Ye, R.; Nishimoto, S.K.; Smith, J.C.; Baudry, J.; Quarles, L.D. Evidence for osteocalcin binding and activation of GPRC6A in β-cells. *Endocrinology* **2016**, *157*, 1866–1880. [CrossRef]
121. Pi, M.; Nishimoto, S.K.; Quarles, L.D. GPRC6A: Jack of all metabolism (or master of none). *Mol. Metab.* **2017**, *6*, 185–193. [CrossRef] [PubMed]
122. Carrelli, A.; Bucovsky, M.; Horst, R.; Cremers, S.; Zhang, C.; Bessler, M.; Schrope, B.; Evanko, J.; Blanco, J.; Silverberg, S.J.; et al. Vitamin D storage in adipose tissue of obese and normal weight women. *J. Bone Miner. Res.* **2017**, *32*, 237–242. [CrossRef]
123. de Pergola, G.; Nitti, A.; Bartolomeo, N.; Gesuita, A.; Giagulli, V.A.; Triggiani, V.; Guastamacchia, E.; Silvestris, F. Possible role of hyperinsulinemia and insulin resistance in lower vitamin D levels in overweight and obese patients. *BioMed Res. Int.* **2013**, *2013*. [CrossRef]
124. Uwitonze, A.M.; Razzaque, M.S. Role of magnesium in vitamin D activation and function. *J. Am. Osteopath. Assoc.* **2018**, *118*, 181–189. [CrossRef] [PubMed]
125. Frost, M.; Abrahamsen, B.; Nielsen, T.L.; Hagen, C.; Andersen, M.; Brixen, K. Vitamin D status and PTH in young men: A cross-sectional study on associations with bone mineral density, body composition and glucose metabolism. *Clin. Endocrinol.* **2010**, *73*, 573–580. [CrossRef]
126. Huang, M.S.; Lu, J.; Ivanov, Y.; Sage, A.P.; Tseng, W.; Demer, L.L.; Tintut, Y. Hyperlipidemia impairs osteoanabolic effects of PTH. *J. Bone Miner. Res.* **2008**, *23*, 1672–1679. [CrossRef] [PubMed]
127. Alemzadeh, R.; Kichler, J. Parathyroid hormone is associated with biomarkers of insulin resistance and inflammation, independent of vitamin D status, in obese adolescent. *Metab. Syndr. Relat. Disord.* **2012**, *10*, 422–429. [CrossRef] [PubMed]
128. Rahimi, Z. Parathyroid hormone, glucose metabolism and diabetes mellitus. *J. Parathyr. Dis.* **2014**, *2*, 55–56.
129. Bilinski, W.J.; Paradowski, P.T.; Sypniewska, G. Bone health and hyperglycemia in pediatric populations. *Crit. Rev. Clin. Lab. Sci.* **2020**, *57*, 444–457. [CrossRef]
130. Raisingani, M.; Preneet, B.; Kohn, B.; Yakar, S. Skeletal growth and bone mineral acquisition in type 1 diabetic children; Abnormalities of the GH/IGF-1 axis. *Growth Horm. IGF Res.* **2017**, *34*, 13–21. [CrossRef]
131. Goran, M.I.; Gower, B.A. Longitudinal study on pubertal insulin resistance. *Diabetes* **2001**, *50*, 2444–2450. [CrossRef] [PubMed]
132. Kelsey, M.M.; Zeitler, P.S. Insulin resistance of puberty. *Curr. Diab. Rep.* **2016**, *16*, 64. [CrossRef]
133. Lieben, L.; Masuyama, R.; Torrekens, S.; Van Looveren, R.; Schrooten, J.; Baatsen, P.; Lafage-Proust, M.H.; Dresselaers, T.; Feng, J.Q.; Bonewald, L.F.; et al. Normocalcemia is maintained in mice under conditions of calcium malabsorption by vitamin D-induced inhibition of bone mineralization. *J. Clin. Investig.* **2012**, *122*, 1803–1815. [CrossRef]

134. Tresguerres, F.G.F.; Torres, J.; López-Quiles, J.; Hernández, G.; Vega, J.A.; Tresguerres, I.F. The Osteocyte: A multifunctional cell within the bone. *Ann. Anat.* **2020**, *227*, 151422. [CrossRef] [PubMed]
135. Bikle, D. Vitamin D: Production, Metabolism, And Mechanisms of Action. Available online: http://www.ncbi.nlm.nih.gov/pubmed/25905172 (accessed on 11 August 2017).
136. Bikle, D.D.; Patzek, S.; Wang, Y. Physiologic and pathophysiologic roles of extra renal CYP27b1: Case report and review. *Bone Rep.* **2018**, *8*, 255–267. [CrossRef] [PubMed]
137. Lian, J.; Stewart, C.; Puchacz, E.; Mackowiak, S.; Shalhoub, V.; Collart, D.; Zambetti, G.; Stein, G. Structure of the rat osteocalcin gene and regulation of vitamin D-dependent expression. *Proc. Natl. Acad. Sci. USA.* **1989**, *86*, 1143–1147. [CrossRef] [PubMed]
138. Pramyothin, P.; Biancuzzo, R.M.; Lu, Z.; Hess, D.T.; Apovian, C.M.; Holick, M.F. Vitamin D in adipose tissue and serum 25-hydroxyvitamin D after Roux-en-Y gastric bypass. *Obesity* **2011**, *19*, 2228–2234. [CrossRef]
139. Landrier, J.-F.; Marcotorchino, J.; Tourniaire, F. Lipophilic micronutrients and adipose tissue biology. *Nutrients* **2012**, *4*, 1622–1649. [CrossRef]
140. Skottke, J.; Gelinsky, M.; Bernhardt, A. In vitro co-culture model of primary human osteoblasts and osteocytes in collagen gels. *Int. J. Mol. Sci.* **2019**, *20*, 1998. [CrossRef]
141. Yamauchi, M.; Yamaguchi, T.; Nawata, K.; Takaoka, S.; Sugimoto, T. Relationships between undercarboxylated osteocalcin and vitamin K intakes, bone turnover, and bone mineral density in healthy women. *Clin. Nutr.* **2010**, *29*, 761–765. [CrossRef]
142. Niemeier, A.; Niedzielska, D.; Secer, R.; Schilling, A.; Merkel, M.; Enrich, C.; Rensen, P.C.N.; Heeren, J. Uptake of postprandial lipoproteins into bone in vivo: Impact on osteoblast function. *Bone* **2008**, *43*, 230–237. [CrossRef]
143. Al-Haideri, M.; Goldberg, I.J.; Galeano, N.F.; Gleeson, A.; Vogel, T.; Gorecki, M.; Sturley, S.L.; Deckelbaum, R.J. Heparan sulfate proteoglycan-mediated uptake of apolipoprotein E- triglyceride-rich lipoprotein particles: A major pathway at physiological particle concentrations. *Biochemistry* **1997**, *36*, 12766–12772. [CrossRef]
144. Laffont, I.; Shuvaev, V.V.; Briand, O.; Lestavel, S.; Barbier, A.; Taniguchi, N.; Fruchart, J.C.; Clavey, V.; Siest, G. Early-glycation of apolipoprotein E: Effect on its binding to LDL receptor, scavenger receptor A and heparan sulfates. *Biochim. Biophys. Acta-Mol. Cell Biol. Lipids* **2002**, *1583*, 99–107. [CrossRef]
145. Bolt, M.J.G.; Liu, W.; Qiao, G.; Kong, J.; Zheng, W.; Krausz, T.; Cs-Szabo, G.; Sitrin, M.D.; Li, Y.C. Critical role of vitamin D in sulfate homeostasis: Regulation of the sodium-sulfate cotransporter by 1,25-dihydroxyvitamin D3. *Am. J. Physiol. Metab.* **2004**, *287*, E744–E749. [CrossRef]
146. Ke, L.; Chan, H.; Chen, C.; Chang, C.; Lu, P.; Chu, C.; Lai, W.; Shin, S.; Liu, F.; Chen, C. Increased APOE glycosylation plays a key role in the atherogenicity of L5 low-density lipoprotein. *FASEB J.* **2020**, *34*, 9802–9813. [CrossRef]
147. An, X.-F.F.; Zhou, L.; Jiang, P.-J.J.; Yan, M.; Huang, Y.-J.J.; Zhang, S.-N.N.; Niu, Y.-F.F.; Ten, S.-C.C.; Yu, J.-Y.Y. Advanced glycation end-products induce heparanase expression in endothelial cells by the receptor for advanced glycation end products and through activation of the FOXO4 transcription factor. *Mol. Cell. Biochem.* **2011**, *354*, 47–55. [CrossRef]
148. Goldberg, R.; Meirovitz, A.; Abecassis, A.; Hermano, E.; Rubinstein, A.M.; Nahmias, D.; Grinshpun, A.; Peretz, T.; Elkin, M. Regulation of heparanase in diabetes-associated pancreatic carcinoma. *Front. Oncol.* **2019**, *9*, 1405. [CrossRef]
149. Maxhimer, J.B.; Somenek, M.; Rao, G.; Pesce, C.E.; Baldwin, D.; Gattuso, P.; Schwartz, M.M.; Lewis, E.J.; Prinz, R.A.; Xu, X. Heparanase-1 gene expression and regulation by high glucose in renal epithelial cells. *Diabetes* **2005**, *54*, 2172–2178. [CrossRef]
150. Brestoff, J.R.; Wilen, C.B.; Moley, J.R.; Li, Y.; Zou, W.; Malvin, N.P.; Rowen, M.N.; Saunders, B.T.; Ma, H.; Mack, M.R.; et al. Intercellular mitochondria transfer to macrophages regulates white adipose tissue homeostasis and is impaired in obesity. *Cell Metab.* **2021**, *33*, 270–282e8. [CrossRef]
151. Oury, F.; Khrimian, L.; Denny, C.A.; Gardin, A.; Chamouni, A.; Goeden, N.; Huang, Y.; Lee, H.; Srinivas, P.; Gao, X.B.; et al. Maternal and offspring pools of osteocalcin influence brain development and functions. *PLoS ONE* **2013**, *8*, e57375. [CrossRef]
152. Khrimian, L.; Obri, A.; Ramos-Brossier, M.; Rousseaud, A.; Moriceau, S.; Nicot, A.S.; Mera, P.; Kosmidis, S.; Karnavas, T.; Saudou, F.; et al. Gpr158 mediates osteocalcin's regulation of cognition. *J. Exp. Med.* **2017**, *214*, 2859–2873. [CrossRef]
153. Norwitz, N.G.; Hu, M.T.; Clarke, K. The mechanisms by which the ketone body D-β-hydroxybutyrate may improve the multiple cellular pathologies of Parkinson's disease. *Front. Nutr.* **2019**, *6*, 63. [CrossRef]
154. Shan, C.; Ghosh, A.; Guo, X.Z.; Wang, S.M.; Hou, Y.F.; Li, S.T.; Liu, J.M. Roles for osteocalcin in brain signalling: Implications in cognition- and motor-related disorders. *Mol. Brain* **2019**, *12*, 1–11. [CrossRef]
155. Yang, H.; Shan, W.; Zhu, F.; Wu, J.; Wang, Q. Ketone bodies in neurological diseases: Focus on neuroprotection and underlying mechanisms. *Front. Neurol.* **2019**, *10*, 585. [CrossRef]
156. Bradburn, S.; Mcphee, J.S.; Bagley, L.; Sipila, S.; Stenroth, L.; Narici, M.V.; Pääsuke, M.; Gapeyeva, H.; Osborne, G.; Sassano, L.; et al. Association between osteocalcin and cognitive performance in healthy older adults. *Age Ageing* **2016**, *45*, 844–849. [CrossRef]
157. Schatz, M.; Saravanan, S.; D'Adesky, N.D.; Bramlett, H.; Perez-Pinzon, M.A.; Raval, A.P. Osteocalcin, Ovarian senescence, and brain health. *Front. Neuroendocrinol.* **2020**, *59*, 100861. [CrossRef]
158. Zhao, Y.; Shen, L.; Ji, H.F. Alzheimer's disease and risk of hip fracture: A meta-analysis study. *Sci. World J.* **2012**, *2012*. [CrossRef]
159. Lyell, V.; Henderson, E.; Devine, M.; Gregson, C. Assessment and management of fracture risk in patients with parkinson's disease. *Age Ageing* **2015**, *44*, 34–41. [CrossRef] [PubMed]
160. Oh, E.S.; Blennow, K.; Bigelow, G.E.; Inouye, S.K.; Marcantonio, E.R.; Neufeld, K.J.; Rosenberg, P.B.; Troncoso, J.C.; Wang, N.-Y.; Zetterberg, H.; et al. Abnormal CSF amyloid-β42 and Tau levels in hip fracture patients without dementia. *PLoS ONE* **2018**, *13*, e0204695. [CrossRef]

161. Kim, S.Y.; Lee, J.K.; Lim, J.-S.; Park, B.; Choi, H.G. Increased risk of dementia after distal radius, hip, and spine fracture. *Medicine* **2020**, *99*, e19048. [CrossRef] [PubMed]
162. Mizokami, A.; Yasutake, Y.; Gao, J.; Matsuda, M.; Takahashi, I.; Takeuchi, H.; Hirata, M. Osteocalcin induces release of glucagon-like peptide-1 and thereby stimulates insulin secretion in mice. *PLoS ONE* **2013**, *8*, 57375. [CrossRef] [PubMed]
163. Drucker, D.J. Mechanisms of action and therapeutic application of glucagon-like peptide-1. *Cell Metab.* **2018**, *27*, 740–756. [CrossRef]
164. Plamboeck, A.; Veedfald, S.; Deacon, C.F.; Hartmann, B.; Wettergren, A.; Svendsen, L.B.; Meisner, S.; Hovendal, C.; Vilsbøll, T.; Knop, F.K.; et al. The effect of exogenous GLP-1 on food intake is lost in male truncally vagotomized subjects with pyloroplasty. *Am. J. Physiol.-Gastrointest. Liver Physiol.* **2013**, *304*, 1117–1127. [CrossRef]
165. Cataldo Bascuñan, L.R.; Lyons, C.; Bennet, H.; Artner, I.; Fex, M. Serotonergic regulation of insulin secretion. *Acta Physiol.* **2019**, *225*, e13101. [CrossRef]
166. Nakamura, M.; Imaoka, M.; Takeda, M. Interaction of bone and brain: Osteocalcin and cognition. *Int. J. Neurosci.* **2020**, 1–9. [CrossRef]
167. Paulmann, N.; Grohmann, M.; Voigt, J.P.; Bert, B.; Vowinckel, J.; Bader, M.; Skelin, M.; Jevšek, M.; Fink, H.; Rupnik, M.; et al. Intracellular serotonin modulates insulin secretion from pancreatic β-cells by protein serotonylation. *PLoS Biol.* **2009**, *7*, 1000229. [CrossRef]
168. Almaça, J.; Molina, J.; Menegaz, D.; Pronin, A.N.; Tamayo, A.; Slepak, V.; Berggren, P.O.; Caicedo, A. Human beta cells produce and release serotonin to inhibit glucagon secretion from alpha cells. *Cell Rep.* **2016**, *17*, 3281–3291. [CrossRef]
169. Kelly, T.; Unwin, D.; Finucane, F. Low-carbohydrate diets in the management of obesity and type 2 diabetes: A review from clinicians using the approach in practice. *Int. J. Environ. Res. Public Health* **2020**, *17*, 2557. [CrossRef]
170. Athinarayanan, S.J.; Adams, R.N.; Hallberg, S.J.; McKenzie, A.L.; Bhanpuri, N.H.; Campbell, W.W.; Volek, J.S.; Phinney, S.D.; McCarter, J.P. Long-term effects of a novel continuous remote care intervention including nutritional ketosis for the management of type 2 diabetes: A 2-year nonrandomized clinical trial. *Front. Endocrinol.* **2019**, *10*, 348. [CrossRef]

biomedicines

Review

Metabolic Phenotypes and Step by Step Evolution of Type 2 Diabetes: A New Paradigm

Isabella D. Cooper [1,*], Kenneth H. Brookler [2], Yvoni Kyriakidou [1], Bradley T. Elliott [1] and Catherine A. P. Crofts [3]

1 Translational Physiology Research Group, School of Life Sciences, University of Westminster, 115 New Cavendish Street, London W1W 6UW, UK; y.kyriakidou@westminster.ac.uk (Y.K.); b.elliott@westminster.ac.uk (B.T.E.)
2 Research Collaborator, Aerospace Medicine and Vestibular Research Laboratory, Mayo Clinic, Scottsdale, AZ 85259, USA; brookler.kenneth@mayo.edu
3 Faculty of Health and Environmental Sciences, School of Public Health and Interdisciplinary Studies, Auckland University of Technology, Auckland 0627, New Zealand; catherine.crofts@aut.ac.nz
* Correspondence: bellamitochondria@gmail.com

Abstract: Unlike bolus insulin secretion mechanisms, basal insulin secretion is poorly understood. It is essential to elucidate these mechanisms in non-hyperinsulinaemia healthy persons. This establishes a baseline for investigation into pathologies where these processes are dysregulated, such as in type 2 diabetes (T2DM), cardiovascular disease (CVD), certain cancers and dementias. Chronic hyperinsulinaemia enforces glucose fueling, depleting the NAD+ dependent antioxidant activity that increases mitochondrial reactive oxygen species (mtROS). Consequently, beta-cell mitochondria increase uncoupling protein expression, which decreases the mitochondrial ATP surge generation capacity, impairing bolus mediated insulin exocytosis. Excessive ROS increases the Drp1:Mfn2 ratio, increasing mitochondrial fission, which increases mtROS; endoplasmic reticulum-stress and impaired calcium homeostasis ensues. Healthy individuals in habitual ketosis have significantly lower glucagon and insulin levels than T2DM individuals. As beta-hydroxybutyrate rises, hepatic gluconeogenesis and glycogenolysis supply extra-hepatic glucose needs, and osteocalcin synthesis/release increases. We propose insulin's primary role is regulating beta-hydroxybutyrate synthesis, while the role of bone regulates glucose uptake sensitivity via osteocalcin. Osteocalcin regulates the alpha-cell glucagon secretory profile via glucagon-like peptide-1 and serotonin, and beta-hydroxybutyrate synthesis via regulating basal insulin levels. Establishing metabolic phenotypes aids in resolving basal insulin secretion regulation, enabling elucidation of the pathological changes that occur and progress into chronic diseases associated with ageing.

Keywords: hyperinsulinaemia; insulin resistance; osteocalcin; beta-hydroxybutyrate; phenotype; stages; serotonin; glucagon-like peptide-1; glucagon; type 2 diabetes; hyperglycaemia

Citation: Cooper, I.D.; Brookler, K.H.; Kyriakidou, Y.; Elliott, B.T.; Crofts, C.A.P. Metabolic Phenotypes and Step by Step Evolution of Type 2 Diabetes: A New Paradigm. *Biomedicines* **2021**, *9*, 800. https://doi.org/10.3390/biomedicines9070800

Academic Editor: Susan J. Burke

Received: 21 May 2021
Accepted: 5 July 2021
Published: 9 July 2021

1. Introduction

The hormone insulin is synthesized and secreted by pancreatic beta cells [1,2]. The commonly accepted principle is that insulin is secreted in a basal/bolus pattern, with the latter predominately released upon rising blood glucose (most likely from a meal) stimulus [3,4]. The mechanism by which bolus insulin is secreted from the pancreas is well established [5]; however, little is known about the mechanisms by which basal insulin is secreted.

Once we understand the processes by which insulin is secreted in both the basal and bolus states in a healthy person, we can begin to unravel the pathologies whereby these processes are dysregulated, such as in type 2 diabetes mellitus (T2DM), cardiovascular disease (CVD), certain cancers and dementias [6].

2. Bolus Insulin Secretion

The commonly accepted premise is that the primary function of insulin secretion is to regulate glucose uptake into muscle cells [7,8]. To better understand basal versus bolus insulin secretion regulation and physiological roles, a firm understanding of how glucose drives rapid insulin exocytosis is warranted.

Glycaemic elevations, such as following an oral carbohydrate bolus, results in rapid entry of glucose into the pancreatic beta cells via GLUT1 and/or GLUT3 glucose transporters (Km = 6 mmol/L and Km = 1–1.4 mmol/L, respectively) (Figure 1) [9–11]. GLUT1 is the predominant glucose transporter in humans, its Km of 6 mmol/L indicates that this transporter is only activated when significantly high levels (above the physiological concentration of 5 mmol/L) of glucose are detected in the blood stream [3]. However, the high glucose affinity of GLUT3, suggests its role for metabolic fuel homeostasis during fasting periods where there would be glucose/carbohydrate deprivation/restriction. Theoretically, in this setting the GLUT3 receptor expression is upregulated.

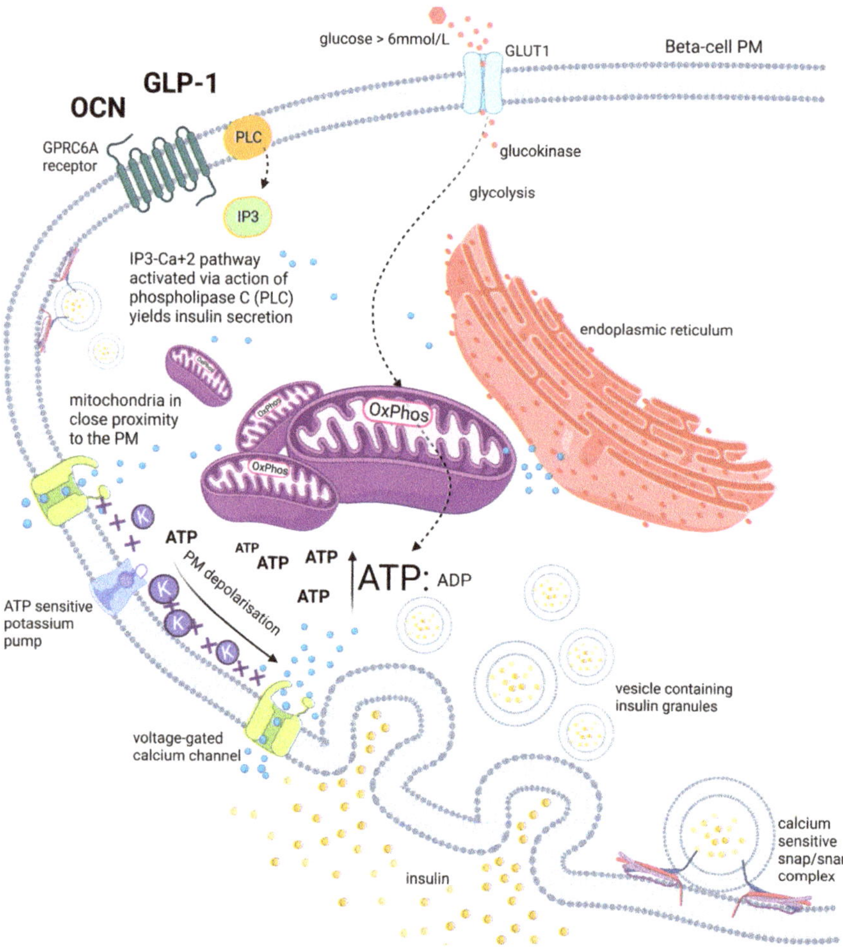

Figure 1. Schematic representation of beta-cell intracellular mechanisms involved in insulin secretion. Adenosine diphosphate (ADP), adenosine triphosphate (ATP), calcium (Ca^{2+}), glucagon-like peptide-1 (GLP-1), glucose transporter 1 (GLUT1), G-protein coupled receptor 6A (GPRC6A), inositol-1,4,5-trisphosphate (IP3), plasma membrane (PM), osteocalcin (OCN), oxidative phosphorylation (OxPhos), phospholipase C (PLC), potassium (K^{+}).

Upon entry into beta cells, glucose is phosphorylated to glucose-6-phosphate by glucokinase (GK), an isozyme of hexokinase [12]. GK has a high Km value = 10 mmol/L, indicating a low affinity for glucose [13,14]. GK is predominantly only expressed in hepatocytes and pancreatic beta cells [13,15,16], and also found to be expressed by cells in the hypothalamus [17]. Glucose is catabolised through the glycolytic pathway to pyruvate, in the process generating reducing equivalents NADH. These enter the mitochondria to undergo a series of redox reactions to yield adenosine triphosphate (ATP) via the electron transport chain (ETC) oxidative phosphorylation (OxPhos) machinery coupled to ATPase. The ETC-OxPhos complexes are situated on the cristae of the inner mitochondrial membrane (IMM) [3]. Beta cells express low levels of lactate dehydrogenase, indicating a "preference" to fully oxidise glucose via OxPhos generating maximal ATP (~36 ATP/glucose molecule), as opposed to via the fermentation pathway (~2 ATP/glucose molecule) [18,19].

As glucose is fully oxidised through OxPhos, a surge of ATP is generated via the mitochondrial high electrochemical-potential gradient inner membrane ($\Delta\Psi m$) [20]. As a result, there is a steep increase in the ATP concentration relative to adenosine diphosphate (ADP) [21]. The mitochondria are situated close to the beta cell plasma membrane (PM) [22]. The surge in ATP affects ATP-sensitive potassium channels (K_{ATP}), causing them to close [23,24]. This depolarises the PM, causing PM voltage gated calcium channels (VGCC) to open [23,25]. The subsequent rapid influx of calcium divalent cations, through P/Q-type Ca^{2+} channels [26], activates the calcium sensitive SNAP/SNARE complexes that hold vesicles containing insulin granules in a "ready" docking position on the cytosolic side of the PM [27]. This induces the rapid exocytosis of insulin, which is referred to as the first phase response [28]. Further insulin is stored in "reserve pool" vesicles and released only in response to fuel secretagogues [27], known as the "second-phase". This second phase of insulin release has an observed delay and insulin nadir after the first phase, and has a lower amplitude and longer direction, but only lasts while the beta cells are stimulated [20].

A whole host of aspects of the glucose-mediated insulin first phase response needs to be closely coordinated in-order to promote exocytosis of a high concentration of insulin granules, outside of the beta-cells basal pulsatile release. These include the expression of GK, nicotinamide adenine dinucleotide (NAD+) availability, and maintaining a high capacity to perform OxPhos that requires a high $\Delta\Psi m$, in-order to generate the necessary surge in ATP concentration to affect the (K_{ATP}) channels. Finally, calcium homeostasis must also be well regulated, as a PM concentration gradient and consequent signal:response ratio are required to elicit the exocytosis response. Therefore, calcium relocation mechanisms must also be effective [29].

3. Basal Insulin

In contrast to the above-described regulation of bolus insulin secretion, the regulation of insulin secretion in the basal phase is not well understood. It could be hypothesised that the same mechanism is used for insulin secretion in the basal state. It is recognised that GLUT3 has a lower Km for glucose (Km = 1–1.4 mmol/L) and in the fasted state GLUT3 may become upregulated, thereby increasing the role of glucose in basal insulin secretion regulation. However, if this were the case, increased insulin release would down regulate beta-hydroxybutyrate (BHB) synthesis. This is not corroborated by the observation of individuals in the fasted state, where a lower basal state of insulin and glucagon, with the presence of BHB, has been observed in people in habitual ketosis [30–32]. Another argument for it not being GLUT3 dominant stimulated in the basal state, is that glucose in the fasted state is at a relatively steady state with low degrees of magnitude in the blood, which, therefore, would not correspond to basal insulin oscillatory patterns.

Basal insulin is recognised to be released from pancreatic beta cells in a pulsatile rhythmic pattern, approximately every 4 min, in addition to circadian and ultradian periodicities (Figure 2A) [20]. Five-to-fifteen-minute fast oscillations modulate the ultradian periodicity, which has a range of 40 to 180 min [7,20,33]. It is believed that the modulatory fast oscillations are influenced by the degree of insulin resistance (IR) within an individual [34].

The bone derived hormone osteocalcin (OCN) activates beta cell calcium signalling via the GPRC6A receptor (Figure 1) and we hypothesise it is the OCN that regulates the oscillatory insulin secretion pattern (Figure 2B) [35].

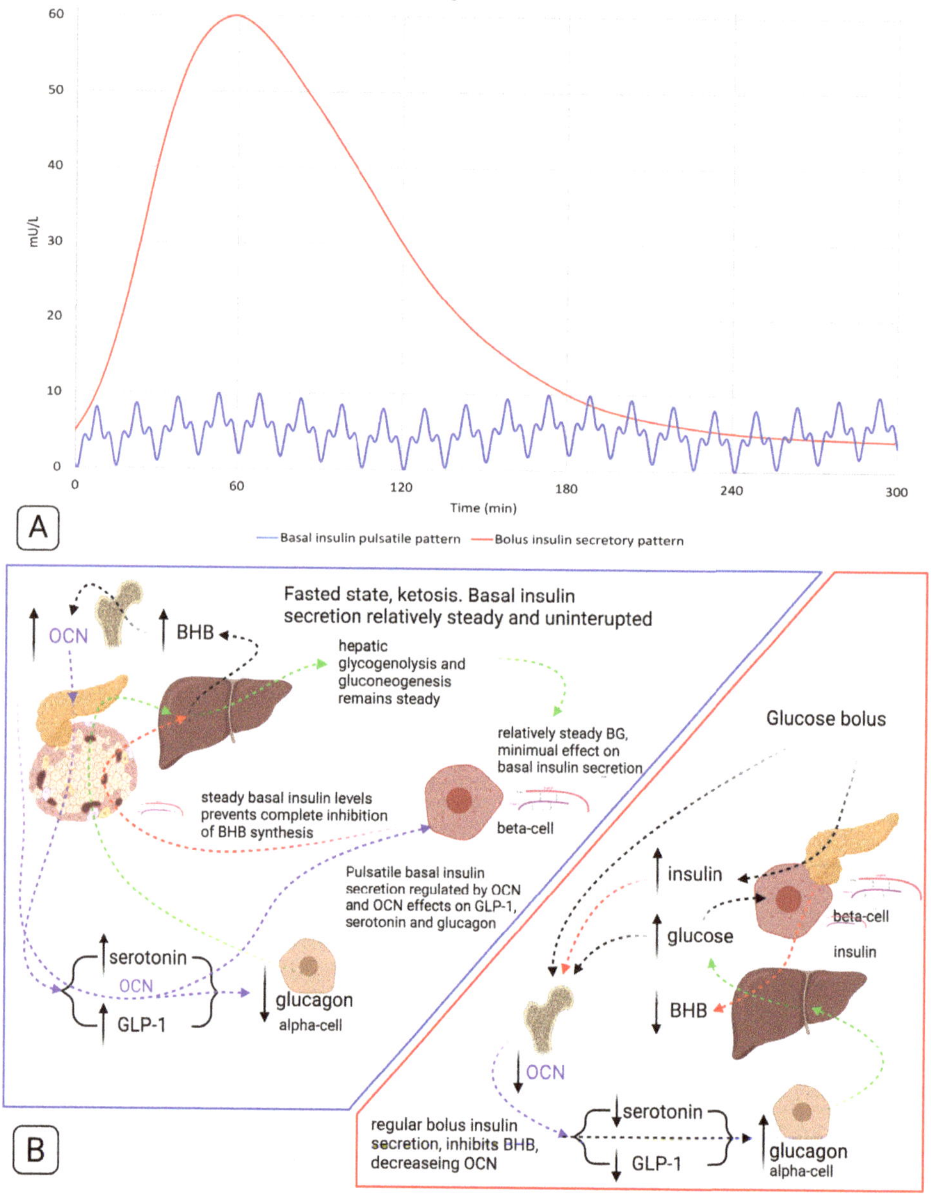

Figure 2. Schematic representations of basal and bolus insulin secretory patterns (**A**) and secretion regulation (**B**). (**A**) The red line conceptually models glucose bolus mediated insulin secretory response pattern (Kraft I) [36], and the blue line conceptually models basal insulin pulsatile secretory pattern, in metabolically healthy individuals [34]. (**B**) Schematic representation of the regulatory cycles of basal and bolus insulin secretion in metabolically healthy, habitually fasted individuals. Beta-hydroxybutyrate (BHB), blood glucose (BG), glucagon-like peptide-1 (GLP-1), osteocalcin (OCN).

The pulsatility pattern likely reduces the risk of negative consequences from potentially downregulating insulin receptors (INSR), which would result in further insulin resistance. Therefore, a pulsatile secretory pattern plausibly is more effective in regulating blood glucose levels [7]. In addition to contributing to sympathovagal balance, there is a tight coupling between the pancreatic ultradian periodicity and the neuroendocrine, cardiovascular and autonomic nervous systems [37].

Additional challenges in the understanding of pancreatic beta cell insulin secretion, include changes to the IMM. Mitochondria are the largest intracellular source of reactive oxygen species (ROS) [38–40], specifically from the activities of the ETC, as electrons "leak" and react with oxygen, forming superoxide [41,42]. Cellular mechanisms to counter ROS are facilitated by anti-oxidant enzymes such as mitochondrial superoxide dismutase (SOD2) and reduced glutathione (GSH) [43]. Both SOD2 and GSH require NAD+ [44]. Glucose oxidation has a greater NAD+ depletion effect over beta-oxidation or ketolysis; as a result, this increases ROS levels via a reduced ability to counter ROS [43,45–47]. Furthermore, excessive insulin signalling increases ROS levels via ceramide synthesis [48], which, in turn, leads to cellular apoptosis. Without sufficient anti-oxidative enzymes to manage the excessive ROS levels, the mitochondrial strategy is to increase the expression of uncoupling proteins (UCP2) in the IMM [20,49]. This causes the uncoupling of proton flow from the higher concentration within the inner membrane space of the mitochondrial double membrane lipid bilayer, into the mitochondrial matrix, without generating ATP. This results in a less hyperpolarised IMM, and thus resultant inability to generate the surge in ATP, leading to a lower ability to trigger rapid insulin release via the calcium dependent route [50].

Careful regulation of calcium is key for mitochondrial activity. Mitochondria are organelles that store calcium, however, not at a significant concentration to effect cytosolic concentrations [51]. The relationship of mitochondria with calcium is two-fold: ATP production and calcium-trafficking or redistribution. Calcium uptake by mitochondria enhances their ATP production potential; however, a fine balance must be struck, as calcium is required, but too much calcium induces apoptosis. In-order to manage this, calcium efflux must be effective to avoid overload [52]. Mitochondrial calcium uptake is mediated by the calcium uniporter channel complexes (MCUC), while mitochondrial calcium extrusion is facilitated by the sodium/calcium exchanger (NCLX). Both the MCUC and NCLX are electrogenically driven by the high $\Delta\Psi m$ [52,53]. Hyperinsulinaemia (HI) increases mtROS production via ceramide synthesis, and NAD+ depletion from concomitant elevated glucose-metabolism, leading to decreased counter-ROS management. Excess ROS generation increases UCP2 in beta cell mitochondria, resulting in a decrease in the $\Delta\Psi m$ [54]. This impairs the mitochondrial MCUC and NCLX dependent uptake and the redistribution of calcium that is required for maximal ATP synthesis [51,55]. The second role of mitochondrial calcium homeostasis is focused on redistribution efforts [53,56].

Mitochondria facilitate the trafficking of cytosolic calcium uptake into the endoplasmic and sarcoplasmic reticulae via Ca^{2+} ATPase (SERCA) pumps [57,58]. The beta cell endoplasmic reticulum (ER) regulates cytosolic calcium partitioning, while ATP dependent SERCA pumps dominate in mediating calcium exocytosis, in-order to "re-set" PM calcium levels. These mechanisms enable the cycle of the calcium-mediated exocytosis of insulin granules packaged in vesicles, docked along the inner PM, to repeat [59]. Roughly 20% of the beta cell sub-plasma membrane is in close proximity with mitochondria, exerting a strong calcium buffering effect [22]. Mitochondrial calcium uptake facilitates the signal:response ratio sensitivity in PM depolarisation and increases the cytosolic calcium concentration [60]. A reduction in the mitochondrial calcium uptake via reduced MCUC activity, leads to an increase in calcium within the PM sub-membrane compartment upon depolarisation [29,55]. This attenuates any increase in cytoplasmic calcium and results in a net reduction in rapid insulin exocytosis. An increase in ROS increases UCP2 expression, which lowers the $\Delta\Psi m$, consequently reducing calcium uptake and efflux [20,49,50,52]. As a result, this not only decreases rapid ATP synthesis potential, but also impairs calcium trafficking to the ER [61].

As a consequence, dysregulated cytosolic calcium levels may reduce/impair the docking "ready–set–go" positions of the exocytosis mediating SNAP/SNARE proteins [62–64]. This impairment results in the disabling of a rapid response from the exocytosis machinery to the required extracellular calcium influx along the PM, of the much-needed steep calcium concentration gradient that is sensitive to the signal:response ratio to elicit rapid insulin release. In short, when the concentration gradient is not steep, the signal is not strong, resulting in a poor response.

Excessive ROS levels, along with chronic insulin signalling, increases the ratio of dynamin-related protein 1 (Drp1) to mitofusin-2 proteins (Mfn2) [65,66]. Drp1 mediates mitochondria fission, while Mfn2 mediates mitochondrial fusion and is required for ER association [51]. When there is an increase in Drp1 relative to Mfn2, there is a net increase in mitochondrial fission [51,66]. This results in a decrease in ER association and OxPhos capacity, and an increase in mtROS production [66,67]. Consequently there is an increase in ER stress, a reduction in ER mediated calcium homeostasis, and a reduction in mitochondrial (mt) OxPhos that is required in order to generate the ATP surge needed for first phase insulin exocytosis [58,68]. It is, therefore, clear that the health of beta cell mitochondria are essential for a functional first phase response to a glucose bolus [54,69,70].

4. Insulin Secretion in the Insulin Resistant/Hyperinsulinaemic State

Having established the processes in the healthy state, the hyperinsulinaemic individual can be considered. However, a number of factors first need to be addressed in the research literature on T2DM and pancreatic beta cells. A large majority of the literature states that in T2DM, there is a significant loss of beta cell mass, and this, consequently, results in insulin insufficiency [71,72]. However, this stage of T2DM is the far end of the condition, where the pathology is entering into the final stages of T2DM-induced pseudo-T1DM, as a result of beta cell "exhaustion", failure and/or increased apoptosis [72]. However, in many people, this is a relative deficiency [73], not an absolute, as examination of the Kraft dataset shows they still produce more insulin than the normoglycaemic/normoinsulinaemia population [36].

A large phase of the pathogenesis of T2DM is the silent normo-glycaemia hyperinsulinaemia phase, often termed (pre-)pre-diabetes. In reality, this phase would be best termed stage-1 T2DM and mildly elevated glycaemia, currently termed pre-diabetes, stage-2 T2DM (Figure 3). Waiting to see hyperglycaemia (HG) (stage-2/3), in-order to diagnose T2DM, is already deep into pathology progression, where chronic excess insulin levels are no longer able to mask the problem. Obese individuals are more likely to have stage-1/2 T2 diabetes and have a higher risk of developing HG-T2DM (phenotype-3 stage 3) than non-obese individuals. An abnormally high percentage of islet tissue, and increased beta cell mass has been found in the pancreas of obese individuals in comparison to lean subjects [74,75]. Given the scale and changes in phenotype along the trajectory pathogenesis of T2DM, it is vital that distinctions are made between the stages. Furthermore, that investigations in biological samples and participants belonging to different pathology stage categories, are not pooled. This is to avoid cancelling out effects/observations between two or more stages. For example, stage-1/2 normo-glycaemia HI T2DM with increased beta cell mass, pooled with stage-3/4 HG-HI T2DM with decreased beta cell mass [73]. In this example, the net pooling results in a cancelling out of any signal. This can lead to the forming of incorrect premises that would contribute to a misinterpretation of results and, in addition, potentially cause the development of in vitro and animal models that do not truly represent the full scope of the disease. Again, this, consequently, increases the risk of producing results that are correct for the experiment only, which is based on a flawed premise, in turn, sending the researcher on a merry-go-round.

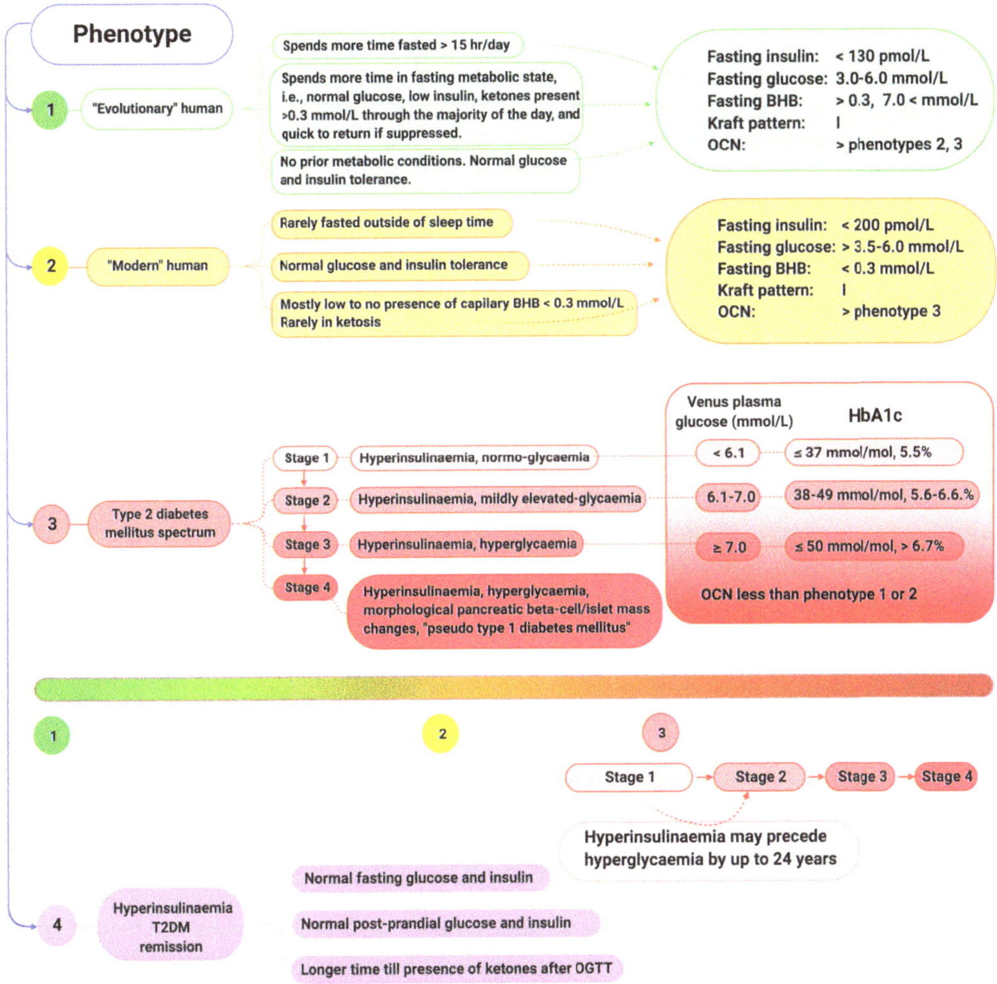

Figure 3. Classification of metabolic phenotypes. Beta-Hydroxybutyrate (BHB), haemoglobin A1c (HbA1c), oral glucose tolerance test (OGTT), osteocalcin (OCN), type 2 diabetes mellitus (T2DM).

5. An Alternative Hypothesis: Insulin's Main Role Is to Regulate Beta-Hydroxybutyrate Synthesis

Let us assume the natural human state is to be in a hunter–gatherer pattern, the equivalent is hypothesised to be found in the early European exploration of the traditional Inuit and the Hadza examples. These societies are characterised by predominantly fasted—metabolic phenotype 1 (Figure 3), often only consuming one meal a day, potentially not having food every day, and most meals are low in digestible carbohydrates [76–78]. In this context, blood glucose levels infrequently rise above 6 mmol/L, only with occasional access to honey or fruit, or a meal containing a high glycogen content, such as liver. Alternatively, blood glucose may surge in response to an acute stress response. Aside from these contexts, blood glucose levels remain relatively constant and may even dip to levels that conventional medical wisdom considers puts the individual at risk of a hypoglycaemic coma [79]. However, it has been demonstrated that humans in nutritional ketosis are able to function comparably, even optimally, when plasma glucose levels are below the standard reference ranges, due to the elevated presence of the ketone body beta-hydroxybutyrate (BHB) [80]. The fasted state, or carbohydrate restriction, induces the metabolic phenotype of ketosis, where plasma insulin is low, glucose is normal to low, and BHB is detectable above 0.5 mmol/L [43]. Just as hyperglycaemia may become pathological, hyperketonaemia may also become pathological, especially when BHB levels exceed 25 mmol/L. Within humans, diabetic ketoacidosis (DKA) pathology is when there are elevated ketones, with hyperglycaemia and decreased bicarbonate levels, resulting in a decrease in blood pH [81–83]. Other common symptoms experienced with DKA include nausea, vomiting, and gastrointestinal symptoms including abdominal pain [84]. Insulin regulates hepatocyte BHB synthesis, therefore hyperinsulinaemic individuals are at a very low risk of developing DKA, unless they are on sodium-glucose co-transporter-2 (SGLT2) inhibitors and simultaneously embark on carbohydrate restriction without adjusting medications [43,85].

If the natural state of humans is to spend more time in the metabolically fasted phenotype of ketosis (phenotype 1), it is plausible that the role of basal pulsatile insulin secretion is to regulate BHB synthesis. Cells that are wholly or substantially glucose dependent do not require insulin to take up glucose [86,87]. In the fasted state, blood glucose is provided from the liver, either from the glycogen stored from a rare glucose loaded meal or, more likely, from gluconeogenesis that both replenishes hepatic glycogen stores as well as providing extra-hepatic tissue needs [88,89]. In this context, hepatocytes are metabolising fatty acids for their own energy provision and in the process synthesise BHB [90]. Hepatocytes are unable to use BHB for fuel [91], instead the BHB is released into the bloodstream, to provide energy and act as a signalling molecule to extra-hepatic tissues, such as the brain, heart, and muscular-skeletal system [46,92]. In the absence of insulin production, BHB synthesis would continue unabated, as seen in type 1 diabetes mellitus (T1DM). However, a small amount of insulin is able to inhibit ketogenesis [93]. The signal for insulin release is required though, this then begs the question, what provides the signal?

It is first important to understand that when humans are in ketosis as their de-facto state, the body's supply of glucose is dependent on hepatic synthesis and provision [30,89]. It is essential that the liver does not respond to insulin's effects on glucose output, even for one meal. This is because, hypothetically, one high carbohydrate meal that would induce a high insulin output, could shut-down hepatic glycogenolysis and gluconeogenesis. This insulin signalling effect may overshoot in duration, resulting in depriving glucose dependent cells of hepatic glucose, fatty acids, and BHB, which are also regulated by insulin [89]. The liver becomes physiologically "IR", which is really a state of adaptive homeostasis. In reality, the liver is not uniformly IR, as it can be seen, a glucose bolus in keto-adapted individuals rapidly curtails BHB synthesis, whilst not inhibiting hepatic glucose output. The hepatic "IR" is selective [94,95].

Individuals in habitual ketosis have significantly lower glucagon and insulin levels than hyperinsulinaemia T2DM patients [31,43,80,96]. Hepatic glucose output is both regulated by glucagon and is likely due to the continual "draw-down" of plasma glucose by extra-hepatic tissues [31,97]. As BHB rises and glucose levels are restricted, this stimulates an increase in osteocalcin (OCN) synthesis and release from the bone by osteoblasts, osteocytes, and osteoclasts [98–100]. OCN significantly increases glucose uptake independent of insulin and, in the fasted state, likely functions as the glucose uptake regulator, rather than insulin [101,102]. However, OCN potentiates insulin's glucose uptake effect, therefore requiring less insulin, which effectively improves insulin sensitivity with regard to glucose uptake [103]. OCN also induces glucagon-like peptide 1 (GLP-1) synthesis. Both OCN and GLP1 signal beta cells to increase insulin synthesis and release [101,104,105]. The resulting effect is that elevated BHB and glucose restriction drives OCN synthesis and release, and OCN increases GLP-1 [105]. Together, OCN and GLP-1 increase insulin secretion [106,107], resulting in downregulating BHB synthesis, which, in turn, down regulates OCN release from the bone, which removes the signal for insulin secretion, and, therefore, insulin levels decrease. This feedback loop effectively regulates BHB synthesis and glucose homeostasis

(Figure 4). It is also intriguing that the half-life of OCN and GLP-1 are both 5 min, and the pulsatile pattern of insulin secretion is 4 to 15 min [20,108,109]. The combined effects of OCN and GLP-1 may enhance the signal for insulin secretion to potentially match or synergise in generating the following feedback cycle: BHB increase → OCN increase (+GLP-1) → insulin release → BHB decrease → OCN decrease → insulin decrease → BHB increase. Furthermore, BHB, together with lactate or low levels of glucose, potentiates the strength of the signal for insulin release, indicating that BHB directly, although not independently, stimulates an insulin response [110].

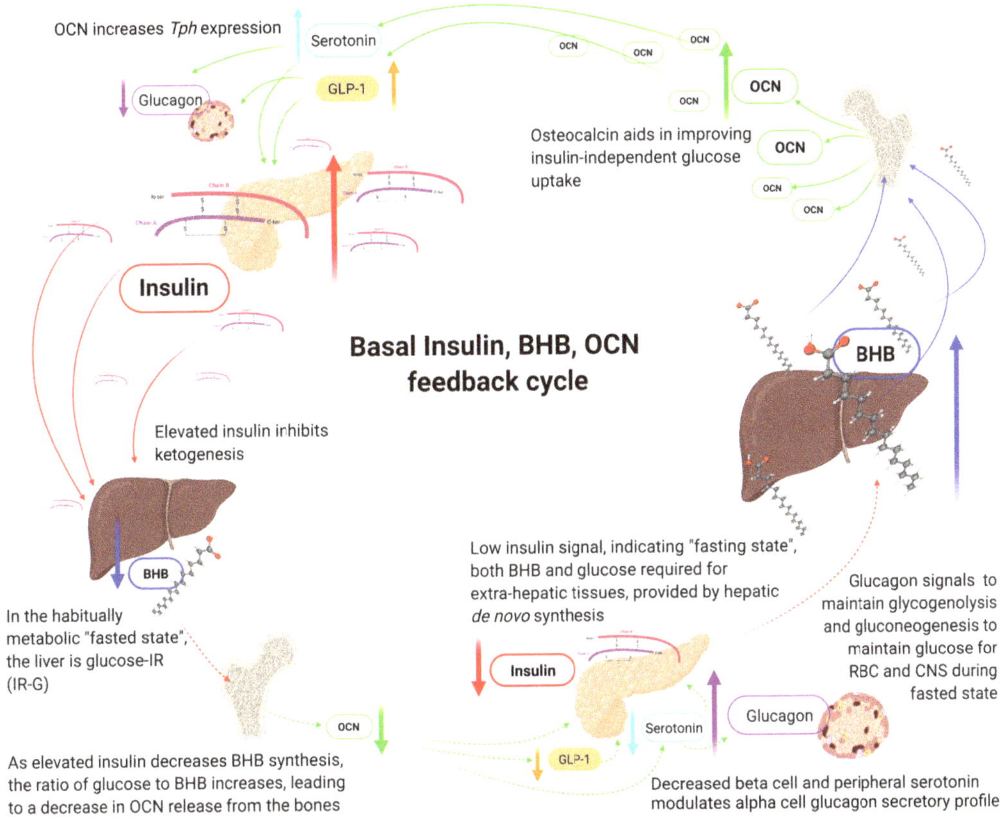

Figure 4. Proposed basal insulin, beta-hydroxybutyrate, osteocalcin feedback cycle in phenotype-1 individuals. Beta-hydroxybutyrate (BHB), central nervous system (CNS), glucagon-like peptide-1 (GLP-1), insulin resistance (IR), glucose-insulin resistance (IR-G), osteocalcin (OCN), red blood cells (RBC), tryptophan hydroxylase (*Tph*).

It would be remiss to not indicate the significant role of the pancreatic alpha cells in glucose and insulin homeostasis. A higher insulin level should predict a lower fasting plasma glucose. However, as we see, fasting glucose is elevated in T2DM, as well as higher glucagon and insulin [36,96]. The question is why would the alpha cells be secreting more glucagon in a higher glucose background? An explanation maybe found in OCNs role in regulating tryptophan hydroxylase (*Tph*) gene expression, the rate limiting enzyme for serotonin (5-hydroxytryptamine, 5-HT) synthesis, an alpha cell glucagon secretory profile modulator [111,112].

Through a series of elegant experiments, Almaça et al.,(2017) demonstrated in vivo and ex vivo that alpha cell secretion of glucagon is modulated by serotonin signalling via the 5-HT1F receptor, causing rapid inhibition of adenylate cyclase, resulting in a decrease in intracellular second messenger cyclic adenosine monophosphate (cAMP). Furthermore, beta cells from people with normal glucose tolerance, phenotypes 1 and 2, produce and secrete serotonin, and alpha cells respond to this serotonin, leading to the modulation of glucagon secretion under different plasma glucose conditions. Almaça et al. showed that when islet serotonin levels are manipulated (serotonin is depleted or inhibited), alpha cells lose their ability to respond in concordance to surrounding glucose levels. This affects glucose homeostasis, and likely plays a negative role in propagating pernicious increases in fasting basal insulin levels, a pattern seen in the progression toward hyperinsulinaemia phenotype-3 stages 1 and 2, and overt T2DM phenotype-3 stage 3.

Serotonin is a strong paracrine regulator of the secretory profile of alpha cells [112]. Higher levels of serotonin decrease alpha cell glucagon secretion when plasma glucose levels are high and also reduces the amount of glucagon secretion at very low glucose levels [112]. This may explain why phenotype-1 (longstanding metabolically flexible habitual ketosis) and phenotype-2 individuals would have lower glucagon and insulin levels, while maintaining lower glucose levels than found in those with hyperinsulinaemia phenotype-3 [30,96].

GLP-1 receptor activation, in vivo, increases serotonin synthesis [113]. Furthermore, serotonin has been found to be synthesised and stored with insulin inside the beta cell insulin secretory beta-granules, and co-released upon glucose stimulation [114]. Beta cell serotonin secretion shares similarity to the secretory pattern of insulin, having a glucose-dependent and pulsatile pattern [112]. GLP-1 is also able to directly suppress glucagon secretion via inhibiting alpha cell P/Q-type-voltage-gated Ca^{2+} channels [26,115].

OCN increases serotonin and GLP-1 synthesis, both of which modulate the alpha cell glucagon secretory profile. OCN levels are significantly lower in insulin resistant and T2DM individuals [99,116–118]. Glucose restriction and lower insulin levels enhance osteoblastogenesis and osteocytogenesis, and OCN synthesis and release. This provides a plausible metabolic and endocrine regulatory feedback cycle (Figure 4), in maintaining glucose homeostasis, whilst also maintaining lower glucagon and insulin levels in phenotype-1 and -2 individuals relative to phenotype-3 [100,119].

Consider an alternative basic premise; humans evolutionarily spent more time in a metabolically fasted state of ketosis than current modern-day humans (phenotype 2 and 3). Then, plausibly, the role of insulin is to regulate BHB synthesis, while the role of bone regulates glucose uptake sensitivity via OCN. OCN then regulates fasting glucose levels via regulating the alpha cell glucagon secretory profile and BHB synthesis via regulating insulin release. An acute "fight or flight" stress response, such as running from danger or running for hunting, would induce a rapid glucocorticoid stimulated hepatic release of glucose that would then signal for a rapid insulin secretion response [120]. The liver is "IR" toward glucose output, to ensure that the glucose may be provided for increased muscle uptake, increased red blood cell (RBC) use for oxygen transport, and to increase the clotting ability in the case of potential physical harm that could cause haemorrhage and subsequent life-threatening hypovolemia.

The acute hyperinsulinaemia that accompanies the acute hyperglycaemia, inhibits anti-coagulation processes via upregulating plasminogen activator type 1 (PAI-1), and thus upregulates blood coagulability [121,122]. It, therefore, stands to reason that, in the fasted state, hepatic glucose output mechanisms would adapt to not respond and, thus, be inhibited by a rapid surge in insulin secretion, because if it did, then the "glucose tap" would be turned off. This would result in sudden life-threatening deprivation of glucose for the RBCs and certain parts of the central nervous system (CNS), which must receive glucose.

It is equally "essential" that the liver becomes IR in the context of hyperinsulinaemia as well. When a significant enough portion of time is spent in the fed state, which includes carbohydrate consumption, rapid insulin secretion is induced; over time, this down regulates the expression of BHB synthesis enzymes [90,93,123,124]. As a result, the return to ketogenesis does not occur within 3 to 5 hours post prandial. Furthermore, chronic hyperinsulinaemia impairs hepatic and extra-hepatic beta-oxidation, driving a greater extra-hepatic reliance on glucose for fuel [125]. Consequently, glucose becomes the essential fuel for the system, a system that does not easily "switch gears" to using fatty acids nor BHB, which is not readily available. Again, in this circumstance, the liver must continue to release glucose even when there is an external influx of exogenous glucose causing a rapid insulin release [94]. It may be the case that the down regulation of the rapid insulin release may not be wholly pathological if one considers it as a means to reduce any excess insulin signal on the liver to inhibit glycogenolysis. However, this is hypothetical and it is more likely that mitochondrial damage results in the pathological changes that impair the first phase rapid insulin response after a glucose bolus in T2DM patients, and this, in turn, contributes to a pathological feedforward progression of increases in fasting basal insulin levels [29,50,55,126].

The first phase rapid insulin response is impaired in hyperglycaemic T2DM patients [5,36], and may be a significant T2DM risk marker for hyperinsulinaemic/normoglycaemic people (phenotype 3, stage 1) [127]. There is a danger in thinking the solution is to find a way to stimulate the rapid exocytosis of a high concentration of insulin to rapidly lower plasma glucose levels. Although, that would result in what would appear to be, better glucose homeostasis. However, it must be remembered that these patients are also insulin resistant and have hyperinsulinaemia. Finding a way to help individuals with T2DM to activate the rapid first phase response, i.e., increased insulin, would likely only potentiate the insulin resistance and further increase their hyperinsulinaemia, which is present for a period of time preceding hyperglycaemia. The better strategy is to understand what causes the damage to the first phase response. Once this is understood, then the logical action is to remove the detrimental upstream causal stimulus, shown successfully by a variety of different carbohydrate and/or calorie restricted processes [78,128–131].

The current paradigm for insulin secretion regulation, for both basal and bolus, is that glucose is the primary stimulus. Our hypothesis challenges this and proposes a new paradigm: that bolus insulin secretion is regulated by glucose stimulus, but basal insulin secretion is regulated by OCN. As people spend more time in a higher glycaemic state, for example decreased fasting, constantly post prandial, this increases the frequency of the requirements of the bolus secretion dominance, with plasma glucose being above 6 mmol/L, which suppresses the role of basal insulin secretion. As a result, reduced OCN-regulated signalling diminishes serotonin regulation on glucagon output and, consequently, fasting glucose rises again, potentiating bolus insulin release.

This proposal, regarding OCN regulating basal insulin release, needs to be thoroughly tested, for example, in phenotype 1 people, who could be proxied by people in habitual ketosis with no prior metabolic health dysregulation. Following an overnight fast, plasma samples for glucose, insulin, BHB, OCN, GLP1, glucagon, and serotonin are collected at least every 5 min for 60 min, to establish basal state insulin secretion. These people are then given a glucose bolus, to establish a phenotype 1 OGTT metabolic profile.

Further investigations into insulin responses upon oral glucose tolerance tests (OGTT), in individuals with phenotype 1, will likely demonstrate healthy first phase rapid insulin secretion responses. These individuals have no prior metabolic health conditions and maintain a longstanding habitual metabolic fasting-mimicking phenotype. In response to an OGTT, their hepatic glucose output will not decrease; however, BHB synthesis will be decreased/inhibited. The return of ketogenesis then marks the decrease in bolus stimulated insulin signalling, while exogenous and endogenous glucose are both put away, out of the bloodstream in a timely manner, somewhat akin to a healthy transient acute stress response. While this may turn out to be true in acute infrequent trials, chronic glucose tolerance tests, in the form of "three square meals a day", likely induce the pathological changes that evolve into the chronic diseases associated with ageing.

6. Clinical Implications

Clinically, this new paradigm suggests that there needs to be an expansion in the concept of good metabolic health to include the presence of BHB. While ketone esters are available as health supplements, future research needs to consider whether there is a difference in clinical outcomes between endogenous and exogenous BHB. Furthermore, we need a better understanding of the metabolic profiles of people representative of the phenotypes, as described above. For example, people with phenotype 1 are likely to have a greater glucose sparing effect, they may have a heightened glycaemic response following a glucose bolus, compared to people with phenotype 2 or 3. Using multiple metabolic markers, especially the combination of glucose, insulin, BHB, and, ideally, OCN, may provide a better understanding of metabolic or endocrine health. Understanding basal insulin regulation enables earlier detection in changes that are associated with progression in the pathological development of hyperinsulinaemia conditions. With this understanding, it becomes clear that tight glycaemic control via medications that directly increase insulin levels, especially the sulphonylureas or exogenous insulin, only further potentiates hyperinsulinaemia and subsequent pathologies [132]. Lifestyle management, especially carbohydrate restriction, [43,128,130] with adequate individualised support should be considered first line treatment for hyperinsulinaemia, with or without hyperglycaemia, as it offers a more effective method in improving glucose homeostasis whilst also decreasing excess insulin exposure.

7. Conclusions

Once the metabolic phenotype is established, this would enable better selection criteria and grouping for participant/tissue sampling/cell culture-media experimentation. Thus, it would enable the avoidance of pooling participants/patients results that lead to signal cancelling effects. This will then aid in the pursuit of resolving the regulation of basal insulin secretion. Understanding the individual phenotype and stage allows for better patient understanding and compliance with clinical nutritional management.

Author Contributions: I.D.C. performed the literature search, wrote the original draft, designed the figures, and reviewed and edited the final manuscript. K.H.B. contributed to writing and reviewed and edited the final manuscript. Y.K. contributed to writing and reviewed and edited the final manuscript. B.T.E. contributed to writing and reviewed and edited the final manuscript. C.A.P.C. contributed to writing and reviewed and edited the final manuscript. All authors have read and agreed to the published version of the manuscript.

Funding: This research received no external funding.

Institutional Review Board Statement: Not applicable.

Informed Consent Statement: Not applicable.

Data Availability Statement: Data sharing not applicable as no datasets were generated and/or analysed for this study.

Acknowledgments: Figures created with BioRender.com (2021).

Conflicts of Interest: No conflict of interest to declare.

References

1. Boland, B.B.; Brown, J.C.; Alarcon, C.; Demozay, D.; Grimsby, J.S.; Rhodes, C.J. β-Cell Control of Insulin Production During Starvation-Refeeding in Male Rats. *Endocrinology* **2017**, *159*, 895–906. [CrossRef]
2. Aspinwall, C.; Lakey, J.R.T.; Kennedy, R.T. Insulin-stimulated Insulin Secretion in Single Pancreatic Beta Cells. *J. Biol. Chem.* **1999**, *274*, 6360–6365. [CrossRef]
3. Keane, K.N.; Newsholme, P. Metabolic Regulation of Insulin Secretion. *Vitam. Horm.* **2014**, *95*, 1–33. [CrossRef] [PubMed]
4. van Vliet, S.; Koh, H.-C.E.; Patterson, B.W.; Yoshino, M.; LaForest, R.; Gropler, R.J.; Klein, S.; Mittendorfer, B. Obesity Is Associated With Increased Basal and Postprandial β-Cell Insulin Secretion Even in the Absence of Insulin Resistance. *Diabetes* **2020**, *69*, 2112–2119. [CrossRef] [PubMed]
5. Rorsman, P.; Braun, M. Regulation of Insulin Secretion in Human Pancreatic Islets. *Annu. Rev. Physiol.* **2013**, *75*, 155–179. [CrossRef] [PubMed]
6. Crofts, C.A.P. Hyperinsulinemia: A unifying theory of chronic disease? *Diabesity* **2015**, *1*, 34. [CrossRef]
7. Seino, S.; Shibasaki, T.; Minami, K. Dynamics of insulin secretion and the clinical implications for obesity and diabetes. *J. Clin. Investig.* **2011**, *121*, 2118–2125. [CrossRef] [PubMed]

8. Rachdaoui, N. Insulin: The Friend and the Foe in the Development of Type 2 Diabetes Mellitus. *Int. J. Mol. Sci.* **2020**, *21*, 1770. [CrossRef]
9. De Vos, A.; Heimberg, H.; Quartier, E.; Huypens, P.; Bouwens, L.; Pipeleers, D.; Schuit, F. Human and rat beta cells differ in glucose transporter but not in glucokinase gene expression. *J. Clin. Investig.* **1995**, *96*, 2489–2495. [CrossRef]
10. McCulloch, L.J.; van de Bunt, M.; Braun, M.; Frayn, K.N.; Clark, A.; Gloyn, A.L. GLUT2 (SLC2A2) is not the principal glucose transporter in human pancreatic beta cells: Implications for understanding genetic association signals at this locus. *Mol. Genet. Metab.* **2011**, *104*, 648–653. [CrossRef]
11. Gould, G.; Holman, G. The glucose transporter family: Structure, function and tissue-specific expression. *Biochem. J.* **1993**, *295*, 329–341. [CrossRef] [PubMed]
12. Cárdenas, M.L.; Cornish-Bowden, A.; Ureta, T. Evolution and regulatory role of the hexokinases. *Biochim. Biophys. Acta (BBA) Mol. Cell Res.* **1998**, *1401*, 242–264. [CrossRef]
13. Gidh-Jain, M.; Takeda, J.; Xu, L.Z.; Lange, A.J.; Vionnet, N.; Stoffel, M.; Froguel, P.; Velho, G.; Sun, F.; Cohen, D. Glucokinase mutations associated with non-insulin-dependent (type 2) diabetes mellitus have decreased enzymatic activity: Implications for structure/function relationships. *Proc. Natl. Acad. Sci. USA* **1993**, *90*, 1932–1936. [CrossRef] [PubMed]
14. Iynedjian, P. Molecular Physiology of Mammalian Glucokinase. *Cell. Mol. Life Sci.* **2009**, *66*, 27–42. [CrossRef]
15. Fajans, S.S.; Bell, G.I.; Polonsky, K.S. Molecular Mechanisms and Clinical Pathophysiology of Maturity-Onset Diabetes of the Young. *N. Engl. J. Med.* **2001**, *345*, 971–980. [CrossRef]
16. Matschinsky, F.M.; Wilson, D.F. The Central Role of Glucokinase in Glucose Homeostasis: A Perspective 50 Years After Demonstrating the Presence of the Enzyme in Islets of Langerhans. *Front. Physiol.* **2019**, *10*, 148. [CrossRef] [PubMed]
17. Alvarez, E.; Martínez, M.D.; Roncero, I.; Chowen, J.A.; Garcia-Cuartero, B.; Gispert, J.D.; Sanz, C.; Vázquez, P.; Maldonado, A.; De Cáceres, J.; et al. The expression of GLP-1 receptor mRNA and protein allows the effect of GLP-1 on glucose metabolism in the human hypothalamus and brainstem. *J. Neurochem.* **2005**, *92*, 798–806. [CrossRef] [PubMed]
18. Sekine, N.; Cirulli, V.; Regazzi, R.; Brown, L.J.; Gine, E.; Tamarit-Rodriguez, J.; Girotti, M.; Marie, S.; Macdonald, M.J.; Wollheim, C.B. Low lactate dehydrogenase and high mitochondrial glycerol phosphate dehydrogenase in pancreatic beta-cells. Potential role in nutrient sensing. *J. Biol. Chem.* **1994**, *269*, 4895–4902. [CrossRef]
19. Alcazar, O.; Tiedge, M.; Lenzen, S. Importance of lactate dehydrogenase for the regulation of glycolytic flux and insulin secretion in insulin-producing cells. *Biochem. J.* **2000**, *352*, 373–380. [CrossRef]
20. Ritzel, R.A.; Michael, D.J.; Butler, P.C. Insulin Secretion. In *Encyclopedia of Hormones*; Elsevier: Amsterdam, The Netherlands, 2003; pp. 384–390.
21. Henquin, J.-C. Triggering and amplifying pathways of regulation of insulin secretion by glucose. *Diabetes* **2000**, *49*, 1751–1760. [CrossRef]
22. Griesche, N.; Sanchez, G.; Hermans, C.; Idevall-Hagren, O. Cortical mitochondria regulate insulin secretion by local Ca2+ buffering in rodent beta cells. *J. Cell Sci.* **2019**, *132*, jcs.228544. [CrossRef]
23. Cook, D.L.; Hales, N. Intracellular ATP directly blocks K+ channels in pancreatic B-cells. *Nature* **1984**, *311*, 271–273. [CrossRef] [PubMed]
24. Tarasov, A.; Dusonchet, J.; Ashcroft, F. Metabolic regulation of the pancreatic β-cell ATP-sensitive K+ channel: A pas de deux. *Diabetes* **2004**, *53*, S113–S122. [CrossRef]
25. Fridlyand, L.E.; Jacobson, D.A.; Philipson, L. Ion channels and regulation of insulin secretion in human β-cells: A computational systems analysis. *Islets* **2013**, *5*, 1–15. [CrossRef]
26. Braun, M.; Ramracheya, R.; Bengtsson, M.; Zhang, Q.; Karanauskaite, J.; Partridge, C.; Johnson, P.R.; Rorsman, P. Voltage-Gated Ion Channels in Human Pancreatic β-Cells: Electrophysiological Characterization and Role in Insulin Secretion. *Diabetes* **2008**, *57*, 1618–1628. [CrossRef]
27. Rorsman, P.; Renström, E. Insulin granule dynamics in pancreatic beta cells. *Diabetologia* **2003**, *46*, 1029–1045. [CrossRef] [PubMed]
28. Wollheim, C.; Sharp, G.W. Regulation of insulin release by calcium. *Physiol. Rev.* **1981**, *61*, 914–973. [CrossRef]
29. Georgiadou, E.; Rutter, G.A. Control by Ca^{2+} of mitochondrial structure and function in pancreatic β-cells. *Cell Calcium* **2020**, *91*, 102282. [CrossRef]
30. Marliss, E.B.; Aoki, T.T.; Unger, R.H.; Soeldner, J.S.; Cahill, G.F. Glucagon levels and metabolic effects in fasting man. *J. Clin. Investig.* **1970**, *49*, 2256–2270. [CrossRef] [PubMed]
31. Færch, K.; Vistisen, D.; Pacini, G.; Torekov, S.; Johansen, N.B.; Witte, D.R.; Jonsson, A.; Pedersen, O.; Hansen, T.; Lauritzen, T.; et al. Insulin Resistance Is Accompanied by Increased Fasting Glucagon and Delayed Glucagon Suppression in Individuals With Normal and Impaired Glucose Regulation. *Diabetes* **2016**, *65*, 3473–3481. [CrossRef] [PubMed]
32. Gershuni, V.M.; Yan, S.L.; Medici, V. Nutritional Ketosis for Weight Management and Reversal of Metabolic Syndrome. *Curr. Nutr. Rep.* **2018**, *7*, 97–106. [CrossRef]
33. Shannahoff-Khalsa, D.S.; Kennedy, B.; Yates, F.E.; Ziegler, M.G. Low-frequency ultradian insulin rhythms are coupled to cardiovascular, autonomic, and neuroendocrine rhythms. *Am. J. Physiol. Regul. Integr. Comp. Physiol.* **1997**, *272*, R962–R968. [CrossRef] [PubMed]
34. Satin, L.S.; Butler, P.C.; Ha, J.; Sherman, A.S. Pulsatile insulin secretion, impaired glucose tolerance and type 2 diabetes. *Mol. Asp. Med.* **2015**, *42*, 61–77. [CrossRef]

35. Franco, M.C.D.; De Leon, R.F.; Villafan-Bernal, J.R. Osteocalcin-GPRC6A: An update of its clinical and biological multi-organic interactions (Review). *Mol. Med. Rep.* **2019**, *19*, 15–22. [CrossRef]
36. Crofts, C.; Schofield, G.; Zinn, C.; Wheldon, M.; Kraft, J. Identifying hyperinsulinaemia in the absence of impaired glucose tolerance: An examination of the Kraft database. *Diabetes Res. Clin. Pract.* **2016**, *118*, 50–57. [CrossRef] [PubMed]
37. Crofts, C.; Neill, A.; Campbell, A.; Bartley, J.; White, D.E. Sleep architecture, insulin resistance and the nasal cycle: Implications for positive airway pressure therapy. *J. Insul. Resist.* **2018**, *3*. [CrossRef]
38. Brand, M.D. Mitochondrial generation of superoxide and hydrogen peroxide as the source of mitochondrial redox signaling. *Free Radic. Biol. Med.* **2016**, *100*, 14–31. [CrossRef]
39. Choi, T.G.; Kim, S.S. Physiological Functions of Mitochondrial Reactive Oxygen Species. *Free Radic. Med. Biol.* **2019**. [CrossRef]
40. Kowalska, M.; Piekut, T.; Prendecki, M.; Sodel, A.; Kozubski, W.; Dorszewska, J. Mitochondrial and Nuclear DNA Oxidative Damage in Physiological and Pathological Aging. *DNA Cell Biol.* **2020**, *39*, 1410–1420. [CrossRef] [PubMed]
41. Murphy, M.P. How mitochondria produce reactive oxygen species. *Biochem. J.* **2009**, *417*, 1–13. [CrossRef] [PubMed]
42. Robb, E.L.; Hall, A.R.; Prime, T.A.; Eaton, S.; Szibor, M.; Viscomi, C.; James, A.M.; Murphy, M.P. Control of mitochondrial superoxide production by reverse electron transport at complex I. *J. Biol. Chem.* **2018**, *293*, 9869–9879. [CrossRef] [PubMed]
43. Cooper, I.D.; Crofts, C.A.P.; DiNicolantonio, J.J.; Malhotra, A.; Elliott, B.; Kyriakidou, Y.; Brookler, K.H. Relationships between hyperinsulinaemia, magnesium, vitamin D, thrombosis and COVID-19: Rationale for clinical management. *Open Hear.* **2020**, *7*, e001356. [CrossRef]
44. Sack, M.N.; Finkel, T. Mitochondrial Metabolism, Sirtuins, and Aging. *Cold Spring Harb. Perspect. Biol.* **2012**, *4*, a013102. [CrossRef] [PubMed]
45. Yu, T.; Robotham, J.L.; Yoon, Y. Increased production of reactive oxygen species in hyperglycemic conditions requires dynamic change of mitochondrial morphology. *Proc. Natl. Acad. Sci. USA* **2006**, *103*, 2653–2658. [CrossRef]
46. Newman, J.C.; Verdin, E. Ketone bodies as signaling metabolites. *Trends Endocrinol. Metab.* **2014**, *25*, 42–52. [CrossRef] [PubMed]
47. Veech, R.L.; Bradshaw, P.C.; Clarke, K.; Curtis, W.; Pawlosky, R.; King, M.T. Ketone bodies mimic the life span extending properties of caloric restriction. *IUBMB Life* **2017**, *69*, 305–314. [CrossRef]
48. Hansen, M.E.; Tippetts, T.S.; Anderson, M.C.; Holub, Z.E.; Moulton, E.R.; Swensen, A.C.; Prince, J.T.; Bikman, B.T. Insulin Increases Ceramide Synthesis in Skeletal Muscle. *J. Diabetes Res.* **2014**, *2014*, 1–9. [CrossRef] [PubMed]
49. Krauss, S.; Zhang, C.-Y.; Scorrano, L.; Dalgaard, L.T.; St-Pierre, J.; Grey, S.T.; Lowell, B.B. Superoxide-mediated activation of uncoupling protein 2 causes pancreatic β cell dysfunction. *J. Clin. Investig.* **2003**, *112*, 1831–1842. [CrossRef]
50. Anello, M.; Lupi, R.; Spampinato, D.; Piro, S.; Masini, M.; Boggi, U.; Del Prato, S.; Rabuazzo, A.M.; Purrello, F.; Marchetti, P. Functional and morphological alterations of mitochondria in pancreatic beta cells from type 2 diabetic patients. *Diabetologia* **2005**, *48*, 282–289. [CrossRef]
51. Kowaltowski, A.J.; Menezes-Filho, S.L.; Assali, E.A.; Gonçalves, I.G.; Cabral-Costa, J.V.; Abreu, P.; Miller, N.; Nolasco, P.; Laurindo, F.R.M.; Bruni-Cardoso, A.; et al. Mitochondrial morphology regulates organellar Ca^{2+} uptake and changes cellular Ca^{2+} homeostasis. *FASEB J.* **2019**, *33*, 13176–13188. [CrossRef]
52. De la Fuente, S.; Lambert, J.; Nichtova, Z.; Sanz, C.F.; Elrod, J.; Sheu, S.-S.; Csordás, G. Spatial Separation of Mitochondrial Calcium Uptake and Extrusion for Energy-Efficient Mitochondrial Calcium Signaling in the Heart. *Cell Rep.* **2018**, *24*, 3099–3107.e4. [CrossRef]
53. Patergnani, S.; Suski, J.M.; Agnoletto, C.; Bononi, A.; Bonora, M.; De Marchi, E.; Giorgi, C.; Marchi, S.; Missiroli, S.; Poletti, F.; et al. Calcium signaling around Mitochondria Associated Membranes (MAMs). *Cell Commun. Signal.* **2011**, *9*, 19. [CrossRef]
54. Maechler, P.; Li, N.; Casimir, M.; Vetterli, L.; Frigerio, F.; Brun, T. Role of Mitochondria in β-Cell Function and Dysfunction. In *Islets of Langerhans*, 2nd ed.; Springer Science and Business Media LLC: Berlin/Heidelberg, Germany, 2015; pp. 633–657. [CrossRef]
55. Georgiadou, E.; Haythorne, E.; Dickerson, M.T.; Lopez-Noriega, L.; Pullen, T.J.; Xavier, G.D.S.; Davis, S.P.X.; Martinez-Sanchez, A.; Semplici, F.; Rizzuto, R.; et al. The pore-forming subunit MCU of the mitochondrial Ca2+ uniporter is required for normal glucose-stimulated insulin secretion in vitro and in vivo in mice. *Diabetologia* **2020**, *63*, 1368–1381. [CrossRef]
56. Contreras, L.; Drago, I.; Zampese, E.; Pozzan, T. Mitochondria: The calcium connection. *Biochim. Biophys. Acta (BBA) Bioenerg.* **2010**, *1797*, 607–618. [CrossRef] [PubMed]
57. Kaufman, R.J.; Malhotra, J.D. Calcium trafficking integrates endoplasmic reticulum function with mitochondrial bioenergetics. *Biochim. Biophys. Acta (BBA) Mol. Cell Res.* **2014**, *1843*, 2233–2239. [CrossRef] [PubMed]
58. Zhang, I.X.; Ren, J.; Vadrevu, S.; Raghavan, M.; Satin, L.S. ER stress increases store-operated Ca2+ entry (SOCE) and augments basal insulin secretion in pancreatic beta cells. *J. Biol. Chem.* **2020**, *295*, 5685–5700. [CrossRef] [PubMed]
59. Chen, L.; Koh, D.-S.; Hille, B. Dynamics of Calcium Clearance in Mouse Pancreatic β-Cells. *Diabetes* **2003**, *52*, 1723–1731. [CrossRef] [PubMed]
60. Rorsman, P.; Ashcroft, F. Pancreatic β-Cell Electrical Activity and Insulin Secretion: Of Mice and Men. *Physiol. Rev.* **2018**, *98*, 117–214. [CrossRef]
61. Lee, K.-U.; Harris, R.A. Mitochondria and Endoplasmic Reticulum in Diabetes and Its Complications. *Exp. Diabetes Res.* **2012**, *2012*, 1–2. [CrossRef] [PubMed]
62. Di Giovanni, J.; Iborra, C.; Maulet, Y.; Lévêque, C.; El Far, O.; Seagar, M. Calcium-dependent Regulation of SNARE-mediated Membrane Fusion by Calmodulin. *J. Biol. Chem.* **2010**, *285*, 23665–23675. [CrossRef]

63. Thurmond, D.C. Regulation of Insulin Action and Insulin Secretion by SNARE-Mediated Vesicle Exocytosis. In *Mechanisms of Insulin Action*; Springer: New York, NY, USA, 2007; pp. 52–70.
64. Bello, O.; Jouannot, O.; Chaudhuri, A.; Stroeva, E.; Coleman, J.; Volynski, K.E.; Rothman, J.E.; Krishnakumar, S.S. Synaptotagmin oligomerization is essential for calcium control of regulated exocytosis. *Proc. Natl. Acad. Sci. USA* **2018**, *115*, E7624–E7631. [CrossRef]
65. Babbar, M.; Sheikh, M.S. Metabolic Stress and Disorders Related to Alterations in Mitochondrial Fission or Fusion. *Mol. Cell. Pharmacol.* **2013**, *5*, 109–133. [CrossRef]
66. Parker, B.A.; Walton, C.M.; Carr, S.T.; Andrus, J.L.; Cheung, E.C.K.; Duplisea, M.J.; Wilson, E.K.; Draney, C.; Lathen, D.R.; Kenner, K.B.; et al. β-Hydroxybutyrate Elicits Favorable Mitochondrial Changes in Skeletal Muscle. *Int. J. Mol. Sci.* **2018**, *19*, 2247. [CrossRef]
67. Gao, J.; Qin, A.; Liu, D.; Ruan, R.; Wang, Q.; Yuan, J.; Cheung, T.S.; Filipovska, A.; Papadimitriou, J.M.; Dai, K.; et al. Endoplasmic reticulum mediates mitochondrial transfer within the osteocyte dendritic network. *Sci. Adv.* **2019**, *5*, eaaw7215. [CrossRef] [PubMed]
68. Rutter, G.A.; Pullen, T.; Hodson, D.; Sanchez, A.M. Pancreatic β-cell identity, glucose sensing and the control of insulin secretion. *Biochem. J.* **2015**, *466*, 203–218. [CrossRef]
69. Ma, Z.A.; Zhao, Z.; Turk, J. Mitochondrial Dysfunction and β-Cell Failure in Type 2 Diabetes Mellitus. *Exp. Diabetes Res.* **2012**, *2012*, 1–11. [CrossRef]
70. Fex, M.; Nicholas, L.M.; Vishnu, N.; Medina, A.; Sharoyko, V.V.; Nicholls, D.G.; Spégel, P.; Mulder, H. The pathogenetic role of β-cell mitochondria in type 2 diabetes. *J. Endocrinol.* **2018**, *236*, R145–R159. [CrossRef]
71. Matveyenko, A.V.; Butler, P.C. Relationship between β-cell mass and diabetes onset. *Diabetes Obes. Metab.* **2008**, *10*, 23–31. [CrossRef] [PubMed]
72. Chen, C.; Cohrs, C.M.; Stertmann, J.; Bozsak, R.; Speier, S. Human beta cell mass and function in diabetes: Recent advances in knowledge and technologies to understand disease pathogenesis. *Mol. Metab.* **2017**, *6*, 943–957. [CrossRef] [PubMed]
73. Ferrannini, E.; Mari, A. β-Cell function in type 2 diabetes. *Metabolism* **2014**, *63*, 1217–1227. [CrossRef]
74. Ogilvie, R.F. The islands of langerhans in 19 cases of obesity. *J. Pathol. Bacteriol.* **1933**, *37*, 473–481. [CrossRef]
75. Cho, J.-H.; Kim, J.-W.; Shin, J.-A.; Shin, J.; Yoon, K.-H. β-cell mass in people with type 2 diabetes. *J. Diabetes Investig.* **2011**, *2*, 6–17. [CrossRef] [PubMed]
76. Pontzer, H.; Wood, B.M.; Raichlen, D.A. Hunter-gatherers as models in public health. *Obes. Rev.* **2018**, *19*, 24–35. [CrossRef] [PubMed]
77. Lougheed, T. The Changing Landscape of Arctic Traditional Food. *Environ. Health Perspect.* **2010**, *118*, A386. [CrossRef]
78. Anton, S.D.; Moehl, K.; Donahoo, W.; Marosi, K.; Lee, S.; Mainous, A.G.; Leeuwenburgh, C.; Mattson, M.P. Flipping the Metabolic Switch: Understanding and Applying the Health Benefits of Fasting. *Obesity* **2018**, *26*, 254–268. [CrossRef]
79. Owen, O.E.; Morgan, A.P.; Kemp, H.G.; Sullivan, J.M.; Herrera, M.G.; Cahill, G.F. Brain Metabolism during Fasting. *J. Clin. Investig.* **1967**, *46*, 1589–1595. [CrossRef]
80. Cahill, G.F.; Herrera, M.G.; Morgan, A.P.; Soeldner, J.S.; Steinke, J.; Levy, P.L.; Reichard, G.A.; Kipnis, D.M. Hormone-fuel interrelationships during fasting. *J. Clin. Investig.* **1966**, *45*, 1751–1769. [CrossRef]
81. Dhatariya, K.; Savage, M.W.; Claydon, A.; The Management of Diabetic Ketoacidosis in Adults *. *Jt. Br. Diabetes Soc. Inpatient Care Gr. NHS* 2021. Available online: https://www.bsped.org.uk/media/1798/bsped-dka-guideline-2020.pdf (accessed on 1 July 2021).
82. Seetho, I.W.; Wilding, J. The clinical management of diabetes mellitus. In *Clinical Biochemistry: Metabolic and Clinical Aspects*, 3rd ed.; Elsevier: Amsterdam, The Netherlands, 2014; pp. 305–332. [CrossRef]
83. Gosmanov, A.R.; Nematollahi, L.R.; Diabetic Ketoacidosis—Symptoms, Diagnosis and Treatment. *BMJ Best Pract.* 2020. Available online: https://bestpractice.bmj.com/topics/en-us/162 (accessed on 4 August 2020).
84. Elzouki, A.-N.; Eledrisi, M.S. Management of diabetic ketoacidosis in adults: A narrative review. *Saudi J. Med. Med Sci.* **2020**, *8*, 165–173. [CrossRef]
85. Min, S.; Oh, T.; Baek, S.-I.; Lee, D.-H.; Kim, K.; Moon, J.; Choi, S.H.; Park, K.; Jang, H.; Lim, S. Degree of ketonaemia and its association with insulin resistance after dapagliflozin treatment in type 2 diabetes. *Diabetes Metab.* **2018**, *44*, 73–76. [CrossRef]
86. Ebeling, P.; Koistinen, H.A.; Koivisto, V.A. Insulin-independent glucose transport regulates insulin sensitivity. *FEBS Lett.* **1998**, *436*, 301–303. [CrossRef]
87. Sowers, J.R.; Frohlich, E.D. Insulin and insulin resistance: Impact on blood pressure and cardiovascular disease. *Med. Clin. N. Am.* **2004**, *88*, 63–82. [CrossRef]
88. Marliss, E.; Aoki, T.; Felig, P.; Pozefsky, T.; Cahill, G. Hormones and substrates in the regulation of gluconeogenesis in fasting man. *Adv. Enzym. Regul.* **1970**, *8*, 3–11. [CrossRef]
89. Robinson, A.M.; Williamson, D.H. Physiological roles of ketone bodies as substrates and signals in mammalian tissues. *Physiol. Rev.* **1980**, *60*, 143–187. [CrossRef]
90. Grabacka, M.; Pierzchalska, M.; Dean, M.; Reiss, K. Regulation of Ketone Body Metabolism and the Role of PPARα. *Int. J. Mol. Sci.* **2016**, *17*, 2093. [CrossRef] [PubMed]
91. Puchalska, P.; Crawford, P.A. Multi-dimensional Roles of Ketone Bodies in Fuel Metabolism, Signaling, and Therapeutics. *Cell Metab.* **2017**, *25*, 262–284. [CrossRef]

92. Newman, J.C.; Verdin, E. β-Hydroxybutyrate: A Signaling Metabolite. *Annu. Rev. Nutr.* **2017**, *37*, 51–76. [CrossRef]
93. Laffel, L. Ketone bodies: A review of physiology, pathophysiology and application of monitoring to diabetes. *Diabetes Metab. Res. Rev.* **1999**, *15*, 412–426. [CrossRef]
94. Grandl, G.; Straub, L.; Rudigier, C.; Arnold, M.; Wueest, S.; Konrad, D.; Wolfrum, C. Short-term feeding of a ketogenic diet induces more severe hepatic insulin resistance than an obesogenic high-fat diet. *J. Physiol.* **2018**, *596*, 4597–4609. [CrossRef]
95. Santoleri, D.; Titchenell, P.M. Resolving the Paradox of Hepatic Insulin Resistance. *CMGH* **2019**, *7*, 447–456. [CrossRef]
96. Reaven, G.M.; Chen, Y.-D.I.; Golay, A.; Swislocki, A.L.M.; Jaspan, J.B. Documentation of Hyperglucagonemia Throughout the Day in Nonobese and Obese Patients with Noninsulin-Dependent Diabetes Mellitus. *J. Clin. Endocrinol. Metab.* **1987**, *64*, 106–110. [CrossRef]
97. Unger, R.H.; Orci, L. Physiology and pathophysiology of glucagon. *Physiol. Rev.* **1976**, *56*, 778–826. [CrossRef] [PubMed]
98. Ferron, M.; Wei, J.; Yoshizawa, T.; Del Fattore, A.; DePinho, R.; Teti, A.M.; Ducy, P.; Karsenty, G. Insulin Signaling in Osteoblasts Integrates Bone Remodeling and Energy Metabolism. *Cell* **2010**, *142*, 296–308. [CrossRef] [PubMed]
99. Saleem, U.; Mosley, J.T.H.; Kullo, I.J. Serum Osteocalcin Is Associated With Measures of Insulin Resistance, Adipokine Levels, and the Presence of Metabolic Syndrome. *Arterioscler. Thromb. Vasc. Biol.* **2010**, *30*, 1474–1478. [CrossRef] [PubMed]
100. Saito, A.; Yoshimura, K.; Miyamoto, Y.; Kaneko, K.; Chikazu, D.; Yamamoto, M.; Kamijo, R. Enhanced and suppressed mineralization by acetoacetate and β-hydroxybutyrate in osteoblast cultures. *Biochem. Biophys. Res. Commun.* **2016**, *473*, 537–544. [CrossRef] [PubMed]
101. Ferron, M.; Hinoi, E.; Karsenty, G.; Ducy, P. Osteocalcin differentially regulates β cell and adipocyte gene expression and affects the development of metabolic diseases in wild-type mice. *Proc. Natl. Acad. Sci. USA* **2008**, *105*, 5266–5270. [CrossRef]
102. Ferron, M.; Lacombe, J. Regulation of energy metabolism by the skeleton: Osteocalcin and beyond. *Arch. Biochem. Biophys.* **2014**, *561*, 137–146. [CrossRef]
103. Ferron, M.; McKee, M.D.; Levine, R.L.; Ducy, P.; Karsenty, G. Intermittent injections of osteocalcin improve glucose metabolism and prevent type 2 diabetes in mice. *Bone* **2012**, *50*, 568–575. [CrossRef] [PubMed]
104. Hinoi, E.; Gao, N.; Jung, D.Y.; Yadav, V.K.; Yoshizawa, T.; Myers, M.G.; Chua, S.C.; Kim, J.; Kaestner, K.H.; Karsenty, G. The sympathetic tone mediates leptin's inhibition of insulin secretion by modulating osteocalcin bioactivity. *J. Cell Biol.* **2008**, *183*, 1235–1242. [CrossRef]
105. Mizokami, A.; Yasutake, Y.; Gao, J.; Matsuda, M.; Takahashi, I.; Takeuchi, H.; Hirata, M. Osteocalcin Induces Release of Glucagon-Like Peptide-1 and Thereby Stimulates Insulin Secretion in Mice. *PLoS ONE* **2013**, *8*, e57375. [CrossRef]
106. Nauck, M.A.; Niedereichholz, U.; Ettler, R.; Holst, J.J.; Ørskov, C.; Ritzel, R.; Schmiegel, W.H. Glucagon-like peptide 1 inhibition of gastric emptying outweighs its insulinotropic effects in healthy humans. *Am. J. Physiol. Endocrinol. Metab.* **1997**, *273*, E981–E988. [CrossRef]
107. Drucker, D.J. Mechanisms of Action and Therapeutic Application of Glucagon-like Peptide-1. *Cell Metab.* **2018**, *27*, 740–756. [CrossRef] [PubMed]
108. Hui, H.; Farilla, L.; Merkel, P.; Perfetti, R. The short half-life of glucagon-like peptide-1 in plasma does not reflect its long-lasting beneficial effects. *Eur. J. Endocrinol.* **2002**, *146*, 863–869. [CrossRef]
109. Brown, J.P.; Albert, C.; Nassar, B.A.; Adachi, J.D.; Cole, D.; Davison, K.S.; Dooley, K.C.; Don-Wauchope, A.; Douville, P.; Hanley, D.A.; et al. Bone turnover markers in the management of postmenopausal osteoporosis. *Clin. Biochem.* **2009**, *42*, 929–942. [CrossRef]
110. Macdonald, M.J.; Longacre, M.J.; Stoker, S.W.; Brown, L.J.; Hasan, N.M.; Kendrick, M.A. Acetoacetate and β-hydroxybutyrate in combination with other metabolites release insulin from INS-1 cells and provide clues about pathways in insulin secretion. *Am. J. Physiol. Physiol.* **2008**, *294*, C442–C450. [CrossRef]
111. Oury, F.; Khrimian, L.; Denny, C.A.; Gardin, A.; Chamouni, A.; Goeden, N.; Huang, Y.-Y.; Lee, H.; Srinivas, P.; Gao, X.-B.; et al. Maternal and Offspring Pools of Osteocalcin Influence Brain Development and Functions. *Cell* **2013**, *155*, 228–241. [CrossRef]
112. Almaça, J.; Molina, J.; Menegaz, D.; Pronin, A.N.; Tamayo, A.; Slepak, V.; Berggren, P.-O.; Caicedo, A. Human Beta Cells Produce and Release Serotonin to Inhibit Glucagon Secretion from Alpha Cells. *Cell Rep.* **2016**, *17*, 3281–3291. [CrossRef]
113. Ebou, M.H.; Singh-Estivalet, A.; Launay, J.-M.; Callebert, J.; Tronche, F.; Ferre, P.; Gautier, J.-F.; Guillemain, G.; Bréant, B.; Blondeau, B.; et al. Glucocorticoids Inhibit Basal and Hormone-Induced Serotonin Synthesis in Pancreatic Beta Cells. *PLoS ONE* **2016**, *11*, e0149343. [CrossRef]
114. Paulmann, N.; Grohmann, M.; Voigt, J.-P.; Bert, B.; Vowinckel, J.; Bader, M.; Skelin, M.; Jevšek, M.; Fink, H.; Rupnik, M.S.; et al. Intracellular Serotonin Modulates Insulin Secretion from Pancreatic β-Cells by Protein Serotonylation. *PLoS Biol.* **2009**, *7*, e1000229. [CrossRef]
115. Ramracheya, R.; Chapman, C.; Chibalina, M.; Dou, H.; Miranda, C.; Gonzalez, A.; Moritoh, Y.; Shigeto, M.; Zhang, Q.; Braun, M.; et al. GLP-1 suppresses glucagon secretion in human pancreatic alpha-cells by inhibition of P/Q-type Ca^{2+} channels. *Physiol. Rep.* **2018**, *6*, e13852. [CrossRef] [PubMed]
116. Razny, U.; Fedak, D.; Kiec-Wilk, B.; Góralska, J.; Gruca, A.; Zdzienicka, A.; Kiec-Klimczak, M.; Solnica, B.; Hubalewska-Dydejczyk, A.; Malczewska-Malec, M. Carboxylated and undercarboxylated osteocalcin in metabolic complications of human obesity and prediabetes. *Diabetes Metab. Res. Rev.* **2016**, *33*, e2862. [CrossRef] [PubMed]

117. Guney, G.; Sener-Simsek, B.; Tokmak, A.; Yucel, A.; Buyukkagnici, U.; Yilmaz, N.; Engin-Ustun, Y.; Ozgu-Erdinc, A.S. Assessment of the Relationship between Serum Vitamin D and Osteocalcin Levels with Metabolic Syndrome in Non-Osteoporotic Postmenopausal Women. *Geburtshilfe Frauenheilkd.* **2019**, *79*, 293–299. [CrossRef]
118. Gallego, B.R.; García-Molina, L.; Cano-Ibáñez, N.; Sanchez-Delgado, G.; Andújar-Vera, F.; García-Fontana, C.; González-Salvatierra, S.; García-Recio, E.; Martínez-Ruiz, V.; Bueno-Cavanillas, A.; et al. Circulating Undercarboxylated Osteocalcin as Estimator of Cardiovascular and Type 2 Diabetes Risk in Metabolic Syndrome Patients. *Sci. Rep.* **2020**, *10*, 1840. [CrossRef]
119. Sánchez-de-Diego, C.; Artigas, N.; Pimenta-Lopes, C.; Valer, J.A.; Torrejon, B.; Gama-Pérez, P.; Villena, J.A.; Garcia-Roves, P.M.; Rosa, J.L.; Ventura, F. Glucose Restriction Promotes Osteocyte Specification by Activating a PGC-1α-Dependent Transcriptional Program. *iScience* **2019**, *15*, 79–94. [CrossRef]
120. Berger, J.M.; Singh, P.; Khrimian, L.; Morgan, D.A.; Chowdhury, S.; Arteaga-Solis, E.; Horvath, T.L.; Domingos, A.; Marsland, A.L.; Yadav, V.K.; et al. Mediation of the Acute Stress Response by the Skeleton. *Cell Metab.* **2019**, *30*, 890–902.e8. [CrossRef] [PubMed]
121. Stegenga, M.E.; Van Der Crabben, S.N.; Levi, M.; De Vos, A.F.; Tanck, M.; Sauerwein, H.P.; Van Der Poll, T. Hyperglycemia Stimulates Coagulation, Whereas Hyperinsulinemia Impairs Fibrinolysis in Healthy Humans. *Diabetes* **2006**, *55*, 1807–1812. [CrossRef]
122. Perkins, J.M.; Joy, N.G.; Tate, D.B.; Davis, S.N. Acute effects of hyperinsulinemia and hyperglycemia on vascular inflammatory biomarkers and endothelial function in overweight and obese humans. *Am. J. Physiol. Metab.* **2015**, *309*, E168–E176. [CrossRef] [PubMed]
123. Somogyi, M. Effects of insulin upon the production of ketone bodies. *J. Biol. Chem.* **1941**, *141*, 219–227. [CrossRef]
124. Wolfrum, C.; Besser, D.; Luca, E.; Stoffel, M. Insulin regulates the activity of forkhead transcription factor Hnf-3β/Foxa-2 by Akt-mediated phosphorylation and nuclear/cytosolic localization. *Proc. Natl. Acad. Sci. USA* **2003**, *100*, 11624–11629. [CrossRef] [PubMed]
125. Veech, R.L. The therapeutic implications of ketone bodies: The effects of ketone bodies in pathological conditions: Ketosis, ketogenic diet, redox states, insulin resistance, and mitochondrial metabolism. *Prostaglandins Leukot. Essent. Fat. Acids* **2004**, *70*, 309–319. [CrossRef]
126. Sampson, M.; Lathen, D.R.; Dallon, B.W.; Draney, C.; Ray, J.D.; Kener, K.B.; Parker, B.A.; Gibbs, J.L.; Gropp, J.S.; Tessem, J.S.; et al. β-Hydroxybutyrate improves β-cell mitochondrial function and survival. *J. Insul. Resist.* **2017**, *1*, 8. [CrossRef]
127. Hayashi, T.; Boyko, E.J.; Sato, K.K.; McNeely, M.J.; Leonetti, D.L.; Kahn, S.E.; Fujimoto, W.Y. Patterns of Insulin Concentration During the OGTT Predict the Risk of Type 2 Diabetes in Japanese Americans. *Diabetes Care* **2013**, *36*, 1229–1235. [CrossRef] [PubMed]
128. Kelly, T.; Unwin, D.; Finucane, F. Low-Carbohydrate Diets in the Management of Obesity and Type 2 Diabetes: A Review from Clinicians Using the Approach in Practice. *Int. J. Environ. Res. Public Health* **2020**, *17*, 2557. [CrossRef] [PubMed]
129. Murdoch, C.; Unwin, D.; Cavan, D.; Cucuzzella, M.; Patel, M. Adapting diabetes medication for low carbohydrate management of type 2 diabetes: A practical guide. *Br. J. Gen. Pract.* **2019**, *69*, 360–361. [CrossRef]
130. Athinarayanan, S.J.; Adams, R.N.; Hallberg, S.J.; McKenzie, A.L.; Bhanpuri, N.H.; Campbell, W.W.; Volek, J.S.; Phinney, S.D.; McCarter, J.P. Long-Term Effects of a Novel Continuous Remote Care Intervention Including Nutritional Ketosis for the Management of Type 2 Diabetes: A 2-Year Non-randomized Clinical Trial. *Front. Endocrinol.* **2019**, *10*, 348. [CrossRef] [PubMed]
131. Al-Mrabeh, A.; Hollingsworth, K.G.; Shaw, J.A.M.; McConnachie, A.; Sattar, N.; Lean, M.E.J.; Taylor, R. 2-year remission of type 2 diabetes and pancreas morphology: A post-hoc analysis of the DiRECT open-label, cluster-randomised trial. *Lancet Diabetes Endocrinol.* **2020**, *8*, 939–948. [CrossRef]
132. Gerstein, H.; Miller, M.E.; Byington, R.P.; Goff, D.C.; Bigger, J.T.; Buse, J.; Cushman, W.C.; Genuth, S.; Ismail-Beigi, F.; Grimm, R.H.; et al. Effects of Intensive Glucose Lowering in Type 2 Diabetes. *N. Engl. J. Med.* **2008**, *358*, 2545–2559. [CrossRef] [PubMed]

biomedicines

MDPI

Review

Modulation of Insulin Sensitivity by Insulin-Degrading Enzyme

Carlos M. González-Casimiro [1], Beatriz Merino [1], Elena Casanueva-Álvarez [1], Tamara Postigo-Casado [1], Patricia Cámara-Torres [1], Cristina M. Fernández-Díaz [2], Malcolm A. Leissring [3], Irene Cózar-Castellano [1,4,*] and Germán Perdomo [1,*]

[1] Instituto de Biología y Genética Molecular, University of Valladolid-CSIC, 47003 Valladolid, Spain; carlosmanuel.gonzalez.casimiro@uva.es (C.M.G.-C.); bmerino@ibgm.uva.es (B.M.); elenacasanueva15@gmail.com (E.C.-Á.); tamara.postigo@uva.es (T.P.-C.); patriciacamaratorres1@gmail.com (P.C.-T.)

[2] Molecular Oncology Group, IMDEA Food Institute, CEI UAM + CSIC, 28049 Madrid, Spain; cristinamaria.fernandez@imdea.org

[3] Institute for Memory Impairments and Neurological Disorders, University of California, Irvine (UCI MIND), Irvine, CA 92697, USA; m.leissring@uci.edu

[4] Centro de Investigación Biomédica en Red de Diabetes y Enfermedades Metabólicas Asociadas (CIBERDEM), 28029 Madrid, Spain

* Correspondence: irene.cozar@uva.es (I.C.-C.); g.perdomo@csic.es (G.P.); Tel.: +34-983-184-805 (G.P.)

Abstract: Insulin-degrading enzyme (IDE) is a highly conserved and ubiquitously expressed metalloprotease that degrades insulin and several other intermediate-size peptides. For many decades, IDE had been assumed to be involved primarily in hepatic insulin clearance, a key process that regulates availability of circulating insulin levels for peripheral tissues. Emerging evidence, however, suggests that IDE has several other important physiological functions relevant to glucose and insulin homeostasis, including the regulation of insulin secretion from pancreatic β-cells. Investigation of mice with tissue-specific genetic deletion of *Ide* in the liver and pancreatic β-cells (L-IDE-KO and B-IDE-KO mice, respectively) has revealed additional roles for IDE in the regulation of hepatic insulin action and sensitivity. In this review, we discuss current knowledge about IDE's function as a regulator of insulin secretion and hepatic insulin sensitivity, both evaluating the classical view of IDE as an insulin protease and also exploring evidence for several non-proteolytic functions. Insulin proteostasis and insulin sensitivity have both been highlighted as targets controlling blood sugar levels in type 2 diabetes, so a clearer understanding the physiological functions of IDE in pancreas and liver could led to the development of novel therapeutics for the treatment of this disease.

Keywords: insulin-degrading enzyme; insulin resistance; pancreas; liver; insulin receptor; glucose transporters

Citation: González-Casimiro, C.M.; Merino, B.; Casanueva-Álvarez, E.; Postigo-Casado, T.; Cámara-Torres, P.; Fernández-Díaz, C.M.; Leissring, M.A.; Cózar-Castellano, I.; Perdomo, G. Modulation of Insulin Sensitivity by Insulin-Degrading Enzyme. *Biomedicines* **2021**, *9*, 86. https://doi.org/10.3390/biomedicines9010086

Received: 8 December 2020
Accepted: 15 January 2021
Published: 17 January 2021

1. Introduction

Insulin-degrading enzyme (IDE; EC 3.4.24.56; a.k.a. insulin protease, insulinase, insulysin, insulin-glucagon protease, neutral thiol protease, metalloendoprotease, amyloid-degrading protease or peroxisomal protease) is a neutral Zn^{2+} metallo-endopeptidase that is ubiquitously expressed in insulin-responsive and non-responsive cells [1,2]. IDE belongs to a distinct superfamily of zinc-metalloproteases (clan M16) sometimes called "inverzincins" because they are characterized by a zinc-binding consensus sequence (HxxEH) that is inverted with respect to the sequence in most conventional metalloprotease (HExxH) [1,3,4]. IDE and its homologues represent an interesting example of convergent evolution; despite independent origins, IDE shares striking structural and functional similarity with conventional metalloproteases (clan M13 family; e.g., thermolysin and neprilysin) [3–5].

IDE and its homologs and paralogs are highly conserved and present in phylogenetically diverse organisms, ranging from viruses to humans [6], highlighting the fact that IDE is a multifunctional protein with proteolytic and non-proteolytic functions (Table 1; Table 2). Human IDE shares significant sequence similarity with orthologs in bacteria and yeast, for example, including the HxxEH zinc-binding motif. The protease has insulin-binding and -degrading activity in *Neurospora crassa*, *Acinetobacter calcoaceticus* and *Escherichia coli* (*E. coli* protease III or pitrilysin) [7–10]. IDE orthologs exhibit a periplasmic localization in *E. coli* [11] and *A. calcoaceticus* [9], whereas the ortholog in *N. crassa* is membrane bound, leading to a proposed function as a putative insulin receptor (IR) [7]. Of interest, the yeast IDE homologues Axl1p and Ste23 are incapable of degrading insulin despite possessing the conserved zinc-binding motif [12]. The proteolytic activity of Ste23 is required, however, for N-terminal cleavage of the pro-a-factor, the precursor of a pheromone (a-factor) involved in the mating response of haploid yeast cells [13]. Interestingly, rat IDE can promote the formation of mature a-factor in vivo, suggesting that the functional conservation of IDE, Axl1p and Ste23 may be not be bidirectionally conserved [12]. Humans and other mammals also express several IDE paralogs (Table 2), including N-arginine dibasic convertase (a.k.a., nardilysin), a cytosolic, secreted and membrane bound peptidase [14]. Of relevance to the present review, nardilysin regulates β-cell function and identity through the transcriptional factor MafA [15] and also prevents the development of diet-induced steatohepatitis and liver fibrogenesis by regulating chronic liver inflammation [16]. Nardilysin has several other roles unrelated to hepatic or pancreatic function, however, such as modulation of thermoregulation [17]. It also mediates cell migration by acting as a specific receptor for heparin-binding epidermal growth factor-like growth [18]. This protein provides a good example of how paralogs (rather than orthologs) can evolve to develop diverse functions.

Table 1. Orthologs of insulin-degrading enzyme (IDE).

UniProtKB Entry	Organism	Gene Name	Protein Name	Protein Length	E-Value *	Identity (%) *	Positives (%) *
Q5UPX9 (YL233_MIMIV)	*Acanthamoeba polyphaga mimivirus (APMV)*	*MIMI_L233*	Putative zinc protease L233	440	1.3×10^{-5}	20.9	40.4
P31828 (PQQL_ECOLI)	*Escherichia coli (strain K12)*	*pqqL*	Probable zinc protease PqqL	931	2.4×10^{-9}	24.6	41.5
P05458 (PTRA_ECOLI)	*Escherichia coli (strain K12)*	*ptrA*	Protease 3 (Pitrilysin)	962	7.1×10^{-6}	27.6	47.6
O22941 (IDE1_ARATH)	*Arabidopsis thaliana (Mouse-ear cress)*	*PXM16*	Insulin-degrading enzyme-like 1, peroxisomal	970	0.0	38.5	57.1
F4J3D9 (IDE2_ARATH)	*Arabidopsis thaliana (Mouse-ear cress)*	*At3g57470*	Insulin-degrading enzyme-like 2	966	0.0	39.4	57.9
Q06010 (STE23_YEAST)	*Saccharomyces cerevisiae (strain ATCC 204508/S288c) (Baker's yeast)*	*STE23*	A-factor-processing enzyme	1027	0.0	38.4	60.2
P40851 (AXL1_YEAST)	*Saccharomyces cerevisiae (strain ATCC 204508/S288c) (Baker's yeast)*	*AXL1*	Putative protease AXL1	1208	9.2×10^{-50}	21.5	41.5
P22817 (IDE_DROME)	*Drosophila melanogaster (Fruit fly)*	*Ide*	Insulin-degrading enzyme	990	0.0	46.6	67.3
Q9JHR7 (IDE_MOUSE)	*Mus musculus (House mouse)*	*Ide*	Insulin-degrading enzyme	1019	0.0	95.0	97.4
P35559 (IDE_RAT)	*Rattus norvegicus (Norwegian rat)*	*Ide*	Insulin-degrading enzyme	1019	0.0	95.5	97.6
Q24K02 (IDE_BOVIN)	*Bos taurus (Bovine)*	*Ide*	Insulin-degrading enzyme	1019	0.0	98.8	99.2
F7EFL5 (F7EFL5_MACMU)	*Macaca mulatta (Rhesus macaque)*	*Ide*	Insulin-degrading enzyme	1019	0.0	99.5	99.7
P14735 (IDE_HUMAN)	*Homo sapiens (Human)*	*Ide*	Insulin-degrading enzyme	1019	0.0	100	100

* BLAST results with P14735 (IDE_HUMAN) as query sequence.

Table 2. Human paralogs of IDE.

UniProtKB Entry	Organism	Gene Name	Protein Name	Protein Length	E-Value *	Identity (%) *	Positives (%) *
Q5JRX3 (PREP_HUMAN)	*Homo sapiens (Human)*	*PITRM1*	Presequence protease, mitochondrial	1037	7.6×10^{-1}	28.9	51.3
Q10713 (MPPA_HUMAN)	*Homo sapiens (Human)*	*PMPCA*	Mitochondrial-processing peptidase subunit alpha	525	1.7×10^{-1}	24.4	45.5
P31930 (QCR1_HUMAN)	*Homo sapiens (Human)*	*UQCRC1*	Cytochrome b-c1 complex subunit 1, mitochondrial	480	1.3×10^{-2}	21.7	40.1
O75439 (MPPB_HUMAN)	*Homo sapiens (Human)*	*PMPCB*	Mitochondrial-processing peptidase subunit beta	489	7.9×10^{-3}	22.8	39.6
O43847 (NRDC_HUMAN)	*Homo sapiens (Human)*	*NRDC*	Nardilysin	1151	8×10^{-174}	34.6	55.6
P14735 (IDE_HUMAN)	*Homo sapiens (Human)*	*Ide*	Insulin-degrading enzyme	1019	0.0	100	100

* BLAST results with P14735 (IDE_HUMAN) as query sequence.

1.1. The Discovery of IDE: An Historical Perspective

More than 70 years ago, Mirsky and Broh–Kahn characterized the existence of a proteolytic activity they named insulinase, which inactivated insulin in rat tissue extracts from liver, kidney and muscle in vitro [19]. The insulinase activity was a mixture of specific and non-specific proteases. Almost four decades of work, within several laboratories, would be conducted on this crude activity before Roth and colleagues eventually cloned the cDNA of *Ide*, revealing that the insulin-degrading activity within these extracts was attributable primarily if not exclusively to the action of just one protease, IDE [20,21]. Work on this extract within the early literature, however, resulted in several notable findings. For instance, Mirsky reported that fasting markedly reduced insulinase activity of rat liver extracts as compared with fasted rats subsequently fed a regular diet, suggesting a relationship between the insulin content of pancreas and hepatic insulinase activity [22]. However, refeeding of fasted rats with a high-carbohydrate diet resulted in a greater increase in insulinase than refeeding with a high-fat diet (HFD) [23]. Interestingly, Mirsky and colleagues were the first to report the existence of a substance that, in vitro, inhibited the insulinase activity in liver extracts, and they also demonstrated the effect of crude insulinase-inhibitor preparations on insulin degradation in vivo [24–26]. Identification of endogenous proteins that interact with and modulate IDE function has been only partially successful. For example, Brush and colleagues partially purified four competitive IDE inhibitors from human serum, whose molecular weights were in a range that includes insulin (Cohn fraction IV) [27]. Ogawa and colleagues purified an endogenous ~14-kDa protein from rat liver that in a competitive manner inhibited insulin binding and insulin degradation by IDE [28]. Other groups partially purified additional competitive [29] and noncompetitive [30] insulinase inhibitors from rat tissues, but their identities remain unknown. However, Saric and colleagues successfully identified ubiquitin as an IDE-interacting protein that inhibited the proteolytic activity of IDE in a reversible, ATP-independent manner [31]. Although the physiological relevance of the non-covalent, and energy-independent interaction between IDE and ubiquitin remains to be established, these findings may have alternative implications, such as the possibility that IDE interacts with ubiquitin-like modifiers. Finally, Mirsky described that the hypoglycemic effect of oral administration of sulfonylureas in diabetic patients was associated with non-competitive inhibition of insulinase in liver [32], leading to speculation that the mechanism of action of sulfonylureas was inhibition of insulin degradation [33].

The pioneering investigation of the insulinase led to the formulation of an innovative concept to explain the etiology of diabetes. At a time when the classical view of diabetes, proposed by von Mehring and Minkowski as well as Banting and Best [34,35], was a decrease in the production of insulin by pancreatic β-cells, Mirsky hypothesized that an increase in the rate of insulin degradation by extrapancreatic tissues, could explain the insulin insufficiency in some diabetic patients [36]. The role of IDE on the pathophysiology of diabetes has evolved over time, and it will be discussed in this review.

1.2. The Function of IDE as a Protease of Insulin

IDE was first characterized by its capacity to degrade insulin into several fragments in vitro, yielding major and minor products. The initial cleavage events occur at the middle of the insulin A and B chains, without specific amino acid requirements, suggesting that substrate recognition by IDE depends on tertiary structure rather than primary amino acid sequence [37–40]. Consistent with this, whereas IDE has a high affinity for insulin (K_m ~0.1 μM), proinsulin is a poor substrate that is hydrolyzed at very slow rates, acting as a competitive inhibitor [41,42]. Likewise, insulin-like growth factor I (IGF-I) and IGF-II are substrates of IDE, being IGF-II degraded more rapidly than IGF-I, but both acting as competitive inhibitors of IDE [42]. Another potent inhibitor of the insulin-degrading activity of IDE is the leader peptide of rat prethiolase B (P27 peptide), present in peroxisomes [43].

For decades, the standard assay for assessing insulin-protease activity was the use of purified or partially purified enzyme preparations of IDE (e.g., from erythrocytes) and [^{125}I]-labelled insulin. Two methods have been used to quantify the degradation of these radiolabeled peptides, the trichloroacetic acid (TCA) precipitation assay and high-performance liquid chromatography (HPLC), the former being the most sensitive [44,45]. Several other IDE activity assays have been developed, such as fluorogenic FRET-based peptide substrates derived from the sequence of bradykinin [46] or other IDE substrates [47]. However, IDE appears to process such short peptides markedly differently from intermediate-sized substrates [48]; for instance, the hydrolysis of such substrates is activated by ATP [49] and other nucleoside polyphosphates [50], certain small molecules [51] and several substrates [52], while these compounds have no effect or actually inhibit the degradation of more physiological substrates. To account for this intriguing substrate specificity of IDE, several substrate-specific degradation assays have been developed for different IDE substrates, including amyloid β-protein (Aβ) [53], glucagon [54] and amylin [55]. Unfortunately, similarly facile assays for insulin degradation have not yet been identified, so insulin degradation is now typically quantified by ELISA or HPLC.

Early studies of the kinetics of insulin degradation suggested that the endosomal apparatus is a physiological site of degradation of internalized insulin in hepatocytes [56–58]. Within the acidic endosomal lumen, the insulin-IR complex dissociates and insulin is degraded by endosomal proteases, whereas the IR is recycled to the plasma membrane of the hepatocyte [38]. Over the past several decades, it has been proposed that endosomal degradation of insulin is mediated by the action of three endosomal proteases: Cathepsin D [59], neutral Arg aminopeptidase [60] and IDE [38]. Cathepsin D has been shown to be responsible for the majority of the proteolytic degradation within endosomes, in general [59,61]. Neutral Arg aminopeptidase is an endosomal Arg convertase involved in the removal of Arg residues from internalized monoarginyl insulin prior endosomal acidification [60]. The action of this protease is highly selective towards [ArgA0]-human insulin peptide, a proinsulin intermediate containing an additional Arg at the amino-terminal insulin A-chain [60]. The role of IDE in endosomal proteolysis of internalized insulin, however, remains controversial, even though many of the primary sites of cleavage of internalized insulin are consistent with those produced by purified IDE [43,61,62]. At an early time, following insulin endocytosis, endosomal proteases account for major degradation products containing an intact A-chain, and cleavages in the B-chain at the Phe^{B24}-Phe^{B25}, Gly^{B23}-Phe^{B24}, Tyr^{B16}-Leu^{B17} and Ala^{B14}-Leu^{B15} peptide bonds [56–58] (Figure 1). At a later stage, insulin is degraded to its constituent amino acids within endosomes and/or lysosomes [38,58].

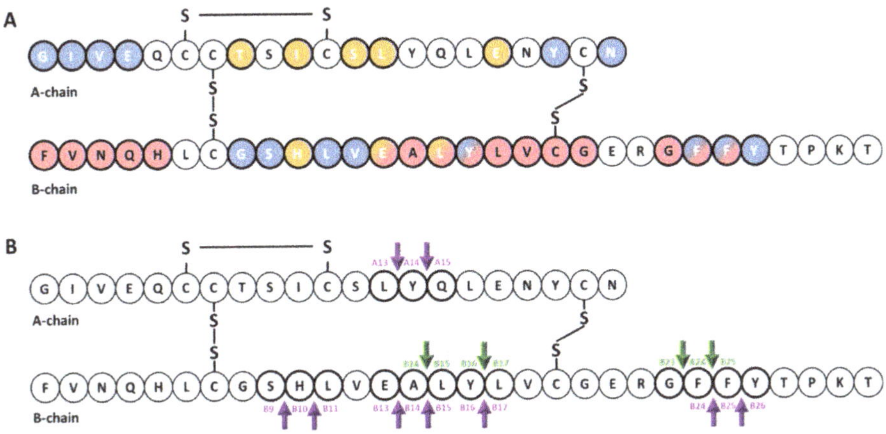

Figure 1. Cartoon illustrating the primary structure and cleavage products of human insulin. (**A**) primary structure of insulin showing amino acids that interact with IDE (red color) [63–65] and with site 1 (blue color) and site 2 (gold color) of the IR [66–69]. (**B**) cleavage products generated by endosomal proteases. At an early time, following insulin endocytosis, endosomal proteases account for major degradation products containing an intact A-chain, and cleavages in the B-chain (green arrows). Purple arrows indicate IDE cleavage sites effected by IDE in vitro [39].

The role of IDE as a protease of insulin has been demonstrated over time in cell extracts and intact cells. Overexpression of human IDE (h*Ide*) in Chinese hamster ovary (CHO) cells [70] and monkey kidney COS cells [71] enhanced the extracellular degradation of exogenously applied insulin. Likewise, cell extracts from NIH3T3 cells overexpressing *Drosphila* IDE (d*Ide*) exhibited enhanced insulin-degrading activity [72]. Based on the use of lysosomotropic agents (which prevent acidification of intracellular lysosomes and endosomes), Kuo and colleagues showed that insulin degradation was affected by intact cells via a d*Ide*-mediated intracellular pathway, independently of the lysosome [71]. In addition, this same team created a stably transfected Ltk^{-} cell line with dexamethasone-inducible overexpression of h*Ide*. IDE expression upon glucocorticoid induction resulted in increased IDE insulin degradation in both cell lysates and intact cells [73].

Consistent with these findings, non-specific inhibitors of IDE decreased intracellular insulin degradation in intact HepG2 cells [74], rat L6 myoblasts [75], and mouse BC3H1 muscle cells [76]. Similarly, monoclonal antibodies against IDE almost completely abolished the insulin-degrading activity of IDE from erythrocytes, and microinjection of these antibodies into HepG2 cells reduced intracellular [^{125}I]-insulin degradation by ~50% [77]. Furthermore, insulin degradation was diminished by ~50% in HepG2 transfected cells with siRNA against h*Ide*, in parallel with a reduction by more than 50% in IDE protein and mRNA levels [78]. Similarly, CRISPR/Cas9 targeting of *Ide* in CHO cells abolished insulin degradation [79]. Finally, cytosolic and membrane fractions of liver from mice with homozygous deletion of the *Ide* gene (IDE-KO mouse) showed impaired [^{125}I]-insulin degradation [80]. Taken together, the above-mentioned studies indicated that increasing IDE activity increases cellular insulin degradation, and conversely decreasing its activity reduces insulin degradation.

Despite the large body of evidence supporting a role for IDE in insulin degradation within cultured cells, IDE resides primarily in the cytosol and does not have a signal peptide, so precisely where IDE interacts with insulin or other substrates remains an unresolved question. On the one hand, Zhao and colleagues found that IDE is exported via an unconventional protein secretion pathway [81]. On the other hand, a recent study by Song and colleagues [82] found only very low levels of IDE secretion from HEK293 and BV2 cells, in quantities never exceeding the non-specific release of other cytosolic proteins such as lactate dehydrogenase.

The involvement of IDE in insulin degradation in vivo is similarly controversial. For instance, on the one hand, Farris and colleagues [80], and Abdul–Hay and colleagues [83] found that IDE-KO mice exhibited significant increases in plasma insulin levels. On the other hand, Miller and colleagues [84] and Steneberg and colleagues [85] reported that plasma insulin levels in IDE-KO mice were unchanged relative to wildtype controls. Moreover, we have shown that mice with liver-specific deletion of *Ide* (L-IDE-KO mice) display normal insulin levels when fed a regular diet [86], but elevated levels when fed a HFD [87], as is discussed in greater detail below.

1.3. Other Proteolytic Functions of IDE

IDE can degrade several other substrates with lower affinity than insulin, including glucagon [88], somatostatin [89], amylin [90], Aβ [91], amyloid precursor protein intracellular domain [92], amyloid Bri and amyloid Dan [93], atrial natriuretic peptide [94], bradykinin and kallidin [95,96], calcitonin and β-endorphin [5], growth hormone-release factor [97], transforming growth factor-α [98], oxidized hemoglobin [99], cytochrome c [100], chemokine ligand (CCL)3 and CCL4 [101] and HIV-p6 protein [102].

A role for IDE in the processing of insulin epitopes for helper T cells has been reported by Semple and colleagues [103,104]. IDE degrades human insulin into peptides that are presented by murine TA3 B-cell antigen-presenting cells to HI/I-Ad-reactive T cells [103]. Of note, IDE is necessary but not sufficient for the recognition of insulin by T cells [103,104]. More controversial is the role of IDE in the proteasome-independent processing of peptides. As shown by Parmentier and colleagues, MAGE-A3, a cytosolic human tumor protein, is degraded by IDE, leading to different sets of antigenic peptides presented by major histocompatibility complex (MHC) class I molecules to cytotoxic T lymphocytes (CTLs) [105]. Immunodepletion of IDE abolished the capacity to produce the antigenic peptide (MAGE-A3$_{168–176}$), whereas expression of recombinant human IDE was able to produce the antigenic peptide [105]. In addition, *Ide* RNAi-treated cells reduced the ability of CTLs to recognize tumor cells [105]. On the other hand, Culina and colleagues, using a number of MHC-I class molecules and a loss-of-IDE-function approach in human cell lines and two different mouse strains (IDE-KO mouse and IDE-KO mouse back-crossed to the non-obese diabetic (NOD) strain), concluded that IDE does not play a general major role in peptide loading to MHC-I molecules [106].

Other non-insulin-related proteolytic functions of IDE include degrading cleaved leader peptides of peroxisomal proteins targeted by the type II motif [43] and, possibly, cleaved mitochondrial targeting sequences, in this case by a mitochondrial form of IDE generated by alternative translation initiation [107]. IDE has also been implicated in the formation and/or degradation of "cryptic" peptides (i.e., hidden peptides derived from proteolytic processing of a substrate with different biochemical functions of parent protein), which is the case for IDE-mediated regulation of cryptic peptides from the neuropeptide FF (NPFF) precursor (pro-NPFF$_A$) [108] as well as Aβ [109]. In addition, IDE has been proposed to mediate the degradation of nociceptin/orphanin 1–16 (OFQ/N), a class of neuropeptides involved in pain transmission. Interestingly, the main hydrolytic peptides of OFQ/N produced by IDE, but not the neuropeptide itself, exhibited inhibitory activity towards IDE-mediated degradation of insulin [110].

An example of multiple catalytic and non-catalytic functions of IDE is its role in binding and degrading viral proteins. For instance, Li and colleagues showed that IDE interacts with the glycoprotein E (gE) from varicella zoster virus (VZV) and proposed that IDE is the cellular receptor for the virus [111]. This group subsequently demonstrated that binding of IDE to the N-terminal domain of gE produced a conformational change, increasing its susceptibility to proteolysis [112]. Berarducci and colleagues found that gE/IDE interaction contributed to skin virulence in vivo [113]. In contrast, the gE/IDE interaction was not necessary for VZV infection of T cells in vivo [113]. On the other hand, IDE is necessary and sufficient for degradation of the mature p6 protein of the human immunodeficiency virus 1 (HIV-1) [102,114]. Of note, p6 is degraded 100-fold more efficiently than insulin [102]. Virus replication was reduced by exogenous insulin or pharmacological inactivation of IDE with the inhibitor 6bK [102,115].

1.4. Non-Proteolytic Functions of IDE

IDE has been reported to directly interact with androgen and glucocorticoid receptors, enhancing specific DNA binding of both receptors [116]. Non-competitive inhibition of IDE's catalytic activity did not block the binding of the androgen receptor to IDE, but competitive inhibition of IDE blocked its binding, suggesting that IDE-binding sites for the receptor and insulin are identical or overlapping [116]. Interestingly, dexamethasone, a synthetic glucocorticoid, significantly reduced insulin binding to IDE without affecting expression levels of the protease in rat hepatoma cells, most likely by inducing a conformational change or blocking insulin-binding sites [117]. Conversely, the steroid's effect was blocked by insulin [117]. The interaction of IDE with androgen and glucocorticoid receptors may be of relevance for steroid hormone action and metabolism, the crosstalk between insulin and steroid hormones, and the pathophysiology of glucocorticoid-mediated insulin resistance [118].

Intriguingly, IDE co-localizes with the ~50-amino acid cytoplasmic tail of the scavenger receptor type A (SR-A), an important domain for SR-A function, in mouse macrophages [119]. The biological significance of this interaction is uncertain, because IDE deficiency in bone-marrow derived macrophages (BMDMs) did not alter protein levels of SR-A or its ability to uptake low-density lipoprotein (LDL)-cholesterol, albeit its deficiency in these cells was associated with higher levels of intermediate density lipoproteins, LDL-cholesterol and accelerated atherogenesis in LDL receptor knockout (*Ldlr*$^{-/-}$) mice [119]. Considering that SR-A participates in multiple cellular processes, including regulation of inflammatory cytokine synthesis through its interaction with TLR4, it is tempting to hypothesize that IDE deficiency in macrophages may cause an inflammatory milieu surrounding the arterial wall, thus contributing to the pathophysiology of atherogenesis.

IDE co-immunoprecipitates with SIRT4, a protein with no histone deacetylase activity but with associated ADP-ribosyl-transferase activity that resides in the mitochondrial matrix [120]. SIRT4 expression is detected in several mouse tissues including liver [121,122], and human pancreatic β-cells [120,122]. SIRT4 depletion in INS1 and MIN6 cells markedly increased insulin secretion without altering basal secretion and intracellular insulin content [120,122]. Conversely, SIRT4 overexpression in INS1 cells suppressed glucose-induced insulin secretion. Of note, insulin secretion stimulated with the secretagogue KCl, which bypasses mitochondrial activation, remained unaltered in SIRT4-depleted INS-1E cells [120]. Once more, the biological significance of this interaction remains undeciphered, because there are no reports on the ADP-ribosylation of IDE, and the mitochondrial function of IDE is not clarified. However, IDE depletion in INS1-E cells using siRNA-IDE and shRNA-IDE significantly decreases glucose-stimulated insulin secretion [123]. The contributions of the cytosolic and mitochondrial forms of IDE in regulating insulin secretion, and the association of the later with SIRT4, warrants further research.

IDE has been identified as an interacting partner of intermediate filaments, one of the three major cytoskeletal components that serve as a scaffold for signaling molecules, modulating their distribution and activity [124]. Specifically, IDE has been shown to interact with disassembled and soluble vimentin/nestin complexes during mitosis [125]. Vimentin plays a dominant role in targeting IDE to the complex and binds to IDE with higher affinity than nestin. Phosphorylation of vimentin is not required for its binding to IDE, but the interaction is enhanced by vimentin phosphorylation at Ser-55. On the other hand, binding of IDE to nestin promotes the disassembly of vimentin intermediate filaments, most likely by rendering the phosphorylated vimentin more accessible for IDE. The binding of IDE to nestin is phosphorylation independent. Interestingly, the binding of nestin or phosphorylated vimentin suppressed by ~2-fold the insulin-degrading activity of IDE but increased its proteolytic activity toward bradykinin [125]. These data suggest that IDE may be involved in regulating the turnover and/or subcellular localization of cytosolic proteins and peptides. In this context, it has to be noted that integrins are a major family of cell adhesion receptors that mediate attachment of cells to the extracellular matrix. Recruitment of integrins and other proteins forms multi-protein complexes on the cytoplasmic face of the membrane named focal adhesions, which allow the anchoring of the actin cytoskeleton to the plasma membrane, providing a linkage between the extracellular environment and the cytoplasm. Integrins do not have intrinsic kinase activity and signaling depends upon the recruitment and activation of focal adhesion kinase (FaK), a cytoplasmic protein tyrosine kinase [126,127]. Significantly, Liu and colleagues identified IDE as a binding partner that interacts with C-terminal domain of FaK [128]. The relevance of the interaction between IDE and FaK in regulating recruitment of cytoskeletal proteins and the assembly of focal adhesions, a process that is important for cell migration, survival and proliferation, remains to be elucidated.

IDE, in a non-proteolytic manner, binds to α-synuclein oligomers leading to the formation of stable and irreversible complexes, precluding amyloid formation [129,130]. α-synuclein is a synaptic signaling protein with three domains—the N-terminus, which interacts with membranes, the amyloidogenic domain and the C-terminus, which is involved in the pathogenesis of Parkinson's disease [131]. Interestingly, the catalytic activity of IDE on a bradykinin-based fluorogenic substrate was increased in the presence of α-synuclein [129]. The interaction between both proteins appears to require electrostatic attraction involving the exosite region of IDE, which is positively charged, and the C-terminus of α-synuclein, which contains many negatively charged amino acids [130]. The role of IDE in the turnover of amyloidogenic proteins and the non-proteolytic prevention of toxic amyloid formation—in the case of α-synuclein via a so-called "dead-end chaperone function"—appears to be important for pancreatic β-cells function. In this connection, Steneberg and colleagues showed that genetic deletion of *Ide* in pancreatic β-cells led to the formation α-synuclein oligomers and fibril accumulation, which was associated with impaired insulin secretion and reduced granule turnover, possibly by disruption of the microtubule network [85].

Sorting nexin 5 (SNX5) is a member of the sorting nexin family that regulates intracellular trafficking and is abundantly expressed in kidney [132,133]. Its expression is reduced in Zucker rats, a model of obesity, hyperinsulinemia and insulin resistance [134]. The *Snx5* gene is located on chromosome 20p, a susceptibility quantitative trait locus for high fasting plasma insulin levels and insulin resistance [135]. Li and colleagues have shown that IDE colocalizes with SNX5 in the brush border membrane of proximal tubules and the luminal side of distal convoluted tubules of human and rat kidneys, in addition to the plasma membrane and perinuclear area of human renal proximal tubule cells (hRPTCs) [134]. Furthermore, exposure of hRPTCs to insulin increased colocalization and co-immunoprecipitation of IDE and SNX5. Interestingly, SNX5-depleted hRPTCs exhibit reduced IDE activity and protein levels, in parallel with decreased expression of the IR and downstream insulin signaling [134,136]. Similarly, renal subcapsular infusion of SNX5-specific siRNA decreased IDE mRNA and protein expression in kidneys of mice [134]. These studies underline the potential significance of renal IDE in the regulation of circulating insulin levels as well as insulin sensitivity in kidneys.

The human retinoblastoma (RB) protein acts as a tumor suppressor that negatively regulates cell cycle progression at the G1/S transition through its interaction with the E2F family of transcription factors [137]. IDE co-purifies with RB on proteasomal preparations of breast cancer and hepatoma cells [138]. Similarly, IDE co-immunoprecipitates with the tumor suppressor phosphatase and tensin homolog deleted on chromosome 10 (PTEN) [139]. IDE accelerates PTEN degradation by SIRT4 in response to nutritional starvation stresses [139]. Although the underlying molecular mechanisms have not been fully elucidated, these findings support a role for IDE in insulin-driven oncogenesis. Likewise, Tundo and colleagues have hypothesized that IDE, in a heat shock protein-like fashion, may be implicated in cell growth regulation and cancer progression [140]. In normal cells (human fibroblasts cell line and human peripheral blood lymphocytes) and malignant cells (human neuroblastoma cell line (SHSY5Y) and human lymphoblastic-like cells line (Jurkat cells)) exposed to heat shock, H_2O_2 and serum starvation, IDE is markedly up-regulated at both protein and mRNA levels. Additionally, delivery of IDE siRNA to SHSY5Y cells led to extensive apoptotic cell death; and administration of ATRA (a vitamin A precursor used in the clinical treatment of neuroblastoma) significantly decreased intracellular IDE content [140].

1.5. Molecular and Biochemical Characteristics of IDE

IDE is synthesized as a single polypeptide with a molecular weight of ~110 kDa by a gene located on human chromosome 10 q23–q25, and mouse chromosome 19, respectively [20,70]. *Ide* coding mutations have been associated with the development of T2DM in the Goto–Kakizaki rat model [141]. Fakhrai–Rad and colleagues identified two missense mutations (H18R and A890V) in IDE that decrease the ability to degrade insulin by 31% in transfected cells. They reported a synergistic effect of the two mutations on insulin degradation, which is somewhat puzzling given that H18R is present within the mitochondrial targeting sequence of IDE produced by alternative translation initiation [107]. Farris and colleagues, by contrast, found that recombinant IDE containing the A980V mutation alone exhibited reduced catalytic efficiency of both insulin and Aβ degradation, suggesting that this mutation is most relevant to proteolytic function [142]. Interestingly, no effect on insulin degradation was seen in cell lysates of *Ide*-transfected COS cells, suggesting that the effect may be dependent upon receptor-mediated internalization of the hormone [141,142].

IDE assembles as a stable homodimer, although it can exist as an equilibrium of monomers, dimers and tetramers [52]. Each monomer is comprised four homologous domains (named 1–4). The first two domains constitute the N-terminal portion (IDE-N), and the last two the C-terminal portion (IDE-C). IDE-N and IDE-C are joined by an extended loop of 28 amino acids. In the human IDE dimer, the interface between the two monomers is formed by 18 residues of domains 3 and 4 (IDE-C) [143]. Interestingly, the crystal structure of rat IDE, obtained by Hersh and colleagues [144], revealed a different homodimer interface than human IDE. In rat IDE, deletion of the last 18 amino acids abolished homodimer formation while simultaneously eliminating allosteric effects reflective of inter-subunit cooperativity [144]. The active site of IDE consists of a catalytic tetramer, $HxxEHx_{76}E$, located inside domain 1, in which two histidine residues (H108 and H112) and a glutamate (E189) coordinate the binding of the Zn^{2+} ion and a second glutamate (E111) plays an essential role in catalysis. Glutamate E111 activates a catalytic water molecule for the nucleophilic attack that mediates peptide hydrolysis [63,145]. Although the catalytic site is entirely inside IDE-N domain, the IDE-C is necessary for correct substrate recognition [146]. Site-directed mutagenesis revealed that mutating IDE H108 (i.e., H108L and H108Q) abolished catalytic activity of the enzyme, but not the ability to bind insulin. Similarly, mutation E111Q abolished proteolytic activity [145,147].

Substrates for IDE are almost exclusively intermediate-size (~20–40 amino acids) peptide substrates, with rare exceptions, such as oxidized hemoglobin [99]. This size preference is the result of the overall structure of IDE, which resembles a clamshell, with two bowl-shaped domains (IDE-C and IDE-N) facing one another, connected by a hinge, and together forming a ~13,000-Å^3 internal chamber (Figure 2). These domains can pivot on the hinge, thus adopting "open" and "closed" conformations. Transition to the open conformation is required for entry of substrates and exit of proteolytic products, and there is strong evidence that transition to the closed conformation is a requirement for proteolytic processing. Consistent with this, site-directed mutagenesis revealed that the complete active site of IDE is in fact bipartite, consisting of residues within both IDE-N and IDE-C [148], a conclusion that is confirmed by numerous crystal structures [149]. Due to the placement of the bipartite active site within the chamber of the closed protease, substrates must be small enough to fit completely within this chamber to be processed. To facilitate binding and subsequent cleavages at the catalytic site, larger substrates interact with an exosite within domain 2 located ~30 Å away from the active-site Zn^{2+}, which anchors the N-terminus of several substrates [50]. Because of the unusual requirement that substrates fit within the internal chamber, the substrate selectivity of IDE is based more on the tertiary structure of substrates than their primary amino acid sequence. IDE shows some preference for cleavage at basic or bulky hydrophobic residues at the P1′site of the target protein [62], but this subsite specificity is not strict, and IDE commonly cleaves at vicinal peptide bonds within substrates [53]. Of note, substrates containing positively charged residues at their C-terminus are poor IDE substrates [63]. Thus glucagon, which lack of positively charged amino acids at the C-terminal is an IDE substrate, but not glucagon-like peptide [63].

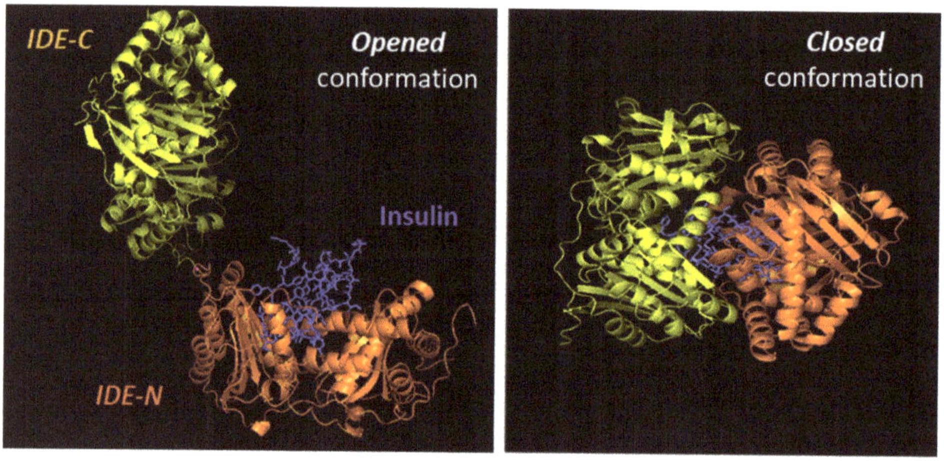

Figure 2. Cartoon illustrating the binding of insulin by IDE. IDE is in an equilibrium between "opened" and "closed" conformational states. In the absence of a substrate (e.g., insulin), IDE is preferentially in the closed conformation. IDE must adopt the open conformation for substrates to enter the internal chamber, whereas the protease must assume the closed conformation for proteolysis to occur. Release of the cleavage products requires a return to the open conformation.

The requirement for a transition between the "closed" and "open" conformations has an additional implication for the activity of IDE. There is extensive hydrogen bonding between the two halves of IDE, creating a "latch" that tends to maintain the protease in the closed conformation [63,150,151]. Consistent with this idea, most crystal structures of IDE, whether empty or occupied by substrate, show the protease in the closed conformation [63]. Notably, mutation of some of the residues mediating this interaction has been shown to activate the protease by as much as 15-fold [63]. It is estimated that in the absence of other factors, ~99% of IDE molecules are normally in the closed conformation (M.L. unpublished observations), suggesting that a significant of latent IDE activity could be untapped, for example, by compounds that disrupt this "latch" [51].

Interestingly, somatostatin, a hormone produced and secreted by the hypothalamus and in the pancreas by δ-cells, that inhibits glucose-stimulated insulin secretion [152], in addition to being a substrate of IDE, also regulates its function. Somatostatin binds to two additional exosites named "somatostatin-binding exosites" which play different roles according to the size of the substrates and its binding mode to the IDE catalytic cleft [95].

As mentioned above, a mitochondrial isoform of IDE was identified by Leissring and colleagues, which is formed by alternative translation initiation [107]. The open reading frames of human, rat, and mouse *Ide* cDNAs contain two in-frame translational codons encoding proteins beginning either at the first (Met^1-IDE) or the 42nd amino acid (Met^{42}-IDE). Met^{42}-IDE (the shorter isoform) is the predominant isoform expressed in tissues and culture cells [107], because the nucleotide sequence surrounding the second initiation codon contains a better Kozak consensus sequence for initiation of translation. Although the Met^1-IDE isoform is predicted to be less efficiently translated, it could nevertheless account for a significant fraction of total cellular IDE [107]. Currently, it is uncertain whether the mitochondrial isoform plays a major role in human disease.

In addition to the two possible translation initiation sites (Met^1-IDE and Met^{42}-IDE), Farris and colleagues identified a novel splice isoform in which exon 15a is replaced by a novel exon, 15b [153]. The resultant variant is widely expressed and present in both cytosol and mitochondria. The 15b-IDE isoform can exist as homodimer or as heterodimer with the 15a isoform. The catalytic efficiency of the 15b-IDE isoform is significantly lower than the 15a-IDE isoform [153].

IDE expression is regulated during cell differentiation and growth. During rat development (6–7 days of age) to adulthood, *Ide* mRNA levels increased in brain, testis and tongue with a concomitant decreased expression in muscle and skin but remained unchanged in other tissues. In the adult rat, *Ide* mRNA is higher in testis, tongue and brain, and lower in spleen, lung, thymus and uterus [2,154]. Interestingly, IDE activity is affected by aging. The highest IDE activity is observed in muscle, liver and kidneys of 4-week-old rats. The IDE activity in muscle and liver at 7 weeks of age is lower than at 4 weeks, with similar activity in kidney. The lowest activity of IDE was observed in muscle, liver and kidneys of 1-year-old rats [155].

1.6. Subcellular Localization of IDE

The subcellular localization of IDE is mainly cytosolic [37,156–158], but it has been reported to be present in several other subcellular compartments, including endosomes [58,159,160], peroxisomes [43], mitochondria [107], plasma membrane [161–165], endoplasmic reticulum [166], exosomes [167], the extracellular space [166] and even in human cerebrospinal fluid [168]. If IDE is primarily cytosolic, its role as an insulin protease seems to be called into question. Two main pathways for insulin internalization have been described. At physiological concentrations, insulin is internalized through an IR-mediated process (see reference [1] for a comprehensive review of IDE's role on insulin uptake and clearance), whereas at higher concentrations a non-receptor mediated uptake internalizes insulin to endosomes [169]. In both pathways, insulin is internalized in endosomes, which begs an important question: How can insulin gain access to the cytosol to be degraded by IDE? Several studies proposed that internalized insulin is probably released to the cytosol by endosomes [170]. The mechanism by which insulin is transported through the membrane of endosomes and enter the cytosol is not well understood, but it has been proposed a two-step process involving acidification of the endosome and unfolding of insulin molecules, which help it pass through the membrane [170]. Modulation of IDE activity may have a significant impact in the accumulation of cytosolic and nuclear insulin or insulin-bound cytoplasmatic proteins [171–174].

An alternative locus for the interaction between IDE and insulin (and other substrates) is the extracellular space. As mentioned, IDE does not have a signal peptide and is not exported via the classical secretion pathway [81]. Many reports indicate that IDE is secreted in significant quantities from various cell types (e.g., [161,166]), but a recent analysis [82] suggests that IDE release from cultured cells might be non-specific. There is a great need for additional research on this topic, as it is of central significance to the functional role of IDE in regulating the levels of insulin and other IDE substrates.

1.7. Transcriptional and Posttranscriptional Regulation of IDE

Although the role of IDE in the regulation of hepatic insulin signaling and glucose homeostasis has been investigated [86,87], the physiological regulation of its expression and activity in hepatocytes remains poorly understood. In human hepatocellular carcinoma HepG2 cells grown in normal glucose medium, exposure to insulin for 24-h did not regulate *Ide* mRNA or protein levels [175]. Likewise, in the presence of high glucose levels, insulin increased expression of *Ide* mRNA, but without changes in levels of IDE [175]. However, insulin increased hepatic IDE activity, but this insulin-mediated effect was abolished in the presence of high glucose levels [175]. The underlying mechanism(s) by which 24-h exposure to insulin regulates IDE activity remains to be deciphered. As mentioned above, human *Ide* mRNA undergoes alternative splicing in exon 15 [153]. Pivovarova and colleagues showed that the relative proportion of the more proteolytically active 15a splice isoform was increased after insulin treatment, independently of glucose levels [175].

Insulin-mediated regulation of IDE has also been investigated in mouse primary hippocampal neurons. Contrary to hepatocytes, exogenous insulin application upregulates IDE protein levels, and this insulin-mediated effect was abolished by inhibition of the insulin-signaling component phosphoinositide 3-kinase (PI3K) [176]. These findings suggest that, within certain cell types, there is a negative feedback mechanism whereby insulin-mediated activation of IR upregulates IDE to prevent chronic activation of the pathway in the presence of high insulin levels.

The effects of glucagon on IDE function in hepatocytes was investigated by Wei and colleagues. In a time-dependent manner, glucagon (100 ng/mL) upregulated IDE protein levels in Hepa 1c1c7 cells. A similar pattern was observed after preincubation of hepatic cells with forskolin (10 μM), an activator of protein kinase A (PKA), suggesting that the glucagon-mediated regulation of IDE proceeds via a cAMP/PKA-dependent pathway [177]. The physiological and pathophysiological relevance of these findings awaits further validation in vivo.

Lin and colleagues demonstrated that Fas-associated protein with death domain (FADD), a classical adaptor in the Fas-FasL signaling pathway, which is phosphorylated in response to members of the tumor necrosis factor receptor family, regulates the expression of IDE at the transcriptional level, without affecting the stability of *Ide* mRNA in HepG2 cells [178]. FADD knockdown in HepG2 cells by siRNA resulted in downregulation of both mRNA and protein levels of IDE without change in IDE mRNA stability. Similar effects on IDE mRNA and protein levels were observed in the liver of mice overexpressing FADD-D (a mimic of constitutively phosphorylated FADD) as well as in primary hepatocytes cultured from these mice, effects that were attributable to reduced stability of IDE protein [178]. Interestingly, in primary hepatocytes from FADD-D mice, nuclear translocation of the transcription factor forkhead box O1 (FoxO1) is enhanced, and the transcriptional activity of the IDE promoter in response to FADD knockdown in HEK293T cells was decreased. Furthermore, the transcriptional activity of the *Ide* promoter was reduced by expressing FoxO1 in HEK293T cells [178]. Altogether, these results point out that FADD phosphorylation may reduce the expression of IDE by promoting the nuclear translocation of FoxO1. The detailed regulatory mechanism by which FADD phosphorylation regulates transcriptional activity of the *Ide* promoter in hepatocytes requires further experimental confirmation. In addition, these findings open an avenue to explore whether the insulin signaling pathway through FoxO1 regulates IDE levels in hepatocytes.

The cannabinoid receptor 1 (CB1) is a seven-transmembrane G protein-coupled receptor present in liver, and its activation by endocannabinoids stimulates lipogenic genes in hepatocytes, leading to increased fatty acid synthesis [179]. Pancellular genetic deletion of CB1 in mice resulted in resistance to diet-induced obesity, and liver-specific deletion in mice fed a HFD showed lower insulin resistance, hyperglycemia and steatosis [179]. In HepG2 cells, the endocannabinoid anandamide, a metabolite of the non-oxidative metabolism of arachidonic acid that is a partial agonist of CB1, causes down-regulation of IDE in a time-dependent manner, in parallel with Ser307 phosphorylation of insulin receptor substrate 1 (IRS1) [180]. Likewise, acute treatment with anandamide or feeding with a HFD reduced hepatic IDE levels, in parallel with insulin resistance in mice. The CB1-mediated regulation of IDE was further corroborated by the finding that hepatic IDE expression is downregulated in mice that overexpress CB1 in hepatocytes (htgCB1 mice) but not in CB1 knockout mice [180]. Of note, htgCB1 mice displayed down-regulation of IDE and its proteolytic activity, but unaltered levels of phosphorylated carcinoembryonic antigen-related cell adhesion molecule 1 (CEACAM1), in parallel with hepatic insulin resistance, lower insulin clearance and moderate hyperinsulinemia [180].

The impact of inflammation mediators on IDE function has also been examined in hepatocytes and pancreatic cells. Interleukin-6 (IL-6) is a pro-inflammatory cytokine that mediates inflammation associated with insulin resistance in the liver and other tissues [181]. In HepG2 cells, exposure to IL-6 increases IDE protein levels [182]. Conversely, in livers of IL-6 knockout mice, which develop glucose intolerance without a change in insulin sensitivity, hepatic insulin clearance is reduced, which is associated with reductions in the levels of IDE mRNA, protein and activity [182]. In addition, IL-6 knockout mice exhibited diminished C-peptide secretion after administration of an intraperitoneal (IP) glucose bolus, and reduced glucose-stimulated insulin secretion in isolated pancreatic islets, leading to lower fasting plasma insulin levels [182]. In the opposite direction of IL-6's effect on hepatic IDE function, exposure to tumor necrosis factor α (TNFα) decreased IDE mRNA and protein levels in Hepa 1c1c7 cells [177]. Thus, additional work is warranted, but these findings suggest IDE may play a role in the mechanisms linking inflammation to the regulation of pancreatic insulin secretion and hepatic insulin resistance.

In addition to the aforementioned, other modulators can regulate IDE protein and activity levels. Thus, ATP and other nucleoside polyphosphates, as well as polyphosphate alone, induce dose-dependent allosteric inhibition of insulin degradation and, concomitantly, activation of short fluorogenic peptide substrates at physiologically relevant concentrations (1–5 mM) [50,183]. Interestingly, the activating effect is stronger for nucleoside triphosphates than nucleoside di- or monophosphates, and is attenuated in the presence of Mg^{2+} [50,184]. Polyphosphate binding has been shown to occur at a specific region within IDE, known as the polyanion-binding domain [144], and appears to mediate activation of the protease towards short substrates by facilitating the transition from the closed state to the open conformation [49,50]. Camberos and colleagues reported that IDE has ATPase activity [185], but this was not confirmed by other investigators (M. Leissring, unpublished observations) and a molecular mechanism for this functionality is not evident from the crystal structures of IDE [63].

Acidic pH also affects the ability of IDE to bind insulin and alter its degradation by inducing dissociation of the oligomerization state into monomeric units. Since IDE is most active at neutral and basic pH, suggest that cellular acidosis may regulate insulin signaling and degradation [186]. Further studies are necessary to understand the impact of clinically relevant diabetic ketoacidosis, and starvation ketoacidosis stimulated by the combination of low insulin and high glucagon on IDE function.

Nitric oxide (NO) production is known to play an important role in permissive regulation of glucose-stimulated insulin secretion, and hepatic insulin resistance [187–189]. In this connection, it is interesting to note that S-nitrosylation of IDE, mediated by S-nitrosoglutathione, a potent physiologically relevant NO donor source in cells, inhibited the proteolytic activity of IDE [96]. Notably, NO donors exert this effect in a non-competitive manner, without affecting insulin binding to IDE or the insulin degradation products it produces [190]. Similarly, Cordes and colleagues showed that NO inhibited the degrading activity of IDE in rat liver homogenates in a dose-dependent manner [191]. Somewhat controversially, Natali and colleagues have postulated that the increased insulin clearance evoked by systemic blockage of endogenous NO synthesis in humans may be accounted for by effects on IDE function in liver [192]; however, the physiological and pathophysiological relevance of this proposed mechanism of IDE regulation on hepatic insulin action and signaling awaits further confirmation.

IDE is vulnerable to oxidative damage in multiple ways. For instance, IDE has been shown to be covalently modified and consequently inactivated by 4-hydroxynonenal (HNE), an oxidative byproduct of lipid metabolism [193,194]. Of note, in the brains of mice, HNE-modified IDE accrues in an age-dependent manner, being particularly abundant in brain regions affecting in Alzheimer's disease [193]. HNE, H_2O_2 and other oxidizing agents act, not only, to inhibit IDE proteolytic activity, but also promote the proteolysis of IDE itself by other proteases [194]. Conversely, dietary vitamin E supplementation, a classical lipophilic antioxidant that protect against free radicals [195], increased IDE mRNA levels in livers of rats, which was associated with improved glucose tolerance and insulin sensitivity [196].

The effects of caloric restriction and exercise on IDE function have been examined in mice, as well. Mice fed a low-protein diet for 14 weeks showed improved glucose tolerance and hepatic insulin sensitivity, in parallel with reduced insulin secretion and clearance, which was associated with ~50% reduction in IDE protein levels in the liver [197]. Likewise, caloric restriction (food restriction of 40%) for 21 days resulted in improved glucose tolerance and insulin sensitivity in rats, which was associated with lower hepatic IDE protein levels [196]. Mice subjected to a single bout of exercise on treadmill for 3-h showed reduced glycemia and insulin sensitivity, in parallel with higher insulin secretion in isolated islets and insulin clearance, which was associated with increased expression of IDE in the liver [198].

Exposure to increasing amounts of free fatty acids (FFA) released from adipose tissue promotes the development of hepatic insulin resistance and impaired glucose-stimulated insulin secretion in pancreatic β-cells [199,200], and some reports suggest that IDE may be either affected by or involved in this phenomenon. Early reports showed that FFA reduced leakage of IDE from isolated rat hepatocytes and inhibited the proteolytic activity of IDE released from adipocytes [201,202]. Svedberg and colleagues investigated the effect of different fatty acids on insulin binding, degradation and action in isolated rat hepatocytes, finding that different fatty acids rapidly decreased insulin binding and degradation in isolated hepatocytes [203]. Furthermore, the effect of FFA was specifically on the rate of insulin receptor internalization and/or recycling [203]. Of note, Hamel and colleagues identified a fatty acid-binding motif within IDE and subsequently examined the effect of FFA and their acyl-coenzyme A thioesters on IDE partially purified from rat livers [204]. They observed that both saturated and unsaturated long-chain FFA, and the corresponding acyl-coenzyme A thioesters, inhibited insulin degradation in a non-competitive manner, but did not inhibit binding of the hormone to IDE [204]. In addition to the effect of FFA on IDE activity, Wei and colleagues investigated the impact of FFA on IDE expression, showing that palmitic acid (300 μM) augmented IDE protein levels in Hepa 1c1c7 cells [177]. These results are not in accordance with effects of FFA on IDE expression in brain, where palmitic acid reduced, while docosahexaenoic increased, IDE protein levels in neuron [205]. Furthermore, palmitic acid attenuated the effect of docosahexaenoic acid in brain [205]. In light of these studies, it is tempting to hypothesize that obese patients are exposed to increasing amounts of saturated FFA (e.g., palmitic acid) released from mesenteric and omental fat via the portal system, which would be predicted to inhibit both proteolytic and non-proteolytic functions of IDE, in turn decreasing insulin clearance, whether directly by reducing IDE levels and/or activity or indirectly via mechanisms involving IR internalization and/or recycling. This mechanism may help explain the insulin resistance and hyperinsulinemia seen in obese patients, but further research is needed to validate this hypothesis.

1.8. Pharmacological Modulation of IDE

The proteolytic activity of IDE in vitro is sensitive to non-specific inhibitors such as EDTA (a metal-chelating agent), 1,10-phenanthroline (a Zn^{2+} chelator), p-hydroxymercuribenzoate and iodoacetate (cysteine proteinase inhibitors), N-ethylmaleimide (NEM) and p-chloromercuribenzoate (sulfhydryl-reactive compounds) and bacitracin (cyclic polypeptides from *B. subtilis* that inhibit bacterial growth) [206–208]. Importantly, IDE is sensitive to inhibition by sulfhydryl-directed reactions, such as alkylation (NEM) and oxidative inactivation (H_2O_2) [145,148,194,209,210]. The multi-functional roles of IDE, and the fact that most existing inhibitors were non-selective, inspired the development of pharmacological inhibitors of IDE. The first potent and selective IDE inhibitor was developed by Leissring and colleagues, who used a rational drug design approach based on analysis of the subsite sequence selectivity of IDE, resulting in Ii1, a highly potent (K_i = 2 nM) peptide hydroxamic acid [211]. Another thiol-targeting IDE inhibitor developed by this group, ML345, is of interest because it selectively targets extracellular IDE while sparing cytosolic

IDE via the formation of a redox-sensitive disulfide bond [209]. More recently, a potent and commercially available IDE inhibitor, 6bK, was developed by Maianti and colleagues [115], which is highly selective because it targets the exosite rather than the active site within IDE. Further, 6bK is notable because it exhibited multiple antidiabetic properties in vivo [115]. A recent drug-repurposing screen conducted by Leroux and colleagues [212] identified an existing drug, ebselen (EB), a synthetic organoselenium compound with antioxidant and anti-inflammatory properties, as a potent pharmacological inhibitor of IDE (apparent IC_{50} against insulin degradation = 14 nM) [212–215]. EB was found to inhibit IDE via unusual mechanisms of action. First, EB was shown to be a reversible inhibitor [212], despite the fact that it is known to covalently modify cysteine residues [216], and despite the fact that it was inactive against a cysteine-free form of IDE [212]. While the reversibility of EB action would suggest the compound does not directly modify thiols, an independent study found that a biotinylated form of EB interacts with IDE in a covalent manner [217]. Second, EB was found to shift the quaternary structure of IDE, destabilizing the homodimer and promoting monomer formation [212]. Interestingly, EB has an insulin-mimetic action, reducing hyperglycemia, enhancing glucose uptake in peripheral tissues [218], restoring glucose-stimulated insulin secretion in pancreatic β-cells [219], and improving hepatic insulin signaling and β-cell survival [220], suggesting that EB-mediated IDE inhibition may be involved in the mechanism of action of this compound. Consistent with the idea that IDE inhibition could potentiate insulin signaling, Leissring and colleagues identified peptidic IDE inhibitors via phage display that promote insulin-dependent collagen production in skin fibroblasts and migration in cultured keratinocytes, suggesting these compounds may have therapeutic value for promoting wound healing [221]. Finally, in another interesting application of IDE inhibitors, Demidowich and colleagues proposed the utility of using bacitracin in blood samples to counteract insulin degradation from IDE released during hemolysis, which complicates interpretation of clinical data [222].

Indirect effects of several diabetes drugs, such as thiazolidinediones, on the levels and functionality of IDE have also been investigated. Pioglitazone, an insulin sensitizer that enhances insulin action and decreases hepatic gluconeogenesis in liver [223], increased IDE mRNA and protein levels in a time-dependent manner in the mouse hepatoma cell line Hepa 1c1c7 [177]. In addition, pioglitazone administration resulted in higher IDE mRNA and protein levels in livers of mice fed a HFD concomitant with improved insulin sensitivity and lower circulating glucose and insulin levels [177]. Troglitazone was the first thiazolidinedione to be used in diabetic patients but was subsequently withdrawn for clinical use due to its hepatotoxicity [224]. Troglitazone administration reduced hepatic triglyceride content and decreased de novo lipogenesis, in parallel with higher insulin clearance in rats fed a high-sucrose diet for two weeks. These metabolic improvements were associated with augmented IDE activity, but similar IDE protein levels in the liver [225].

2. Studies of Global Manipulation of *Ide* on Insulin and Glucose Tolerance In Vivo

2.1. Effects of Pancellular Genetic Deletion of Ide on Insulin and Glucose Tolerance In Vivo

As an initial approach to help elucidate the function of IDE in vivo, several studies were conducted in mice with pancellular deletion of IDE (IDE-KO mice). Farris and colleagues, the first team to investigate diabetes-related endpoints in this mouse model [80], found that 6-month-old IDE-KO mice exhibited elevated fasting plasma glucose and insulin levels along with profound glucose intolerance [80]. Abdul–Hay and colleagues subsequently conducted a longitudinal characterization of IDE-KO mice, conducting glucose and insulin tolerance tests at 2, 4 and 6 months of age [83]. Whereas fasting plasma insulin levels were found to be elevated at all ages in IDE-KO mice, glucose and insulin tolerance transitioned from being modestly improved relative to wildtype controls at 2 months of age to profound glucose and insulin intolerance at 6 months of age [83]. The age-dependent emergence of the diabetic phenotype led Abdul–Hay and colleagues to propose that it arose as a consequence of chronic hyperinsulinemia, reporting that IR levels in muscle, adipose and liver tissue decreased as a function of age [83]. Consistent with the reduction in IR levels, primary adipocytes harvested from 6-month-old IDE-KO mice also showed functional impairments in insulin-stimulated glucose uptake [83]. Finally, IDE-KO mice were also characterized by Steneberg and colleagues [85]. Consistent with previous reports, Abdul–Hay and colleagues found that IDE-KO mice fed a normal diet exhibit pronounced glucose intolerance that increased in severity in an age-dependent manner [85]. Unlike previous studies, however, Steneberg and colleagues conducted a detailed examination of glucose-stimulated insulin secretion, finding that plasma insulin levels were markedly reduced in IDE-KO mice after glucose challenge [85]. Subsequent ex vivo and in vitro analyses yielded confirmatory evidence that insulin secretion was indeed impaired by deletion of *Ide* [85], as discussed in greater detail in Section 3. Collectively, these contradictory studies raise far more questions than they answer, suggesting that analysis of mice with global deletion of IDE might not be the most helpful approach to elucidating the precise role(s) of IDE in regulating insulin and glucose homeostasis.

Given the large number of substrates processed by IDE, its widespread expression in various tissues and subcellular organelles, as well as the various proteolytic and non-proteolytic functions of IDE and its homologs and paralogs, pancellular deletion of IDE might be expected to result in a large number of diverse phenotypes. It is therefore of some note that IDE-KO mice have so far been shown to exhibit very few phenotypic differences unrelated to metabolism. Apart from elevations in Aβ levels in brain already discussed, IDE-KO mice were recently shown to exhibit reduced sperm quality associated with substantial decreases in testes weight and seminiferous tubule diameter [226]. While interesting, these phenotypic changes may be secondary to the metabolic disturbances in IDE-KO mice, since reduced fertility and impaired sperm quality is a known consequence of diabetes [227]. Given that testes express transcripts for both insulin [228] and *Ide* (particularly the 15b splice isoform) [153], further research on this topic is warranted.

2.2. *Effects of Pharmacological Inhibition of IDE on Insulin and Glucose Tolerance In Vivo*

Based on the ability of IDE to avidly degrade insulin, investigators have, for more than a half century, postulated that pharmacological inhibition of this protease could augment insulin action by prolonging the half-life of circulating insulin, thereby improving glucose homeostasis in diabetic patients [229]. Stimulated in part by this rationale, several pharmacological inhibitors of IDE have been developed [38,115,211,230–234]. IDE inhibitors are highly desirable from an experimental perspective, as well. For instance, unlike gene knockout approaches, pharmacological inhibition of IDE allows the effects of acute inactivation of IDE to be investigated, obviating compensatory changes that might accrue over longer time frames in knockout mice, while also theoretically permitting the experimental separation of proteolytic and non-proteolytic functions of IDE. Although several highly potent and selective compounds have been generated, these compounds have exhibited mixed effects on glucose tolerance and circulating insulin levels when tested in vivo. In line with the long-predicted outcome, Maianti and colleagues found that the IDE inhibitor 6bK improved oral glucose tolerance and insulin tolerance in normal and diabetic mice, with no effect evident in IDE-KO mice [115]. Intriguingly, however, 6bK dramatically worsened IP glucose tolerance 1 h after administration [115], a distinction that was attributed to the "incretin effect" and the associated involvement of different hormones [235]. Consistent with the latter idea, 6bK produced increases in insulin, amylin and glucagon, albeit with different temporal profiles [115]. Another mouse study involved the use of the dual-exosite targeting IDE inhibitor, NTE-1, developed by investigators at Eli Lilly [233]. As was true for 6bK, NTE-1 administration also improved the glucose excursion in oral glucose tolerance tests in diet-induced obese mice [233]. However, insulin tolerance was unchanged, and no increases in insulin levels were observed; moreover, euglycemic clamping studies revealed no changes in insulin responsiveness [233]. Interestingly, the researchers found no effect on exogenous insulin clearance in HEK cells treated with the related inhibitor, NTE-2, whether or not these cells expressed the IR; similarly, no effect of IDE overexpression or downregulation was observed in the latter cells [233]. Notably, however, NTE-1 treatment did lead to increases in circulating levels of amylin in vivo [233]. Yet another IDE inhibitor, BDM44768, developed by Deprez–Poulain and colleagues, was also evaluated in mice [234]. Mice treated with BDM44768 exhibited modestly improved insulin tolerance along with modestly increased plasma insulin levels after IP insulin administration [234]. Strikingly, however, BDM44768 administration resulted in a significant worsening of oral and IP glucose tolerance in both wildtype B6 mice and non-obese diabetic mice but not B6 or NOD mice lacking IDE, suggesting the effect was indeed dependent on IDE [234]. Of note, plasma insulin levels were increased after glucose challenge in NOD but not wildtype mice treated with BDM44768 [234]. Moreover, hepatic gluconeogenesis was not altered by BDM44768 as assessed by pyruvate tolerance testing [234]. Significantly, BDM44768 increased in a dose-dependent manner the amount of insulin secreted by isolated islets obtained from wildtype, but not IDE-null, mice [234]. The ratio of insulin secreted in response to high versus low levels of glucose was not altered, however, leading the investigators to conclude that BDM44768 does not influence the responsiveness of β-cells to glucose [234].

Taken together, these in vivo pharmacological studies paint a complicated picture, with some supporting a role for IDE in regulating insulin levels and others confuting this idea. Similarly, some IDE inhibitors exhibited antidiabetic properties in oral glucose tolerance tests, but the opposite was observed for other inhibitors or in other tests of glucose and insulin tolerance. It seems evident that the picture is so complicated in large part because IDE degrades multiple hormones with overlapping and/or contradictory effects on blood glucose. In light of this, it is relevant to note that Maianti and colleagues recently developed an insulin-selective IDE inhibitor [236]. Future studies with this inhibitor hold great promise for disentangling the role of IDE in insulin homeostasis independent of its effects on amylin, glucagon and other substrates.

3. Role of IDE in Pancreatic β-Cells

3.1. IDE Protein Expression in Pancreatic β-Cells

Despite being a widely studied cell type in T2DM, the pancreatic β-cell and precisely how it malfunctions in disease, remains incompletely understood. Several pathogenic mechanisms have been proposed, including cell death [237,238], de-differentiation [239,240], trans-differentiation [241,242] and loss of cell identity [243,244] have been proposed. Understanding the molecular mechanisms underlying these processes will help to develop new strategies to recover β-cell function in T2DM.

With respect to IDE, several genome-wide association studies conducted on large populations have independently identified genetic variations in and around the *Hhex/Ide* locus that are associated with T2DM incidence, decreased insulin secretion, and differential β-cell glucose sensitivity in response to an oral glucose challenge [141,198,245–247]. However, until recently the biology of IDE in the β-cell remained largely unknown. For the first time, in 2018, Fernández–Díaz and colleagues investigated the expression pattern of IDE in pancreatic islet cells [157]. We showed that IDE is differentially expressed in different pancreatic islet cells, being expressed to substantially higher levels in pancreatic α-cells relative to β-cells and all other islet cell types. This finding suggests that it may be relevant to investigate the role of this protease in glucagon-producing cells in future research [157].

There is some evidence suggesting a potential role for IDE in T2DM-associated β-cell dysfunction, although the literature on certain topics is contradictory [85,248–251]. The abundance of contradictory findings likely reflects the multifactorial role that IDE plays in different tissues and cell types, highlighting the need for cell-specific manipulations to deepen our understanding of IDE's various functions. Regarding pancreatic β-cells, by western blotting, Steneberg and colleagues showed that IDE protein levels are diminished by 40% in whole islets from T2DM donors compared to controls [85], a finding that was later corroborated by Fernández–Díaz and colleagues via immunostaining [157]. Furthermore, IDE expression levels were modulated in T2DM patients, depending on the treatment they received. IDE appeared decreased in diabetics treated with oral hypoglycemic agents, which was in line with Steneberg and colleagues' observations in whole pancreatic islets from diabetic donors (whose treatment was unspecified) [85]. On the other hand, insulin-treated patients, showed increased IDE protein levels in pancreatic β-cells relative to patients treated with oral hypoglycemic agents, pointing to an upregulation of IDE under high insulin conditions [157]. This hypothesis was confirmed in multiple experimental paradigms, including preclinical murine models with hyperinsulinemia (*db/db* mice and high fat-fed mice) as well as in in vitro experiments using insulin administration to INS1E cells, as well as to rodent and human pancreatic islets. These consistent findings strongly support the notion that IDE levels in pancreatic β-cells increase in response to insulin exposure [157], in agreement with observations in other cell types, including hepatocytes [175] and primary hippocampal neurons [176].

Why precisely IDE levels increase in response to elevated insulin is unknown, but we can speculate that this may occur as a counter-regulatory adaptative mechanism for clearing excessive insulin and thereby restoring homeostasis. This idea is supported by the observation that insulin that is not cleared by liver and kidney is ultimately removed by other insulin-sensitive cells [37]. Taken together, these studies reveal that IDE expression in pancreatic β-cells is remarkably plastic, varying in response to different metabolic and hormonal milieus to preserve β-cell function.

3.2. *Effects of Genetic Deletion of Ide on Insulin Secretion In Vitro*

The main function of a pancreatic β-cell is the production and secretion of insulin in response to increased circulating glucose levels. To test how IDE is involved in this process, Fernandez–Diaz and colleagues treated both rat and human islets with the IDE inhibitor NTE-2 [123,233]. The result was clear: Transient inhibition of IDE led to an abolition of glucose-stimulated insulin secretion (GSIS), thus demonstrating the crucial role of IDE in β-cell function. This pattern of results was also obtained in INS1E cells, an immortalized β-cell line, following transient downregulation of *Ide* by RNA interference. Of note, these results are consistent with the impaired release of insulin associated with genetic variations around the *Hhex/Ide* chromosomal locus in humans [247,252] and with Steneberg and colleagues' findings, who reported deficient insulin secretion from islets isolated from IDE-KO mice [85].

Confirming these observations, shRNA-mediated silencing of *Ide* in immortalized β-cells (INS1E-shRNA-IDE cells) resulted in decreased insulin secretion in response to glucose [123]. These investigators observed increased intracellular insulin content in INS1E-shRNA-IDE cells and electron microscopy images revealed elevated numbers of insulin granules in the cytoplasm, pointing to a delayed movement of the granules through the cytoplasm during insulin secretion. This mechanism would be in agreement with previous observations showing that decreased insulin release caused by IDE ablation is related to impaired polymerization of α-synuclein, a protein involved in the reorganization of cellular microtubules [85]. One of the limiting steps during insulin secretion is the correct organization of microtubules: These allow insulin granules to travel through cytoplasm to reach the cell membrane and release insulin to the medium in response to increased extracellular glucose levels [253]. Thus, the role of IDE in regulating glucose-stimulated insulin secretion in β-cells appears to be related to this non-proteolytic function rather that to its degradative capacity.

On the other hand, because loss of IDE function in β-cells resulted in intracellular accumulation of insulin, it is plausible to propose a proteolytic role for IDE on regulating insulin levels in mature-beta granules. To maintain insulin stores at optimal levels, mature-insulin granules of the pancreatic β-cells gain access to a lysosomal compartment by crinophagy and/or autophagy where the secretory granule content is degraded [254–256]. If IDE could gain access to mature granules at neutral and/or basic pH conditions, the proteolytic activity of IDE might conceivably regulate the insulin content in mature secretory granules. Whether this speculative mechanism is operative in β-cells remains unclear, however, and more studies are warranted.

3.3. Impaired Insulin Secretion and β-Cell Immaturity in the B-IDE-KO Mouse

As discussed, to assess whether IDE might be a valid therapeutic target for T2DM treatment, many authors have investigated the effects of pancellular *Ide* deletion in mice models [38,80,83,85,115]. The results obtained are somewhat contradictory, but most reports indicate that genetic deletion of *Ide* results in deleterious metabolic effects, such as impaired glucose homeostasis. However, it is difficult to draw conclusions about IDE's specific function in β-cells in IDE-KO mice due to potential secondary or compensatory effects of loss of IDE function occurring in other tissues, such as liver. To overcome this problem, and thereby help elucidate the specific role of IDE in β-cells, Fernandez–Diaz and colleagues generated β-cell-specific *Ide* knock-out mice (B-IDE-KO) [123].

In contrast to the previously reported phenotype of IDE-KO mice, B-IDE-KO mice did not fully recapitulate an impairment in glucose homeostasis. Surprisingly, B-IDE-KO mice showed no changes in fasting or non-fasting plasma glucose levels. Glucose homeostasis measured by IP glucose tolerance tests was normal, but after 6 months of age, mice showed glucose levels significantly increased 15 min after glucose challenge, pointing to modest glucose intolerance. Unexpectedly, plasma C-peptide levels were increased in B-IDE-KO as compared to wildtype mice, indicating perturbed regulation of insulin secretion. This observation could have two explanations: Either β-cell mass was increased by *Ide* deletion or β-cells were hypersecreting insulin. Further experiments showed that β-cell area and islet number were normal, and the elevation in C-peptide levels was instead found to be attributable to constitutive insulin secretion. Specifically, B-IDE-KO isolated islets showed impairments in GSIS in the presence of high glucose, but continued to secrete insulin even in the presence of low glucose levels. This impairment in GSIS agrees with the findings of Steneberg and colleagues in islets isolated from IDE-KO mice [85] and with previous in vitro studies using NTE-2 inhibitor and INS1E-shRNA-IDE in β-cells [123]. Interestingly, B-IDE-KO mice exhibited decreased levels of the glucose transporter GLUT2 and increased levels of GLUT1 in the plasma membrane of β-cells. These findings are in agreement with the constitutive secretion seen in B-IDE-KO islets. On the one hand, GLUT1 is operative at low glucose concentrations (1–3 mM) [257], so the elevated levels of GLUT1 would promote insulin secretion at low glucose concentrations. On the other hand, GLUT2 is maximally functional at high glucose concentrations (15–20 mM), so the decreased levels in B-IDE-KO β-cells would account for the impaired response to high glucose [257–261].

The aforementioned perturbations to GLUT1 and GLUT2 levels also suggest a plausible explanation for the elevated fasting plasma insulin levels reported in IDE-KO mice studies [80,83], which was previously attributed to decreased hepatic catabolism of insulin. Because liver-specific deletion of IDE (L-IDE-KO mouse) showed normal plasma insulin levels [86], the hyperinsulinemia seen in the IDE-KO might be due to alterations in the GLUT1/GLUT2 ratio in pancreatic β-cells. In this context, we note that insulin in plasma from wildtype mice would be vulnerable to degradation by any IDE released by hemolysis, which would be absent in plasma from IDE-KO mice, which could conceivably account for the apparent hyperinsulinemia in IDE-KO mice [222]. Ideally, quantification of plasma C-peptide and insulin levels in IDE-KO mice would help to clarify if the hyperinsulinemia seen in IDE-KO mice is related to pancreatic β-cell function or hepatic insulin clearance.

Livers isolated from B-IDE-KO also display increased expression of the hepatic gluconeogenic genes phosphoenolpyruvate carboxykinase (*Pck1*) and glucose-6-phosphatase (*G6pc*), suggesting hepatic insulin resistance, possibly caused by increased flux of insulin through portal vein to the liver. Hepatic insulin resistance could also be the cause of the mild glucose intolerance observed in B-IDE-KO mice. Taken together, these results implicate IDE as a protein that mediates crosstalk between liver and pancreas to maintain insulin and glucose homeostasis.

Constitutive insulin secretion has been reported as a hallmark of β-cell dysfunction and immaturity [262–265]. In embryonic and neonatal stages, β-cells secrete insulin in a constitutively manner [262–265]. Extending this line of thinking, Fernandez–Diaz and colleagues speculated that the expression of GLUT1 is increased in embryonic tissues to help meet the demand of energy necessary for the rapid growth of fetal cells [260]. In addition, it is noteworthy that GLUT2 has been reported to be decreased in the plasma membrane of immature and dysfunctional β-cells [264–267]. Together, these findings are consistent with the conclusion that genetic deletion of *Ide* disrupts the maturation of β-cells, leaving them in a premature metabolic state.

Islets isolated from B-IDE-KO mice also showed an increase in the secretion of proinsulin in parallel with decreased *Pcsk1/3* expression levels, leading to the secretion of immature insulin granules into the extracellular space, which is another characteristic of immature β-cells [264]. In addition, B-IDE-KO islets showed decreased expression of key genes such as *Ins2* and *Ucn3*, which are necessary for the correct maturation and function of β-cells [262].

In summary, the B-IDE-KO mouse model uncovered an unexpected new IDE function in regulating pancreatic β-cell maturation. These results are consistent with the finding that, during rat development, IDE is differentially expressed in different tissues and in an age-dependent manner [2,75].

4. Role of IDE in Liver

4.1. Metabolic Phenotype of the L-IDE-KO Mouse

Historically, the study of IDE, and speculation about its proposed function, has been focused on the liver, but most studies were limited to the analysis of post-mortem tissue or cultured hepatocytes. The study of IDE-KO mice offered new possibilities for elucidating IDE's function in liver; however, as discussed above, the various studies of these animals have yielded conflicting results. Moreover, by virtue of the sheer complexity of the underlying endocrinology, and the many potential proteolytic and nonproteolytic functions of IDE, the study of animals with pancellular deletion of IDE is of limited value—and may in fact yield confounding results. To investigate the role of IDE in liver in a more focused manner, our group generated a mouse line with selective ablation of *Ide* exclusively in hepatocytes, known as the L-IDE-KO model [86].

L-IDE-KO mice exhibit higher fasting and non-fasting glucose levels, glucose intolerance, and insulin resistance, despite normal plasma insulin levels [86]. Of note, plasma levels of other IDE substrates—including glucagon, amylin and Aβ—remained unchanged in L-IDE-KO mice [86]. Contrary to historical predictions about IDE's function in liver, clearance of exogenously administered insulin was unaltered by hepatic ablation of *Ide*. Moreover, assessment of β-cell function (insulin and C-peptide plasma levels), and histomorphological analyses of pancreas (β-cell mass, β-cell area, number of islets, and mean islets size) revealed that β-cell function and mass in L-IDE-KO mice were similar to wildtype controls [86]. These findings indicate that hepatic *Ide* ablation causes insulin resistance independently of any effect on circulating insulin levels, suggesting that the hyperinsulinemia observed in IDE-KO mice emerged as a secondary compensatory response to systemic insulin resistance.

Another highly intriguing observation made in L-IDE-KO mice involved carcinoembryonic antigen-related cell adhesion molecule 1, a substrate of the IR in liver, that upregulates receptor-mediated insulin endocytosis and degradation in a phosphorylation-dependent manner [268]. As expected, administration of insulin to wildtype mice resulted in robust phosphorylation of CEACAM1 [86]. In marked contrast, however, insulin administration to L-IDE-KO mice resulted in no detectable phosphorylation of CEACAM1, despite unchanged CEACAM1 protein levels [86]. The potential consequences of these observations for insulin resistance are considered in greater detail, below.

To delve more deeply into IDE's role in the pathogenesis of hepatic insulin action, Merino and colleagues fed L-IDE-KO mice a Western HFD (35% carbohydrates and 45% fat) [87]. As was true for animals fed a regular diet [86], L-IDE-KO fed a HFD mice exhibited insulin resistance and glucose intolerance [87]. Unlike the regular diet [86], however, feeding a HFD to L-IDE-KO mice triggered elevated fasting and non-fasting plasma insulin levels, but normal glucagon levels, relative to control mice fed a HFD [87]. The hyperinsulinemia in HFD-fed L-IDE-KO mice could theoretically be attributable to reduced hepatic insulin extraction and/or enhanced β-cell function and mass. Similar to the case with a regular diet [86], however, hepatic insulin clearance and β-cell mass remained unchanged in L-IDE-KO mice fed a HFD [87]. On the other hand, β-cell function was improved, most likely as a compensatory response to insulin resistance triggered by loss of hepatic IDE function [87].

From a mechanistic point of view, the hepatic insulin resistance present in L-IDE-KO mice appears to be related to diminished insulin action in the liver under regular and HFD feeding [86,87]. Hepatic ablation of IDE causes a reduction in IR protein levels and insulin-mediated phosphorylation of IR, leading to lower AKT (protein kinase B) activation and aberrant nuclear distribution of FoxO1, which in turn enhances expression of gluconeogenic genes (*G6p6* and *Pck1*). Interestingly, hepatic *Ide* ablation altered IR levels post-translationally, as IR mRNA levels were unaffected, and did not reduce protein or mRNA levels of the insulin-like growth factor-1 receptor, which exhibits 70% homology to IR and shares some insulin-responsive signaling pathways [86,87].

The effects on IR regulation seen in L-IDE-KO mice fed a regular diet (which exhibit normoinsulinemia) resemble those seen in pancellular IDE-KO mice (which exhibit hyperinsulinemia). In pathophysiological conditions, such as obesity and T2DM, it is well-documented that cellular IR levels decrease [269–272] and, moreover, that hyperinsulinemia is associated with accelerated IR degradation [273]. The rate of IR degradation, in particular, is an important factor for controlling the receptor levels in hepatocytes and, hence, their sensitivity to insulin. Because insulin binding to its receptor initiates insulin action, it is apparent that a decrease in IR levels could lead to insulin resistance, but this relationship is not always so clear due to the "spare receptor" concept [274], which is based on the observation that a maximal insulin effect is achieved at an insulin concentration that occupies less than the total number of cellular receptors (10% in adipocytes [274] and 20% in skeletal muscle [275]). Therefore, at any given point in time, the cellular response to increasing insulin levels increases linearly with receptor occupancy, and the maximal biological response occurs when a particular number of receptors are occupied. Beyond this point, increased occupancy of receptors by insulin does not lead to further increases in biological action of the hormone, because events downstream of IR binding become the rate-limiting steps. Thus, the predicted consequence of a progressive loss of IRs on insulin action would be no change in maximal insulin response as long as enough receptors are present, albeit with more insulin required to achieve the same response. However, if the progressive loss of IR reaches a critical threshold (<10–20% of total), the dose-response to insulin and maximal insulin response will diminish drastically. In L-IDE-KO mice fed a regular diet, IR levels are reduced by ~30%, hypothetically leaving sufficient receptors (~70%) above the critical threshold for maximal insulin response [87]. However, because there is no increase in plasma insulin levels, maximal response to the hormone is not achieved, resulting in insulin resistance. On the other hand, L-IDE-KO mice fed a HFD also show a ~30% reduction in IR levels, but also exhibit hyperinsulinemia, which would theoretically permit maximal insulin action. In this case, however, insulin action is instead blocked downstream of the IR, as evidenced by 75% reduction in intracellular AKT levels response to insulin, leading to insulin resistance by this alternative mechanism [87].

Considering that IDE exhibits numerous non-proteolytic functions, such as regulating cytoskeletal components, protein turnover, and/or subcellular localization of proteins [119,125,129,130,134,136], it is tempting to hypothesize that IDE may regulate intracellular trafficking of the IR independently of its protease activity. We recently proposed a coordinated model of IR trafficking and insulin metabolism by CEACAM1- and IDE-dependent pathways [1]. In this model, CEACAM1 promotes the targeting of IR for degradation in response to insulin, and the main effect of IDE would be on IR recycling to the plasma membrane, an important step in insulin retro-endocytosis. Our analysis of L-IDE-KO mice has generated several important findings in support of this hypothesis. As mentioned, hepatic ablation of IDE reduces both the levels and phosphorylation of the

IR [86,87]; moreover, depletion of IDE completely abrogated insulin-induced phosphorylation of CEACAM1 on the membrane of hepatocytes [86]. Finally, as considered in greater detail below, Merino and colleagues found that hepatic overexpression of IDE leads to co-immunoprecipitation with the IR in response to insulin [87].

While this model is compelling, we cannot exclude the possibility that, in addition to the non-proteolytical action of IDE, its proteolytic function may regulate insulin action and/or the fraction of bound insulin available for internalization in hepatocytes. The first and the rate-limiting step for insulin action and internalization is its binding to the IR at the plasma membrane. From this point, two cellular processes for insulin metabolism have been proposed [276]. On one hand, insulin degradation has been proposed to occur at the cellular membrane, which does not involve internalization of insulin bound to its receptor. This pathway is bacitracin-sensitive and may therefore involve IDE and/or the glutathione-insulin transhydrogenase, accounting for half of the cellular insulin degraded [277–279]. A second pathway requires internalization of the complex IR-insulin into coated pits and the formation of cytoplasmic endosomes. A fraction of the internalized insulin is recycled to the membrane and released intact to the extracellular space (the retroendocytotic pathway), whereas the remained insulin is degraded in endosomes (the degradative pathway) [280]. Acidification of the interior of endocytotic vesicles due to proton pumps facilitates dissociation of the insulin bound to its receptor and degradation of the hormone, most likely by the aspartyl protease cathepsin D [59]. Because IDE activity is pH-dependent, being most active at pH 8.5 [281], it has been suggested that IDE can degrade the B-chain of receptor-bound insulin in light endosomes (early endosomes) prior to endosomal acidification (late endosomes) [57,160,282,283]. In any case, the extent to which proteolytic activity of IDE participates in the regulation of insulin action in hepatocytes, if at all, remains to be determined.

4.2. Metabolic Phenotype of Hepatic IDE Gain of Function in Mice

Merino and colleagues also examined the consequence of a gain-of-function manipulation to hepatic IDE in vivo [87]. To that end, an adenovirus IDE expression construct was administered to mice, resulting in ~4-fold increase in liver IDE levels. In mice fed a HFD, hepatic IDE overexpression improved glucose tolerance and insulin sensitivity independently of changes in body weight or food intake [87]. Moreover, plasma insulin and C-peptides levels, but not glucagon, were reduced by hepatic IDE overexpression [87]. Although the reduction in plasma insulin levels might theoretically be explained by increased hepatic insulin clearance by IDE, insulin clearance was found to be unaltered by hepatic IDE, suggesting instead that as a consequence of improved insulin sensitivity the pancreas reduced insulin production and secretion to meet the demand for the hormone in peripheral tissues. This study is the first proof-of-principle demonstration that augmenting hepatic IDE function in liver can partially revert insulin resistance and glucose intolerance in a preclinical mouse model of obesity and diabetes. The opposing effects of loss versus gain of IDE function on insulin levels and glucose tolerance are consistent with a role for IDE in promoting insulin sensitivity in liver of diet-induced obese mice. In a similar way, Leissring and colleagues demonstrated that transgenic upregulation of IDE in neurons, significantly reduces brain Aβ levels, reduced amyloid plaque formation, and rescued the premature lethality present in amyloid precursor protein transgenic mice [284]. Because reduced IDE function has been implicated in the pathogenesis of both T2DM and Alzheimer disease [142], these studies lend support the notion that pharmacological upregulation of IDE might represent viable therapeutic strategies for the treatment of both diseases.

Interestingly, both gain and loss of IDE function in mice fed a HFD resulted in reductions in total IR protein levels [87]. We hypothesized that this occurs because, on the one hand, depleting IDE reduces IR recycling, while on the other, IDE overexpression speeds up IR turnover. In support of this notion, Li and colleagues showed that insulin increased colocalization and co-immunoprecipitation of IDE and SNX5 in plasma membrane of kidney cells (an important organ for systemic circulation insulin clearance and insulin-mediated gluconeogenesis), whereas loss of SNX5 function led to reduced IDE protein and activity, in parallel with decreased expression of the IR and downstream insulin signaling [134]. Further studies are necessary to more fully delineate the molecular mechanisms by which IDE regulates hepatic IR protein levels as well as the physiological and pathophysiological relevance.

In addition to regulation of IR levels, manipulation of hepatic IDE also alters glucose transporters levels. Thus, loss of IDE function in HFD fed mice resulted in a two-fold increase in GLUT2 protein levels, with a reciprocal two-fold reduction of GLUT2 protein levels in mice overexpressing IDE. In addition, hepatic IDE gain-of-function resulted in a two-fold increase in GLUT1 protein levels, therefore altering the hepatic GLUT1/GLUT2 ratio [87]. Taken together, these findings lend support to the notion that IDE forms complexes with membrane proteins to regulate the intracellular trafficking of the IR independently of its proteolytic function.

4.3. *Novel Insights into the Etiology and Pathophysiology of Hepatic Insulin Resistance: Lessons from Knockout Mouse Models*

Over the past several decades, the study of proteins involved in the regulation of the insulin signaling pathway, and liver knockout mouse models in particular, have generated novel insights into the etiology and pathophysiology of hepatic insulin resistance (Table 3). For most of them, a defect in the insulin signaling pathway translated to hepatic insulin resistance and hyperinsulinemia, with the exception of hepatic ablation of FoxO1, which exhibits normoinsulinemia and heightened insulin sensitivity [285], and hepatic Akt2, which showed indistinguishable insulin sensitivity as compared to control mice [286,287].

A signature of the L-IDE-KO mouse model is the presence of hepatic insulin resistance without associated hyperinsulinemia, leading to augmented blood glucose excursions under normal conditions. Conversely, in the setting of obesity induced by a HFD, loss of IDE function exacerbates hyperinsulinemia and worsens glucose intolerance in wildtype mice fed a HFD [86,87]. Applied to the etiology of T2DM, these observations suggest that loss of IDE function represents an early step of the development of T2DM in healthy individuals, triggering impaired glucose tolerance and/or impaired fasting glycemia, before beginning the compensatory hyperinsulinemic phase. On the other hand, for overweight or obese patients with impaired glucose intolerance or fasting glycemia, loss of IDE function would accelerate the compensatory hyperinsulinemia and eventually facilitate the onset of T2DM. More studies are necessary to demonstrate the cause-effect relationship between hepatic IDE function and the onset of T2DM in lean and overweight/obese patients.

Because L-IDE-KO mice were generated using the Cre/loxP system harboring a null allele in their germline, the metabolic phenotype of these mice is related to pre- or postnatal *Ide* deficiency, and metabolic adaptations may arise across lifespan. Thus, the development of an inducible L-IDE-KO mouse line would be valuable, as it would permit the analysis of disruption to IDE function occurring in adult mice, in the absence of disruptions to IDE function during development. An inducible L-IDE-KO model of study would also help elucidate the impact of IDE function on insulin sensitivity and glucose homeostasis before and after the onset of diabetes in mouse models of T2DM such as the *db/db*.

By way of conclusion, to our best knowledge, the L-IDE-KO is one of the few knockout mice models of hepatic insulin resistance in which insulin clearance has been assessed in normal and HFD feeding. So far, the liver CEACAM1 knockout mice provide an in vivo proof of the key role of impaired hepatic insulin clearance and hyperinsulinemia in the pathogenesis of secondary hepatic insulin resistance. Considering the importance of insulin clearance for the regulation of circulating insulin levels, it would be of interest to investigate how IDE levels and activity are impacted in mouse models with liver-specific deletion of, for example, IR, IRS1/2, PI3K and rictor, which could help to elucidate the mechanistic basis for the hyperinsulinemia occurring in these models (Table 3), and its impact on IDE levels and activity.

Table 3. Knockout mice models of hepatic insulin resistance.

Mouse Model	Genetic Background	Target Protein	Target Tissue	Metabolic Phenotype	Insulin Resistance	Insulin Clearance	Refs.
db/*db*	C57BLKs/J	OB-R	Spontaneous mutation of the leptin receptor	Hyperinsulinemia, hyperglycemia, higher body weight	+	n.d.	[288,289]
ob/*ob*	C57BLKs/J	Leptin	Recessive mutation of leptin	Hyperinsulinemia, hyperglycemia, higher body weight	+	n.d.	[289,290]
L-IDE-KO	C57BL/6J	IDE	Liver	Normoinsulinemia, higher glucose levels, similar body weight	+	=	[86,87]
LIRKO	Mixed genetic background	IR	Liver	Hyperinsulinemia, hyperglycemia, similar body weight	+	Lower	[291,292]
LIrs1KO	Mixed genetic background	IRS1	Liver	Hyperinsulinemia, euglycemia, similar body weight	+ (*)	n.d.	[293]
LIrs2KO	Mixed genetic background	IRS2	Liver	Hyperinsulinemia, euglycemia, similar body weight	+ ($)	n.d.	[293]
L-p110-αKO	Mixed genetic background	PI3K catalytic subunit p110-α	Liver	Hyperinsulinemia, hyperglycemia, increased fat mass	+	n.d.	[294]
L-p110βKO	Mixed genetic background	PI3K catalytic subunit p110-β	Liver	Hyperinsulinemia, similar blood glucose levels	+	n.d.	[295]
L-*Pdk1*KO	Mixed genetic background	PDK1	Liver	Hyperinsulinemia, hyperglycemia, similar body weight	+	n.d.	[296]
L-Akt2	C57BL/6J	AKT2	Liver	Normoinsulinemia, euglycemia, similar body weight	-	n.d.	[286,287]
L1KO(l-FoxO1)	C57BL/6J	FoxO1	Liver	Normoinsulinemia, lower plasma glucose, similar body weight	-	n.d.	[285,297]
LiRiKO	C57BL/6J	Rictor	Liver	Hyperinsulinemia, hyperglycemia, similar body weight	+	n.d.	[298]
L-SACC1 (*AlbCre* + $Cc1^{fl/fl}$)	C57BL/6J	CEACAM1	Liver	Hyperinsulinemia, hyperglycemia, increased fat mass	+	Lower	[299]

* Insulin resistance after refeeding; $ Insulin resistance during fasting; n.d., not determined; + presence of insulin resistance; - no presence of insulin resistance; = no change.

5. Concluding Remarks

More than 70 years ago, IDE was first identified as the protease that predominantly degrades insulin. This finding immediately suggested a major role for this protein in the regulation of insulin homeostasis via hepatic insulin clearance. As revealed by this comprehensive review of numerous aspects of IDE biology, with an emphasis on its role in the regulation of insulin secretion and insulin resistance, the biology of IDE has proven to be considerably more complex. Knockout mouse models have demonstrated that the physiological processes regulated by IDE are much broader than expected and, in particular, strongly implicate non-proteolytic functions of this enzyme. These mouse models reveal that neither loss nor gain of hepatic IDE function affected plasma insulin levels or insulin clearance, with the important caveat that hepatic insulin clearance in L-IDE-KO mice has not yet been evaluated by hyperinsulinemic-euglycemic clamping. Furthermore, in B-IDE-KO mice, loss of IDE function alters the expression of key genes necessary for correct maturation of β-cells, leading to the secretion of immature granules and constitutive insulin secretion independent of glucose levels. These findings underscore the importance of tissue-specific knockout mouse models for unravelling the IDE's roles in regulating insulin metabolism and action (Figure 3).

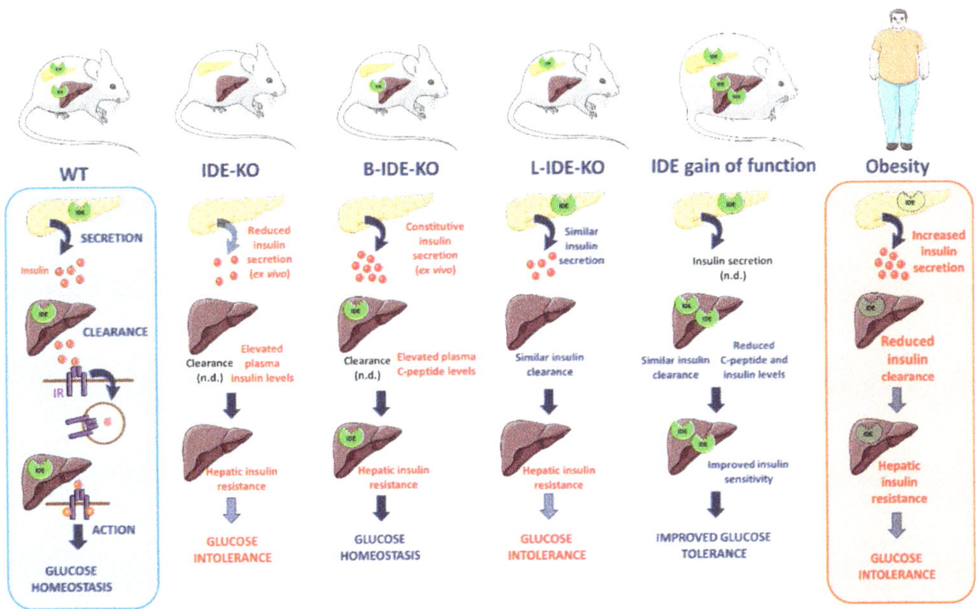

Figure 3. IDE mouse models for the study of insulin proteostasis and insulin sensitivity. In wildtype mice (WT), glucose homeostasis is regulated by insulin secretion out of the pancreas and clearance in the liver. In the fed state, high glucose levels stimulate pancreatic β-cells insulin secretion into portal vein, which is extracted by an insulin receptor-mediated process in hepatic cells. In parallel, insulin promotes glucose utilization and suppresses glucose production in hepatocytes. Pancellular genetic deletion of *Ide* (IDE-KO mice) causes hyperinsulinemia, hepatic insulin resistance, and glucose intolerance, but isolated islets exhibit reduced insulin secretion. Genetic deletion of *Ide* in pancreatic β-cells (B-IDE-KO mice) is associated with elevated plasma C-peptide levels, most likely due to constitutive insulin secretion, leading to hepatic insulin resistance, albeit normal glucose tolerance. Genetic deletion of *Ide* in hepatocytes (L-IDE-KO mice) results in hepatic insulin resistance and glucose intolerance, without altering insulin secretion and clearance. Conversely, IDE overexpression in liver improves hepatic insulin resistance and glucose intolerance, without altering insulin clearance in diet-induced obese mice. Finally, IDE levels are reduced in pancreatic β-cells and the liver of obese patients, which associates with hyperinsulinemia, reduced hepatic insulin clearance, hepatic insulin resistance and glucose intolerance. Each one of the IDE mouse models display hallmarks of the metabolic alterations seen in the setting of obesity. This figure was created using Servier Medical Art (available at https://smart.servier.com/). n.d., not determined.

The L-IDE-KO and B-IDE-KO mouse models have also helped highlight the notion that IDE may participate in the crosstalk between the liver and β-cells. Thus, hepatic loss of IDE function in the setting of diet-induced obesity enhanced β-cells function, leading to increased insulin secretion and hyperinsulinemia to help counteract hepatic insulin resistance. On the other hand, loss of IDE function in β-cells increased expression of hepatic gluconeogenic genes as a result of hepatic insulin resistance, most likely due to increased flux of insulin through the portal vein to the liver. Finally, hepatic IDE overexpression in diet-induced obese mice improves insulin sensitivity and decreases circulating insulin levels. Further studies will be needed to more rigorously assess this idea, but it seems evident that altering IDE function in liver or pancreas reciprocally modifies insulin action and secretion in these tissues.

The less prominent but equally important aspects of IDE function that reside beyond of its proteolytic effect on insulin have been noted for decades, but remained poorly clarified. Numerous studies have found that IDE interacts directly or indirectly with proteins not related to insulin metabolism, such as transcriptional factors, cell receptors, and cytoskeleton. Here, we present evidence that IDE regulates, in both liver and pancreas, glucose transporters (GLUT1 and GLUT2) and hepatic IR levels, suggesting a non-proteolytic role of IDE in the regulation of intracellular trafficking of proteins involved in the regulation of insulin sensitivity and glucose tolerance. The precise molecular and biochemical mechanism(s) by which IDE regulates intracellular trafficking in response to insulin, and its relevance for insulin sensitivity, remains to be deciphered, but is a very attractive idea that warrants additional research.

Research during the coming years may provide answers as to several outstanding questions. Is pharmacological inhibition of IDE a viable approach to the treatment of T2DM? Does the proteolytic activity of IDE play any role in hepatic insulin clearance? How do perturbations to IDE alter insulin sensitivity? Does IDE physically interact with different components of intracellular insulin signaling pathways? How do perturbations of IDE affect pancreatic function and islet maturity? The answers to these and other questions will facilitate our understanding of the etiology and molecular pathogenesis of T2DM and, hopefully, might stimulate the development of novel therapeutic approaches to treating, preventing or potentially reversing this increasingly common disease.

Author Contributions: Conceptualization, I.C.-C. and G.P.; writing—original draft preparation, C.M.G.-C., B.M., E.C.-Á.; T.P.-C., P.C.-T., C.M.F.-D., M.A.L.; writing—review and editing, I.C.-C., M.A.L., and G.P.; supervision, I.C.-C. and G.P.; funding acquisition, G.P. and I.C.-C. All authors have read and agreed to the published version of the manuscript.

Funding: This research was funded by the Ministerio de Economía, Industria y Competitividad, grant numbers SAF2016-77871-C2-1-R to I.C.-C.; SAF2016-77871-C2-2-R to G.P.; Ministerio de Ciencia e Innovación PID2019-110496RB-C21 to I.C.-C.; PID2019-110496RB-C22 to G.P.; European Foundation for the Study of Diabetes (European Diabetes Research Programme on New Targets for Type 2 Diabetes supported by MSD-2017) to I.C.-C. and G.P.; and the US National Institutes of Health (GM115617) to M.A.L. The project leading to these results has received funding from "la Caixa" Foundation, under agreement LCF/PR/PR18/51130007 to G.P., C.M.G.-C. and E.C.-Á. were supported by fellowships from the Junta de Castilla y León and the European Social Fund (ORDER EDU/574/2018 and ORDER EDU/556/2019, respectively).

Conflicts of Interest: The authors declare no conflict of interest. The funders had no role in the design of the study; in the collection, analyses, or interpretation of data; in the writing of the manuscript, or in the decision to publish the results.

References

1. Najjar, S.M.; Perdomo, G. Hepatic insulin clearance: Mechanism and physiology. *Physiology* **2019**, *34*, 198–215. [CrossRef]
2. Kuo, W.L.; Montag, A.G.; Rosner, M.R. Insulin-degrading enzyme is differentially expressed and developmentally regulated in various rat tissues. *Endocrinology* **1993**, *132*, 604–611. [CrossRef] [PubMed]
3. Hooper, N.M. Families of zinc metalloproteases. *FEBS Lett.* **1994**, *354*, 1–6. [CrossRef]

4. Rawlings, N.D.; Morton, F.R.; Kok, C.Y.; Kong, J.; Barrett, A.J. MEROPS: The peptidase database. *Nucleic Acids Res.* **2008**, *36*, D320–D325. [CrossRef] [PubMed]
5. Fernandez-Gamba, A.; Leal, M.C.; Morelli, L.; Castaño, E.M. Insulin-degrading enzyme: Structure-function relationship and its possible roles in health and disease. *Curr. Pharm. Des.* **2009**, *15*, 3644–3655. [CrossRef]
6. Rawlings, N.D.; Barrett, A.J. Homologues of insulinase, a new superfamily of metalloendopeptidases. *Biochem. J.* **1991**, *275 Pt 2*, 389–391. [CrossRef]
7. Kole, H.K.; Muthukumar, G.; Lenard, J. Purification and properties of a membrane-bound insulin binding protein, a putative receptor, from Neurospora crassa. *Biochemistry* **1991**, *30*, 682–688. [CrossRef]
8. Kole, H.K.; Smith, D.R.; Lenard, J. Characterization and partial purification of an insulinase from Neurospora crassa. *Arch. Biochem. Biophys.* **1992**, *297*, 199–204. [CrossRef]
9. Fricke, B.; Betz, R.; Friebe, S. A periplasmic insulin-cleaving proteinase (ICP) from Acinetobacter calcoaceticus sharing properties with protease III from Escherichia coli and IDE from eucaryotes. *J. Basic Microbiol.* **1995**, *35*, 21–31. [CrossRef]
10. Cheng, Y.S.; Zipser, D. Purification and characterization of protease III from Escherichia coli. *J. Biol. Chem.* **1979**, *254*, 4698–4706.
11. Dykstra, C.C.; Kushner, S.R. Physical characterization of the cloned protease III gene from Escherichia coli K-12. *J. Bacteriol.* **1985**, *163*, 1055–1059. [CrossRef] [PubMed]
12. Kim, S.; Lapham, A.N.; Freedman, C.G.; Reed, T.L.; Schmidt, W.K. Yeast as a tractable genetic system for functional studies of the insulin-degrading enzyme. *J. Biol. Chem.* **2005**, *280*, 27481–27490. [CrossRef] [PubMed]
13. Adames, N.; Blundell, K.; Ashby, M.N.; Boone, C. Role of yeast insulin-degrading enzyme homologs in propheromone processing and bud site selection. *Science* **1995**, *270*, 464–467. [CrossRef] [PubMed]
14. Cohen, P.; Pierotti, A.R.; Chesneau, V.; Foulon, T.; Prat, A. N-arginine dibasic convertase. *Methods Enzymol.* **1995**, *248*, 703–716. [CrossRef] [PubMed]
15. Nishi, K.; Sato, Y.; Ohno, M.; Hiraoka, Y.; Saijo, S.; Sakamoto, J.; Chen, P.M.; Morita, Y.; Matsuda, S.; Iwasaki, K.; et al. Nardilysin Is Required for Maintaining Pancreatic β-Cell Function. *Diabetes* **2016**, *65*, 3015–3027. [CrossRef]
16. Ishizu-Higashi, S.; Seno, H.; Nishi, E.; Matsumoto, Y.; Ikuta, K.; Tsuda, M.; Kimura, Y.; Takada, Y.; Kimura, Y.; Nakanishi, Y.; et al. Deletion of nardilysin prevents the development of steatohepatitis and liver fibrotic changes. *PLoS ONE* **2014**, *9*, e98017. [CrossRef]
17. Hiraoka, Y.; Matsuoka, T.; Ohno, M.; Nakamura, K.; Saijo, S.; Matsumura, S.; Nishi, K.; Sakamoto, J.; Chen, P.M.; Inoue, K.; et al. Critical roles of nardilysin in the maintenance of body temperature homoeostasis. *Nat. Commun.* **2014**, *5*, 3224. [CrossRef]
18. Nishi, E.; Prat, A.; Hospital, V.; Elenius, K.; Klagsbrun, M. N-arginine dibasic convertase is a specific receptor for heparin-binding EGF-like growth factor that mediates cell migration. *EMBO J.* **2001**, *20*, 3342–3350. [CrossRef]
19. Mirsky, I.A.; Broh-Kahn, R.H. The inactivation of insulin by tissue extracts; the distribution and properties of insulin inactivating extracts. *Arch. Biochem.* **1949**, *20*, 1–9.
20. Affholter, J.A.; Fried, V.A.; Roth, R.A. Human insulin-degrading enzyme shares structural and functional homologies with E. coli protease III. *Science* **1988**, *242*, 1415–1418. [CrossRef]
21. Duckworth, W.C.; Hamel, F.G.; Bennett, R.; Ryan, M.P.; Roth, R.A. Human red blood cell insulin-degrading enzyme and rat skeletal muscle insulin protease share antigenic sites and generate identical products from insulin. *J. Biol. Chem.* **1990**, *265*, 2984–2987. [CrossRef]
22. Broh-Kahn, R.H.; Mirsky, I.A. The inactivation of insulin by tissue extracts; the effect of fasting on the insulinase content of rat liver. *Arch. Biochem.* **1949**, *20*, 10–14. [PubMed]
23. Broh-Kahn, R.H.; Simkin, B.; Mirsky, A. The inactivation of insulin by tissue extracts. V. The effect of the composition of the diet on the restoration of the liver insulinase activity of the fasted rat. *Arch. Biochem.* **1950**, *27*, 174–184. [PubMed]
24. Mirsky, I.A.; Simkin, B.; Broh-Kahn, R.H. The inactivation of insulin by tissue extracts. VI. The existence, distribution and properties of an insulinase inhibitor. *Arch. Biochem.* **1950**, *28*, 415–423.
25. Mirsky, I.A.; Perisutti, G.; Diengott, D. Effect of insulinase-inhibitor on destruction of insulin by intact mouse. *Proc. Soc. Exp. Biol. Med.* **1955**, *88*, 76–78. [CrossRef]
26. Mirsky, I.A.; Perisutti, G.; Jinks, R. The destruction of insulin by intact mice. *Endocrinology* **1955**, *56*, 484–488. [CrossRef]
27. Brush, J.S.; Shah, R.J. Purification and characterization of inhibitors of insulin specific protease in human serum. *Biochem. Biophys. Res. Commun.* **1973**, *53*, 894–903. [CrossRef]
28. Ogawa, W.; Shii, K.; Yonezawa, K.; Baba, S.; Yokono, K. Affinity purification of insulin-degrading enzyme and its endogenous inhibitor from rat liver. *J. Biol. Chem.* **1992**, *267*, 1310–1316. [CrossRef]
29. McKenzie, R.A.; Burghen, G.A. Partial purification and characterization of insulin protease and its intracellular inhibitor from rat liver. *Arch. Biochem. Biophys.* **1984**, *229*, 604–611. [CrossRef]
30. Ryan, M.P.; Duckworth, W.C. Partial characterization of an endogenous inhibitor of a calcium-dependent form of insulin protease. *Biochem. Biophys. Res. Commun.* **1983**, *116*, 195–203. [CrossRef]
31. Saric, T.; Müller, D.; Seitz, H.J.; Pavelic, K. Non-covalent interaction of ubiquitin with insulin-degrading enzyme. *Mol. Cell. Endocrinol.* **2003**, *204*, 11–20. [CrossRef]
32. Mirsky, I.A. The hypoglycemic action of insulinase-inhibitors by mouth in patients with diabetes mellitus. *Trans. Assoc. Am. Physicians* **1956**, *69*, 262–275. [PubMed]
33. News of Science. *Science* **1956**, *123*, 258–262. [CrossRef] [PubMed]

34. Mering, J.; Minkowski, O. Diabetes mellitus nach Pankreasexstirpation. *Arch. Exp. Pathol. Pharmakol.* **1890**, *26*, 371–387. [CrossRef]
35. Banting, F.G.; Best, C.H.; Collip, J.B.; Campbell, W.R.; Fletcher, A.A. Pancreatic Extracts in the Treatment of Diabetes Mellitus. *Can. Med. Assoc. J.* **1922**, *12*, 141–146.
36. Mirsky, I.A. Insulinase. *Diabetes* **1957**, *6*, 448–449. [CrossRef]
37. Duckworth, W.C. Insulin degradation: Mechanisms, products, and significance. *Endocr. Rev.* **1988**, *9*, 319–345. [CrossRef]
38. Duckworth, W.C.; Bennett, R.G.; Hamel, F.G. Insulin degradation: Progress and potential. *Endocr. Rev.* **1998**, *19*, 608–624. [CrossRef]
39. Manolopoulou, M.; Guo, Q.; Malito, E.; Schilling, A.B.; Tang, W.J. Molecular basis of catalytic chamber-assisted unfolding and cleavage of human insulin by human insulin-degrading enzyme. *J. Biol. Chem.* **2009**, *284*, 14177–14188. [CrossRef]
40. Grasso, G.; Rizzarelli, E.; Spoto, G. AP/MALDI-MS complete characterization of the proteolytic fragments produced by the interaction of insulin degrading enzyme with bovine insulin. *J. Mass Spectrom.* **2007**, *42*, 1590–1598. [CrossRef]
41. Duckworth, W.C.; Heinemann, M.A.; Kitabchi, A.E. Purification of insulin-specific protease by affinity chromatography. *Proc. Natl. Acad. Sci. USA* **1972**, *69*, 3698–3702. [CrossRef] [PubMed]
42. Misbin, R.I.; Almira, E.C.; Duckworth, W.C.; Mehl, T.D. Inhibition of insulin degradation by insulin-like growth factors. *Endocrinology* **1983**, *113*, 1525–1527. [CrossRef] [PubMed]
43. Authier, F.; Bergeron, J.J.; Ou, W.J.; Rachubinski, R.A.; Posner, B.I.; Walton, P.A. Degradation of the cleaved leader peptide of thiolase by a peroxisomal proteinase. *Proc. Natl. Acad. Sci. USA* **1995**, *92*, 3859–3863. [CrossRef]
44. Frank, B.H.; Peavy, D.E.; Hooker, C.S.; Duckworth, W.C. Receptor binding properties of monoiodotyrosyl insulin isomers purified by high performance liquid chromatography. *Diabetes* **1983**, *32*, 705–711. [CrossRef] [PubMed]
45. Ryan, M.P.; Duckworth, W.C. Insulin degradation: Assays and enzymes. In *The Insulin Receptor*; Kahn, C., Harrison, L., Eds.; Alan R Liss Inc.: New York, NY, USA, 1988; Volume 1, pp. 29–57.
46. Song, E.S.; Mukherjee, A.; Juliano, M.A.; Pyrek, J.S.; Goodman, J.P., Jr.; Juliano, L.; Hersh, L.B. Analysis of the subsite specificity of rat insulysin using fluorogenic peptide substrates. *J. Biol. Chem.* **2001**, *276*, 1152–1155. [CrossRef] [PubMed]
47. Fosam, A.; Sikder, S.; Abel, B.S.; Tella, S.H.; Walter, M.F.; Mari, A.; Muniyappa, R. Reduced Insulin Clearance and Insulin-Degrading Enzyme Activity Contribute to Hyperinsulinemia in African Americans. *J. Clin. Endocrinol. Metab.* **2020**, *105*, e1835–e1846. [CrossRef] [PubMed]
48. Song, E.S.; Hersh, L.B. Insulysin: An allosteric enzyme as a target for Alzheimer's disease. *J. Mol. Neurosci.* **2005**, *25*, 201–206. [CrossRef]
49. Song, E.S.; Juliano, M.A.; Juliano, L.; Fried, M.G.; Wagner, S.L.; Hersh, L.B. ATP effects on insulin-degrading enzyme are mediated primarily through its triphosphate moiety. *J. Biol. Chem.* **2004**, *279*, 54216–54220. [CrossRef]
50. Im, H.; Manolopoulou, M.; Malito, E.; Shen, Y.; Zhao, J.; Neant-Fery, M.; Sun, C.Y.; Meredith, S.C.; Sisodia, S.S.; Leissring, M.A.; et al. Structure of substrate-free human insulin-degrading enzyme (IDE) and biophysical analysis of ATP-induced conformational switch of IDE. *J. Biol. Chem.* **2007**, *282*, 25453–25463. [CrossRef]
51. Cabrol, C.; Huzarska, M.A.; Dinolfo, C.; Rodriguez, M.C.; Reinstatler, L.; Ni, J.; Yeh, L.A.; Cuny, G.D.; Stein, R.L.; Selkoe, D.J.; et al. Small-molecule activators of insulin-degrading enzyme discovered through high-throughput compound screening. *PLoS ONE* **2009**, *4*, e5274. [CrossRef]
52. Song, E.S.; Juliano, M.A.; Juliano, L.; Hersh, L.B. Substrate activation of insulin-degrading enzyme (insulysin). A potential target for drug development. *J. Biol. Chem.* **2003**, *278*, 49789–49794. [CrossRef] [PubMed]
53. Leissring, M.A.; Lu, A.; Condron, M.M.; Teplow, D.B.; Stein, R.L.; Farris, W.; Selkoe, D.J. Kinetics of amyloid beta-protein degradation determined by novel fluorescence- and fluorescence polarization-based assays. *J. Biol. Chem.* **2003**, *278*, 37314–37320. [CrossRef] [PubMed]
54. Suire, C.N.; Lane, S.; Leissring, M.A. Development and Characterization of Quantitative, High-Throughput-Compatible Assays for Proteolytic Degradation of Glucagon. *SLAS Discov.* **2018**, *23*, 1060–1069. [CrossRef] [PubMed]
55. Suire, C.N.; Brizuela, M.K.; Leissring, M.A. Quantitative, High-Throughput Assays for Proteolytic Degradation of Amylin. *Methods Protoc.* **2020**, *3*, 81. [CrossRef]
56. Clot, J.P.; Janicot, M.; Fouque, F.; Desbuquois, B.; Haumont, P.Y.; Lederer, F. Characterization of insulin degradation products generated in liver endosomes: In vivo and in vitro studies. *Mol. Cell. Endocrinol.* **1990**, *72*, 175–185. [CrossRef]
57. Hamel, F.G.; Posner, B.I.; Bergeron, J.J.; Frank, B.H.; Duckworth, W.C. Isolation of insulin degradation products from endosomes derived from intact rat liver. *J. Biol. Chem.* **1988**, *263*, 6703–6708. [CrossRef]
58. Seabright, P.J.; Smith, G.D. The characterization of endosomal insulin degradation intermediates and their sequence of production. *Biochem. J.* **1996**, *320 Pt 3*, 947–956. [CrossRef]
59. Authier, F.; Metioui, M.; Fabrega, S.; Kouach, M.; Briand, G. Endosomal proteolysis of internalized insulin at the C-terminal region of the B chain by cathepsin D. *J. Biol. Chem.* **2002**, *277*, 9437–9446. [CrossRef]
60. Kouach, M.; Desbuquois, B.; Authier, F. Endosomal proteolysis of internalised [ArgA0]-human insulin at neutral pH generates the mature insulin peptide in rat liver in vivo. *Diabetologia* **2009**, *52*, 2621–2632. [CrossRef]
61. Authier, F.; Posner, B.I.; Bergeron, J.J. Endosomal proteolysis of internalized proteins. *FEBS Lett.* **1996**, *389*, 55–60. [CrossRef]
62. Authier, F.; Posner, B.I.; Bergeron, J.J. Insulin-degrading enzyme. *Clin. Investig. Med.* **1996**, *19*, 149–160.
63. Shen, Y.; Joachimiak, A.; Rosner, M.R.; Tang, W.J. Structures of human insulin-degrading enzyme reveal a new substrate recognition mechanism. *Nature* **2006**, *443*, 870–874. [CrossRef] [PubMed]

64. Stefanidis, L.; Fusco, N.D.; Cooper, S.E.; Smith-Carpenter, J.E.; Alper, B.J. Molecular Determinants of Substrate Specificity in Human Insulin-Degrading Enzyme. *Biochemistry* **2018**, *57*, 4903–4914. [CrossRef] [PubMed]
65. Affholter, J.A.; Cascieri, M.A.; Bayne, M.L.; Brange, J.; Casaretto, M.; Roth, R.A. Identification of residues in the insulin molecule important for binding to insulin-degrading enzyme. *Biochemistry* **1990**, *29*, 7727–7733. [CrossRef] [PubMed]
66. Schäffer, L. A model for insulin binding to the insulin receptor. *Eur. J. Biochem.* **1994**, *221*, 1127–1132. [CrossRef]
67. Kristensen, C.; Kjeldsen, T.; Wiberg, F.C.; Schäffer, L.; Hach, M.; Havelund, S.; Bass, J.; Steiner, D.F.; Andersen, A.S. Alanine scanning mutagenesis of insulin. *J. Biol. Chem.* **1997**, *272*, 12978–12983. [CrossRef]
68. De Meyts, P. Insulin/receptor binding: The last piece of the puzzle? What recent progress on the structure of the insulin/receptor complex tells us (or not) about negative cooperativity and activation. *BioEssays* **2015**, *37*, 389–397. [CrossRef]
69. Macháčková, K.; Mlčochová, K.; Potalitsyn, P.; Hanková, K.; Socha, O.; Buděšínský, M.; Muždalo, A.; Lepšík, M.; Černeková, M.; Radosavljević, J.; et al. Mutations at hypothetical binding site 2 in insulin and insulin-like growth factors 1 and 2 result in receptor- and hormone-specific responses. *J. Biol. Chem.* **2019**, *294*, 17371–17382. [CrossRef]
70. Affholter, J.A.; Hsieh, C.L.; Francke, U.; Roth, R.A. Insulin-degrading enzyme: Stable expression of the human complementary DNA, characterization of its protein product, and chromosomal mapping of the human and mouse genes. *Mol. Endocrinol.* **1990**, *4*, 1125–1135. [CrossRef]
71. Kuo, W.L.; Gehm, B.D.; Rosner, M.R. Regulation of insulin degradation: Expression of an evolutionarily conserved insulin-degrading enzyme increases degradation via an intracellular pathway. *Mol. Endocrinol.* **1991**, *5*, 1467–1476. [CrossRef]
72. Kuo, W.L.; Gehm, B.D.; Rosner, M.R. Cloning and expression of the cDNA for a Drosophila insulin-degrading enzyme. *Mol. Endocrinol.* **1990**, *4*, 1580–1591. [CrossRef] [PubMed]
73. Kuo, W.L.; Gehm, B.D.; Rosner, M.R.; Li, W.; Keller, G. Inducible expression and cellular localization of insulin-degrading enzyme in a stably transfected cell line. *J. Biol. Chem.* **1994**, *269*, 22599–22606. [PubMed]
74. Gehm, B.D.; Rosner, M.R. Regulation of insulin, epidermal growth factor, and transforming growth factor-alpha levels by growth factor-degrading enzymes. *Endocrinology* **1991**, *128*, 1603–1610. [CrossRef] [PubMed]
75. Kayalar, C.; Wong, W.T. Metalloendoprotease inhibitors which block the differentiation of L6 myoblasts inhibit insulin degradation by the endogenous insulin-degrading enzyme. *J. Biol. Chem.* **1989**, *264*, 8928–8934. [CrossRef]
76. Kayalar, C.; Wong, W.T.; Hendrickson, L. Differentiation of BC3H1 and primary skeletal muscle cells and the activity of their endogenous insulin-degrading enzyme are inhibited by the same metalloendoprotease inhibitors. *J. Cell. Biochem.* **1990**, *44*, 137–151. [CrossRef]
77. Shii, K.; Roth, R.A. Inhibition of insulin degradation by hepatoma cells after microinjection of monoclonal antibodies to a specific cytosolic protease. *Proc. Natl. Acad. Sci. USA* **1986**, *83*, 4147–4151. [CrossRef]
78. Fawcett, J.; Permana, P.A.; Levy, J.L.; Duckworth, W.C. Regulation of protein degradation by insulin-degrading enzyme: Analysis by small interfering RNA-mediated gene silencing. *Arch. Biochem. Biophys.* **2007**, *468*, 128–133. [CrossRef]
79. Louie, S.; Lakkyreddy, J.; Castellano, B.M.; Haley, B.; Dang, A.N.; Lam, C.; Tang, D.; Lang, S.; Snedecor, B.; Misaghi, S. Insulin Degrading Enzyme (IDE) Expressed by Chinese Hamster Ovary (CHO) Cells Is Responsible for Degradation of Insulin in Culture Media. *J. Biotechnol.* **2020**. [CrossRef]
80. Farris, W.; Mansourian, S.; Chang, Y.; Lindsley, L.; Eckman, E.A.; Frosch, M.P.; Eckman, C.B.; Tanzi, R.E.; Selkoe, D.J.; Guenette, S. Insulin-degrading enzyme regulates the levels of insulin, amyloid beta-protein, and the beta-amyloid precursor protein intracellular domain in vivo. *Proc. Natl. Acad. Sci. USA* **2003**, *100*, 4162–4167. [CrossRef]
81. Zhao, J.; Li, L.; Leissring, M.A. Insulin-degrading enzyme is exported via an unconventional protein secretion pathway. *Mol. Neurodegener.* **2009**, *4*, 4. [CrossRef]
82. Song, E.S.; Rodgers, D.W.; Hersh, L.B. Insulin-degrading enzyme is not secreted from cultured cells. *Sci. Rep.* **2018**, *8*, 2335. [CrossRef] [PubMed]
83. Abdul-Hay, S.O.; Kang, D.; McBride, M.; Li, L.; Zhao, J.; Leissring, M.A. Deletion of insulin-degrading enzyme elicits antipodal, age-dependent effects on glucose and insulin tolerance. *PLoS ONE* **2011**, *6*, e20818. [CrossRef] [PubMed]
84. Miller, B.C.; Eckman, E.A.; Sambamurti, K.; Dobbs, N.; Chow, K.M.; Eckman, C.B.; Hersh, L.B.; Thiele, D.L. Amyloid-beta peptide levels in brain are inversely correlated with insulysin activity levels in vivo. *Proc. Natl. Acad. Sci. USA* **2003**, *100*, 6221–6226. [CrossRef] [PubMed]
85. Steneberg, P.; Bernardo, L.; Edfalk, S.; Lundberg, L.; Backlund, F.; Ostenson, C.G.; Edlund, H. The type 2 diabetes-associated gene ide is required for insulin secretion and suppression of alpha-synuclein levels in beta-cells. *Diabetes* **2013**, *62*, 2004–2014. [CrossRef] [PubMed]
86. Villa-Pérez, P.; Merino, B.; Fernandez-Diaz, C.M.; Cidad, P.; Lobaton, C.D.; Moreno, A.; Muturi, H.T.; Ghadieh, H.E.; Najjar, S.M.; Leissring, M.A.; et al. Liver-specific ablation of insulin-degrading enzyme causes hepatic insulin resistance and glucose intolerance, without affecting insulin clearance in mice. *Metab. Clin. Exp.* **2018**, *88*, 1–11. [CrossRef]
87. Merino, B.; Fernández-Díaz, C.M.; Parrado-Fernández, C.; González-Casimiro, C.M.; Postigo-Casado, T.; Lobaton, C.D.; Leissring, M.A.; Cózar-Castellano, I.; Perdomo, G. Hepatic insulin-degrading enzyme regulates glucose and insulin homeostasis in diet-induced obese mice. *Metab. Clin. Exp.* **2020**, *113*, 154352. [CrossRef]
88. Duckworth, W.C. Insulin and glucagon degradation by the kidney. I. Subcellular distribution under different assay condition. *Biochim. Biophys. Acta* **1976**, *437*, 518–530. [CrossRef]

89. Ciaccio, C.; Tundo, G.R.; Grasso, G.; Spoto, G.; Marasco, D.; Ruvo, M.; Gioia, M.; Rizzarelli, E.; Coletta, M. Somatostatin: A novel substrate and a modulator of insulin-degrading enzyme activity. *J. Mol. Biol.* **2009**, *385*, 1556–1567. [CrossRef]
90. Bennett, R.G.; Duckworth, W.C.; Hamel, F.G. Degradation of amylin by insulin-degrading enzyme. *J. Biol. Chem.* **2000**, *275*, 36621–36625. [CrossRef]
91. Kurochkin, I.V.; Goto, S. Alzheimer's beta-amyloid peptide specifically interacts with and is degraded by insulin degrading enzyme. *FEBS Lett.* **1994**, *345*, 33–37. [CrossRef]
92. Edbauer, D.; Willem, M.; Lammich, S.; Steiner, H.; Haass, C. Insulin-degrading enzyme rapidly removes the beta-amyloid precursor protein intracellular domain (AICD). *J. Biol. Chem.* **2002**, *277*, 13389–13393. [CrossRef] [PubMed]
93. Morelli, L.; Llovera, R.E.; Alonso, L.G.; Frangione, B.; de Prat-Gay, G.; Ghiso, J.; Castano, E.M. Insulin-degrading enzyme degrades amyloid peptides associated with British and Danish familial dementia. *Biochem. Biophys. Res. Commun.* **2005**, *332*, 808–816. [CrossRef] [PubMed]
94. Muller, D.; Schulze, C.; Baumeister, H.; Buck, F.; Richter, D. Rat insulin-degrading enzyme: Cleavage pattern of the natriuretic peptide hormones ANP, BNP, and CNP revealed by HPLC and mass spectrometry. *Biochemistry* **1992**, *31*, 11138–11143. [CrossRef] [PubMed]
95. Tundo, G.R.; Di Muzio, E.; Ciaccio, C.; Sbardella, D.; Di Pierro, D.; Polticelli, F.; Coletta, M.; Marini, S. Multiple allosteric sites are involved in the modulation of insulin-degrading-enzyme activity by somatostatin. *FEBS J.* **2016**, *283*, 3755–3770. [CrossRef] [PubMed]
96. Malito, E.; Ralat, L.A.; Manolopoulou, M.; Tsay, J.L.; Wadlington, N.L.; Tang, W.J. Molecular bases for the recognition of short peptide substrates and cysteine-directed modifications of human insulin-degrading enzyme. *Biochemistry* **2008**, *47*, 12822–12834. [CrossRef] [PubMed]
97. Safavi, A.; Miller, B.C.; Cottam, L.; Hersh, L.B. Identification of gamma-endorphin-generating enzyme as insulin-degrading enzyme. *Biochemistry* **1996**, *35*, 14318–14325. [CrossRef] [PubMed]
98. Garcia, J.V.; Gehm, B.D.; Rosner, M.R. An evolutionarily conserved enzyme degrades transforming growth factor-alpha as well as insulin. *J. Cell Biol.* **1989**, *109*, 1301–1307. [CrossRef]
99. Fagan, J.M.; Waxman, L. Purification of a protease in red blood cells that degrades oxidatively damaged haemoglobin. *Biochem. J.* **1991**, *277 Pt 3*, 779–786. [CrossRef]
100. Werlen, R.C.; Offord, R.E.; Rose, K. Preparation and characterization of novel substrates of insulin proteinase (EC 3.4.99.45). *Biochem. J.* **1994**, *302 Pt 3*, 907–911. [CrossRef]
101. Liang, W.G.; Ren, M.; Zhao, F.; Tang, W.J. Structures of human CCL18, CCL3, and CCL4 reveal molecular determinants for quaternary structures and sensitivity to insulin-degrading enzyme. *J. Mol. Biol.* **2015**, *427*, 1345–1358. [CrossRef]
102. Hahn, F.; Schmalen, A.; Setz, C.; Friedrich, M.; Schlosser, S.; Kolle, J.; Spranger, R.; Rauch, P.; Fraedrich, K.; Reif, T.; et al. Proteolysis of mature HIV-1 p6 Gag protein by the insulin-degrading enzyme (IDE) regulates virus replication in an Env-dependent manner. *PLoS ONE* **2017**, *12*, e0174254. [CrossRef] [PubMed]
103. Semple, J.W.; Lang, Y.; Speck, E.R.; Delovitch, T.L. Processing and presentation of insulin. III. Insulin degrading enzyme: A neutral metalloendoproteinase that is non-homologous to classical endoproteinases mediates the processing of insulin epitopes for helper T cells. *Int. Immunol.* **1992**, *4*, 1161–1167. [CrossRef] [PubMed]
104. Semple, J.W.; Ellis, J.; Delovitch, T.L. Processing and presentation of insulin. II. Evidence for intracellular, plasma membrane-associated and extracellular degradation of human insulin by antigen-presenting B cells. *J. Immunol.* **1989**, *142*, 4184–4193. [PubMed]
105. Parmentier, N.; Stroobant, V.; Colau, D.; de Diesbach, P.; Morel, S.; Chapiro, J.; van Endert, P.; Van den Eynde, B.J. Production of an antigenic peptide by insulin-degrading enzyme. *Nat. Immunol.* **2010**, *11*, 449–454. [CrossRef] [PubMed]
106. Culina, S.; Mauvais, F.X.; Hsu, H.T.; Burgevin, A.; Guenette, S.; Moser, A.; van Endert, P. No major role for insulin-degrading enzyme in antigen presentation by MHC molecules. *PLoS ONE* **2014**, *9*, e88365. [CrossRef]
107. Leissring, M.A.; Farris, W.; Wu, X.; Christodoulou, D.C.; Haigis, M.C.; Guarente, L.; Selkoe, D.J. Alternative translation initiation generates a novel isoform of insulin-degrading enzyme targeted to mitochondria. *Biochem. J.* **2004**, *383*, 439–446. [CrossRef]
108. Grasso, G.; Mielczarek, P.; Niedziolka, M.; Silberring, J. Metabolism of cryptic peptides derived from neuropeptide FF precursors: The involvement of insulin-degrading enzyme. *Int. J. Mol. Sci.* **2014**, *15*, 16787–16799. [CrossRef]
109. Kummer, M.P.; Heneka, M.T. Truncated and modified amyloid-beta species. *Alzheimer's Res. Ther.* **2014**, *6*, 28. [CrossRef]
110. Zingale, G.A.; Bellia, F.; Ahmed, I.M.M.; Mielczarek, P.; Silberring, J.; Grasso, G. IDE Degrades Nociceptin/Orphanin FQ through an Insulin Regulated Mechanism. *Int. J. Mol. Sci.* **2019**, *20*, 4447. [CrossRef]
111. Li, Q.; Ali, M.A.; Cohen, J.I. Insulin degrading enzyme is a cellular receptor mediating varicella-zoster virus infection and cell-to-cell spread. *Cell* **2006**, *127*, 305–316. [CrossRef]
112. Li, Q.; Ali, M.A.; Wang, K.; Sayre, D.; Hamel, F.G.; Fischer, E.R.; Bennett, R.G.; Cohen, J.I. Insulin degrading enzyme induces a conformational change in varicella-zoster virus gE, and enhances virus infectivity and stability. *PLoS ONE* **2010**, *5*, e11327. [CrossRef] [PubMed]
113. Berarducci, B.; Rajamani, J.; Zerboni, L.; Che, X.; Sommer, M.; Arvin, A.M. Functions of the unique N-terminal region of glycoprotein E in the pathogenesis of varicella-zoster virus infection. *Proc. Natl. Acad. Sci. USA* **2010**, *107*, 282–287. [CrossRef] [PubMed]

114. Schmalen, A.; Karius-Fischer, J.; Rauch, P.; Setz, C.; Korn, K.; Henklein, P.; Fossen, T.; Schubert, U. The N-Terminus of the HIV-1 p6 Gag Protein Regulates Susceptibility to Degradation by IDE. *Viruses* **2018**, *10*, 710. [CrossRef] [PubMed]
115. Maianti, J.P.; McFedries, A.; Foda, Z.H.; Kleiner, R.E.; Du, X.Q.; Leissring, M.A.; Tang, W.J.; Charron, M.J.; Seeliger, M.A.; Saghatelian, A.; et al. Anti-diabetic activity of insulin-degrading enzyme inhibitors mediated by multiple hormones. *Nature* **2014**, *511*, 94–98. [CrossRef] [PubMed]
116. Kupfer, S.R.; Wilson, E.M.; French, F.S. Androgen and glucocorticoid receptors interact with insulin degrading enzyme. *J. Biol. Chem.* **1994**, *269*, 20622–20628.
117. Harada, S.; Smith, R.M.; Hu, D.Q.; Jarett, L. Dexamethasone inhibits insulin binding to insulin-degrading enzyme and cytosolic insulin-binding protein p82. *Biochem. Biophys. Res. Commun.* **1996**, *218*, 154–158. [CrossRef]
118. Protzek, A.O.; Rezende, L.F.; Costa-Junior, J.M.; Ferreira, S.M.; Cappelli, A.P.; de Paula, F.M.; de Souza, J.C.; Kurauti, M.A.; Carneiro, E.M.; Rafacho, A.; et al. Hyperinsulinemia caused by dexamethasone treatment is associated with reduced insulin clearance and lower hepatic activity of insulin-degrading enzyme. *J. Steroid Biochem. Mol. Biol.* **2016**, *155*, 1–8. [CrossRef]
119. Caravaggio, J.W.; Hasu, M.; MacLaren, R.; Thabet, M.; Raizman, J.E.; Veinot, J.P.; Marcel, Y.L.; Milne, R.W.; Whitman, S.C. Insulin-degrading enzyme deficiency in bone marrow cells increases atherosclerosis in LDL receptor-deficient mice. *Cardiovasc. Pathol.* **2013**, *22*, 458–464. [CrossRef]
120. Ahuja, N.; Schwer, B.; Carobbio, S.; Waltregny, D.; North, B.J.; Castronovo, V.; Maechler, P.; Verdin, E. Regulation of insulin secretion by SIRT4, a mitochondrial ADP-ribosyltransferase. *J. Biol. Chem.* **2007**, *282*, 33583–33592. [CrossRef]
121. Shi, T.; Wang, F.; Stieren, E.; Tong, Q. SIRT3, a mitochondrial sirtuin deacetylase, regulates mitochondrial function and thermogenesis in brown adipocytes. *J. Biol. Chem.* **2005**, *280*, 13560–13567. [CrossRef]
122. Haigis, M.C.; Mostoslavsky, R.; Haigis, K.M.; Fahie, K.; Christodoulou, D.C.; Murphy, A.J.; Valenzuela, D.M.; Yancopoulos, G.D.; Karow, M.; Blander, G.; et al. SIRT4 inhibits glutamate dehydrogenase and opposes the effects of calorie restriction in pancreatic beta cells. *Cell* **2006**, *126*, 941–954. [CrossRef] [PubMed]
123. Fernández-Díaz, C.M.; Merino, B.; López-Acosta, J.F.; Cidad, P.; de la Fuente, M.A.; Lobaton, C.D.; Moreno, A.; Leissring, M.A.; Perdomo, G.; Cózar-Castellano, I. Pancreatic beta-cell-specific deletion of insulin-degrading enzyme leads to dysregulated insulin secretion and beta-cell functional immaturity. *Am. J. Physiol. Endocrinol. Metab.* **2019**, *317*, E805–E819. [CrossRef] [PubMed]
124. Etienne-Manneville, S. Cytoplasmic Intermediate Filaments in Cell Biology. *Annu. Rev. Cell Dev. Biol.* **2018**, *34*, 1–28. [CrossRef] [PubMed]
125. Chou, Y.H.; Kuo, W.L.; Rosner, M.R.; Tang, W.J.; Goldman, R.D. Structural changes in intermediate filament networks alter the activity of insulin-degrading enzyme. *FASEB J.* **2009**, *23*, 3734–3742. [CrossRef] [PubMed]
126. Kleinschmidt, E.G.; Schlaepfer, D.D. Focal adhesion kinase signaling in unexpected places. *Curr. Opin. Cell Biol.* **2017**, *45*, 24–30. [CrossRef] [PubMed]
127. Murphy, J.M.; Jeong, K.; Lim, S.S. FAK Family Kinases in Vascular Diseases. *Int. J. Mol. Sci.* **2020**, *21*, 3630. [CrossRef] [PubMed]
128. Liu, Y.; Loijens, J.C.; Martin, K.H.; Karginov, A.V.; Parsons, J.T. The association of ASAP1, an ADP ribosylation factor-GTPase activating protein, with focal adhesion kinase contributes to the process of focal adhesion assembly. *Mol. Biol. Cell* **2002**, *13*, 2147–2156. [CrossRef]
129. Sharma, S.K.; Chorell, E.; Steneberg, P.; Vernersson-Lindahl, E.; Edlund, H.; Wittung-Stafshede, P. Insulin-degrading enzyme prevents alpha-synuclein fibril formation in a nonproteolytical manner. *Sci. Rep.* **2015**, *5*, 12531. [CrossRef]
130. Sharma, S.K.; Chorell, E.; Wittung-Stafshede, P. Insulin-degrading enzyme is activated by the C-terminus of alpha-synuclein. *Biochem. Biophys. Res. Commun.* **2015**, *466*, 192–195. [CrossRef]
131. Winner, B.; Jappelli, R.; Maji, S.K.; Desplats, P.A.; Boyer, L.; Aigner, S.; Hetzer, C.; Loher, T.; Vilar, M.; Campioni, S.; et al. In vivo demonstration that alpha-synuclein oligomers are toxic. *Proc. Natl. Acad. Sci. USA* **2011**, *108*, 4194–4199. [CrossRef]
132. Cullen, P.J.; Korswagen, H.C. Sorting nexins provide diversity for retromer-dependent trafficking events. *Nat. Cell Biol.* **2011**, *14*, 29–37. [CrossRef] [PubMed]
133. Johannes, L.; Wunder, C. The SNXy flavours of endosomal sorting. *Nat. Cell Biol.* **2011**, *13*, 884–886. [CrossRef] [PubMed]
134. Li, F.; Yang, J.; Villar, V.A.M.; Asico, L.D.; Ma, X.; Armando, I.; Sanada, H.; Yoneda, M.; Felder, R.A.; Jose, P.A.; et al. Loss of renal SNX5 results in impaired IDE activity and insulin resistance in mice. *Diabetologia* **2017**. [CrossRef]
135. Chiu, Y.F.; Chuang, L.M.; Hsiao, C.F.; Hung, Y.J.; Lin, M.W.; Chen, Y.T.; Grove, J.; Jorgenson, E.; Quertermous, T.; Risch, N.; et al. An autosomal genome-wide scan for loci linked to pre-diabetic phenotypes in nondiabetic Chinese subjects from the Stanford Asia-Pacific Program of Hypertension and Insulin Resistance Family Study. *Diabetes* **2005**, *54*, 1200–1206. [CrossRef] [PubMed]
136. Li, F.; Yang, J.; Jones, J.E.; Villar, V.A.; Yu, P.; Armando, I.; Felder, R.A.; Jose, P.A. Sorting nexin 5 and dopamine d1 receptor regulate the expression of the insulin receptor in human renal proximal tubule cells. *Endocrinology* **2015**, *156*, 2211–2221. [CrossRef]
137. Rubin, S.M.; Sage, J.; Skotheim, J.M. Integrating Old and New Paradigms of G1/S Control. *Mol. Cell* **2020**, *80*, 183–192. [CrossRef]
138. Radulescu, R.T.; Duckworth, W.C.; Levy, J.L.; Fawcett, J. Retinoblastoma protein co-purifies with proteasomal insulin-degrading enzyme: Implications for cell proliferation control. *Biochem. Biophys. Res. Commun.* **2010**, *395*, 196–199. [CrossRef]
139. Liu, M.; Wang, Z.; Ren, M.; Yang, X.; Liu, B.; Qi, H.; Yu, M.; Song, S.; Chen, S.; Liu, L.; et al. SIRT4 regulates PTEN stability through IDE in response to cellular stresses. *FASEB J.* **2019**, *33*, 5535–5547. [CrossRef]
140. Tundo, G.R.; Sbardella, D.; Ciaccio, C.; Bianculli, A.; Orlandi, A.; Desimio, M.G.; Arcuri, G.; Coletta, M.; Marini, S. Insulin-degrading enzyme (IDE): A novel heat shock-like protein. *J. Biol. Chem.* **2013**, *288*, 2281–2289. [CrossRef]

141. Fakhrai-Rad, H.; Nikoshkov, A.; Kamel, A.; Fernström, M.; Zierath, J.R.; Norgren, S.; Luthman, H.; Galli, J. Insulin-degrading enzyme identified as a candidate diabetes susceptibility gene in GK rats. *Hum. Mol. Genet.* **2000**, *9*, 2149–2158. [CrossRef]
142. Farris, W.; Mansourian, S.; Leissring, M.A.; Eckman, E.A.; Bertram, L.; Eckman, C.B.; Tanzi, R.E.; Selkoe, D.J. Partial loss-of-function mutations in insulin-degrading enzyme that induce diabetes also impair degradation of amyloid beta-protein. *Am. J. Pathol.* **2004**, *164*, 1425–1434. [CrossRef]
143. Song, E.S.; Rodgers, D.W.; Hersh, L.B. A monomeric variant of insulin degrading enzyme (IDE) loses its regulatory properties. *PLoS ONE* **2010**, *5*, e9719. [CrossRef] [PubMed]
144. Noinaj, N.; Bhasin, S.K.; Song, E.S.; Scoggin, K.E.; Juliano, M.A.; Juliano, L.; Hersh, L.B.; Rodgers, D.W. Identification of the allosteric regulatory site of insulysin. *PLoS ONE* **2011**, *6*, e20864. [CrossRef] [PubMed]
145. Perlman, R.K.; Gehm, B.D.; Kuo, W.L.; Rosner, M.R. Functional analysis of conserved residues in the active site of insulin-degrading enzyme. *J. Biol. Chem.* **1993**, *268*, 21538–21544. [CrossRef]
146. Li, P.; Kuo, W.L.; Yousef, M.; Rosner, M.R.; Tang, W.J. The C-terminal domain of human insulin degrading enzyme is required for dimerization and substrate recognition. *Biochem. Biophys. Res. Commun.* **2006**, *343*, 1032–1037. [CrossRef]
147. Gehm, B.D.; Kuo, W.L.; Perlman, R.K.; Rosner, M.R. Mutations in a zinc-binding domain of human insulin-degrading enzyme eliminate catalytic activity but not insulin binding. *J. Biol. Chem.* **1993**, *268*, 7943–7948. [CrossRef]
148. Neant-Fery, M.; Garcia-Ordoñez, R.D.; Logan, T.P.; Selkoe, D.J.; Li, L.; Reinstatler, L.; Leissring, M.A. Molecular basis for the thiol sensitivity of insulin-degrading enzyme. *Proc. Natl. Acad. Sci. USA* **2008**, *105*, 9582–9587. [CrossRef]
149. Tang, W.J. Targeting Insulin-Degrading Enzyme to Treat Type 2 Diabetes Mellitus. *Trends Endocrinol. Metab.* **2016**, *27*, 24–34. [CrossRef]
150. Leissring, M.A.; Selkoe, D.J. Structural biology: Enzyme target to latch on to. *Nature* **2006**, *443*, 761–762. [CrossRef]
151. McCord, L.A.; Liang, W.G.; Dowdell, E.; Kalas, V.; Hoey, R.J.; Koide, A.; Koide, S.; Tang, W.J. Conformational states and recognition of amyloidogenic peptides of human insulin-degrading enzyme. *Proc. Natl. Acad. Sci. USA* **2013**, *110*, 13827–13832. [CrossRef]
152. Huising, M.O. Paracrine regulation of insulin secretion. *Diabetologia* **2020**, *63*, 2057–2063. [CrossRef] [PubMed]
153. Farris, W.; Leissring, M.A.; Hemming, M.L.; Chang, A.Y.; Selkoe, D.J. Alternative splicing of human insulin-degrading enzyme yields a novel isoform with a decreased ability to degrade insulin and amyloid beta-protein. *Biochemistry* **2005**, *44*, 6513–6525. [CrossRef] [PubMed]
154. Baumeister, H.; Müller, D.; Rehbein, M.; Richter, D. The rat insulin-degrading enzyme. Molecular cloning and characterization of tissue-specific transcripts. *FEBS Lett.* **1993**, *317*, 250–254. [CrossRef]
155. Runyan, K.; Duckworth, W.C.; Kitabchi, A.E.; Huff, G. The effect of age on insulin-degrading activity in rat tissue. *Diabetes* **1979**, *28*, 324–325. [CrossRef] [PubMed]
156. Sudoh, S.; Frosch, M.P.; Wolf, B.A. Differential effects of proteases involved in intracellular degradation of amyloid beta-protein between detergent-soluble and -insoluble pools in CHO-695 cells. *Biochemistry* **2002**, *41*, 1091–1099. [CrossRef] [PubMed]
157. Fernández-Díaz, C.M.; Escobar-Curbelo, L.; López-Acosta, J.F.; Lobaton, C.D.; Moreno, A.; Sanz-Ortega, J.; Perdomo, G.; Cózar-Castellano, I. Insulin degrading enzyme is up-regulated in pancreatic beta cells by insulin treatment. *Histol. Histopathol.* **2018**, *33*, 1167–1180. [CrossRef]
158. Akiyama, H.; Shii, K.; Yokono, K.; Yonezawa, K.; Sato, S.; Watanabe, K.; Baba, S. Cellular localization of insulin-degrading enzyme in rat liver using monoclonal antibodies specific for this enzyme. *Biochem. Biophys. Res. Commun.* **1988**, *155*, 914–922. [CrossRef]
159. Song, E.S.; Jang, H.; Guo, H.F.; Juliano, M.A.; Juliano, L.; Morris, A.J.; Galperin, E.; Rodgers, D.W.; Hersh, L.B. Inositol phosphates and phosphoinositides activate insulin-degrading enzyme, while phosphoinositides also mediate binding to endosomes. *Proc. Natl. Acad. Sci. USA* **2017**, *114*, E2826–E2835. [CrossRef]
160. Hamel, F.G.; Mahoney, M.J.; Duckworth, W.C. Degradation of intraendosomal insulin by insulin-degrading enzyme without acidification. *Diabetes* **1991**, *40*, 436–443. [CrossRef]
161. Vekrellis, K.; Ye, Z.; Qiu, W.Q.; Walsh, D.; Hartley, D.; Chesneau, V.; Rosner, M.R.; Selkoe, D.J. Neurons regulate extracellular levels of amyloid beta-protein via proteolysis by insulin-degrading enzyme. *J. Neurosci.* **2000**, *20*, 1657–1665. [CrossRef]
162. Duckworth, W.C. Insulin degradation by liver cell membranes. *Endocrinology* **1979**, *104*, 1758–1764. [CrossRef] [PubMed]
163. Yokono, K.; Imamura, Y.; Sakai, H.; Baba, S. Insulin-degrading activity of plasma membranes from rat skeletal muscle: Its isolation, characterization, and biologic significance. *Diabetes* **1979**, *28*, 810–817. [CrossRef] [PubMed]
164. Yokono, K.; Roth, R.A.; Baba, S. Identification of insulin-degrading enzyme on the surface of cultured human lymphocytes, rat hepatoma cells, and primary cultures of rat hepatocytes. *Endocrinology* **1982**, *111*, 1102–1108. [CrossRef] [PubMed]
165. Goldfine, I.D.; Williams, J.A.; Bailey, A.C.; Wong, K.Y.; Iwamoto, Y.; Yokono, K.; Baba, S.; Roth, R.A. Degradation of insulin by isolated mouse pancreatic acini. Evidence for cell surface protease activity. *Diabetes* **1984**, *33*, 64–72. [CrossRef] [PubMed]
166. Qiu, W.Q.; Walsh, D.M.; Ye, Z.; Vekrellis, K.; Zhang, J.; Podlisny, M.B.; Rosner, M.R.; Safavi, A.; Hersh, L.B.; Selkoe, D.J. Insulin-degrading enzyme regulates extracellular levels of amyloid beta-protein by degradation. *J. Biol. Chem.* **1998**, *273*, 32730–32738. [CrossRef] [PubMed]
167. Sanderson, R.D.; Bandari, S.K.; Vlodavsky, I. Proteases and glycosidases on the surface of exosomes: Newly discovered mechanisms for extracellular remodeling. *Matrix Biol.* **2019**, *75–76*, 160–169. [CrossRef] [PubMed]
168. Qiu, W.Q.; Folstein, M.F. Insulin, insulin-degrading enzyme and amyloid-beta peptide in Alzheimer's disease: Review and hypothesis. *Neurobiol. Aging* **2006**, *27*, 190–198. [CrossRef]

169. Harada, S.; Loten, E.G.; Smith, R.M.; Jarett, L. Nonreceptor mediated nuclear accumulation of insulin in H35 rat hepatoma cells. *J. Cell. Physiol.* **1992**, *153*, 607–613. [CrossRef]
170. Harada, S.; Smith, R.M.; Jarett, L. Mechanisms of nuclear translocation of insulin. *Cell Biochem. Biophys.* **1999**, *31*, 307–319. [CrossRef]
171. Harada, S.; Smith, R.M.; Smith, J.A.; Jarett, L. Inhibition of insulin-degrading enzyme increases translocation of insulin to the nucleus in H35 rat hepatoma cells: Evidence of a cytosolic pathway. *Endocrinology* **1993**, *132*, 2293–2298. [CrossRef]
172. Shah, N.; Zhang, S.; Harada, S.; Smith, R.M.; Jarett, L. Electron microscopic visualization of insulin translocation into the cytoplasm and nuclei of intact H35 hepatoma cells using covalently linked Nanogold-insulin. *Endocrinology* **1995**, *136*, 2825–2835. [CrossRef] [PubMed]
173. Harada, S.; Smith, R.M.; Jarett, L. 1,10-Phenanthroline increases nuclear accumulation of insulin in response to inhibiting insulin degradation but has a biphasic effect on insulin's ability to increase mRNA levels. *DNA Cell Biol.* **1994**, *13*, 487–493. [CrossRef] [PubMed]
174. Harada, S.; Smith, R.M.; Smith, J.A.; Shah, N.; Jarett, L. Demonstration of specific insulin binding to cytosolic proteins in H35 hepatoma cells, rat liver and skeletal muscle. *Biochem. J.* **1995**, *306 Pt 1*, 21–28. [CrossRef]
175. Pivovarova, O.; Gogebakan, O.; Pfeiffer, A.F.; Rudovich, N. Glucose inhibits the insulin-induced activation of the insulin-degrading enzyme in HepG2 cells. *Diabetologia* **2009**, *52*, 1656–1664. [CrossRef] [PubMed]
176. Zhao, L.; Teter, B.; Morihara, T.; Lim, G.P.; Ambegaokar, S.S.; Ubeda, O.J.; Frautschy, S.A.; Cole, G.M. Insulin-degrading enzyme as a downstream target of insulin receptor signaling cascade: Implications for Alzheimer's disease intervention. *J. Neurosci.* **2004**, *24*, 11120–11126. [CrossRef] [PubMed]
177. Wei, X.; Ke, B.; Zhao, Z.; Ye, X.; Gao, Z.; Ye, J. Regulation of insulin degrading enzyme activity by obesity-associated factors and pioglitazone in liver of diet-induced obese mice. *PLoS ONE* **2014**, *9*, e95399. [CrossRef] [PubMed]
178. Lin, Y.; Liu, J.; Chen, J.; Yao, C.; Yang, Y.; Wang, J.; Zhuang, H.; Hua, Z.C. FADD Phosphorylation Modulates Blood Glucose Levels by Decreasing the Expression of Insulin-Degrading Enzyme. *Mol. Cells* **2020**, *43*, 373–383. [CrossRef]
179. de Kloet, A.D.; Woods, S.C. Minireview: Endocannabinoids and their receptors as targets for obesity therapy. *Endocrinology* **2009**, *150*, 2531–2536. [CrossRef]
180. Liu, J.; Zhou, L.; Xiong, K.; Godlewski, G.; Mukhopadhyay, B.; Tam, J.; Yin, S.; Gao, P.; Shan, X.; Pickel, J.; et al. Hepatic cannabinoid receptor-1 mediates diet-induced insulin resistance via inhibition of insulin signaling and clearance in mice. *Gastroenterology* **2012**, *142*, 1218–1228. [CrossRef]
181. Pal, M.; Febbraio, M.A.; Whitham, M. From cytokine to myokine: The emerging role of interleukin-6 in metabolic regulation. *Immunol. Cell Biol.* **2014**, *92*, 331–339. [CrossRef]
182. Kurauti, M.A.; Costa-Junior, J.M.; Ferreira, S.M.; Santos, G.J.; Sponton, C.H.G.; Carneiro, E.M.; Telles, G.D.; Chacon-Mikahil, M.P.T.; Cavaglieri, C.R.; Rezende, L.F.; et al. Interleukin-6 increases the expression and activity of insulin-degrading enzyme. *Sci. Rep.* **2017**, *7*, 46750. [CrossRef] [PubMed]
183. Camberos, M.C.; Pérez, A.A.; Udrisar, D.P.; Wanderley, M.I.; Cresto, J.C. ATP inhibits insulin-degrading enzyme activity. *Exp. Biol. Med.* **2001**, *226*, 334–341. [CrossRef] [PubMed]
184. Ivancic, V.A.; Krasinski, C.A.; Zheng, Q.; Meservier, R.J.; Spratt, D.E.; Lazo, N.D. Enzyme kinetics from circular dichroism of insulin reveals mechanistic insights into the regulation of insulin-degrading enzyme. *Biosci. Rep.* **2018**, *38*. [CrossRef] [PubMed]
185. Camberos, M.C.; Cresto, J.C. Insulin-degrading enzyme hydrolyzes ATP. *Exp. Biol. Med.* **2007**, *232*, 281–292.
186. Grasso, G.; Satriano, C.; Milardi, D. A neglected modulator of insulin-degrading enzyme activity and conformation: The pH. *Biophys. Chem.* **2015**, *203–204*, 33–40. [CrossRef]
187. Beffy, P.; Lajoix, A.D.; Masiello, P.; Dietz, S.; Péraldi-Roux, S.; Chardès, T.; Ribes, G.; Gross, R. A constitutive nitric oxide synthase modulates insulin secretion in the INS-1 cell line. *Mol. Cell. Endocrinol.* **2001**, *183*, 41–48. [CrossRef]
188. Fujimoto, M.; Shimizu, N.; Kunii, K.; Martyn, J.A.; Ueki, K.; Kaneki, M. A role for iNOS in fasting hyperglycemia and impaired insulin signaling in the liver of obese diabetic mice. *Diabetes* **2005**, *54*, 1340–1348. [CrossRef]
189. Natali, A.; Santini, E.; Delbarba, A.; Baldi, S.; Venturi, E.; Tulipani, A.; Nisoli, E.; Ferrannini, E. Effects of short and prolonged mild intracellular nitric oxide manipulations on various aspects of insulin secretion in INS-1E β-cells. *Exp. Clin. Endocrinol. Diabetes* **2012**, *120*, 210–216. [CrossRef]
190. Cordes, C.M.; Bennett, R.G.; Siford, G.L.; Hamel, F.G. Nitric oxide inhibits insulin-degrading enzyme activity and function through S-nitrosylation. *Biochem. Pharmacol.* **2009**, *77*, 1064–1073. [CrossRef]
191. Cordes, C.M.; Bennett, R.G.; Siford, G.L.; Hamel, F.G. Redox regulation of insulin degradation by insulin-degrading enzyme. *PLoS ONE* **2011**, *6*, e18138. [CrossRef]
192. Natali, A.; Ribeiro, R.; Baldi, S.; Tulipani, A.; Rossi, M.; Venturi, E.; Mari, A.; Macedo, M.P.; Ferrannini, E. Systemic inhibition of nitric oxide synthesis in non-diabetic individuals produces a significant deterioration in glucose tolerance by increasing insulin clearance and inhibiting insulin secretion. *Diabetologia* **2013**, *56*, 1183–1191. [CrossRef] [PubMed]
193. Caccamo, A.; Oddo, S.; Sugarman, M.C.; Akbari, Y.; LaFerla, F.M. Age- and region-dependent alterations in Abeta-degrading enzymes: Implications for Abeta-induced disorders. *Neurobiol. Aging* **2005**, *26*, 645–654. [CrossRef] [PubMed]
194. Shinall, H.; Song, E.S.; Hersh, L.B. Susceptibility of amyloid beta peptide degrading enzymes to oxidative damage: A potential Alzheimer's disease spiral. *Biochemistry* **2005**, *44*, 15345–15350. [CrossRef]

195. Minamiyama, Y.; Takemura, S.; Bito, Y.; Shinkawa, H.; Tsukioka, T.; Nakahira, A.; Suehiro, S.; Okada, S. Supplementation of alpha-tocopherol improves cardiovascular risk factors via the insulin signalling pathway and reduction of mitochondrial reactive oxygen species in type II diabetic rats. *Free Radic. Res.* **2008**, *42*, 261–271. [CrossRef] [PubMed]
196. Venturini, P.R.; Thomazini, B.F.; Oliveira, C.A.; Alves, A.A.; Camargo, T.F.; Domingues, C.E.C.; Barbosa-Sampaio, H.C.L.; do Amaral, M.E.C. Vitamin E supplementation and caloric restriction promotes regulation of insulin secretion and glycemic homeostasis by different mechanisms in rats. *Biochem. Cell Biol.* **2018**, *96*, 777–785. [CrossRef]
197. Rezende, L.F.; Camargo, R.L.; Branco, R.C.; Cappelli, A.P.; Boschero, A.C.; Carneiro, E.M. Reduced insulin clearance and lower insulin-degrading enzyme expression in the liver might contribute to the thrifty phenotype of protein-restricted mice. *Br. J. Nutr.* **2014**, *112*, 900–907. [CrossRef]
198. Kurauti, M.A.; Freitas-Dias, R.; Ferreira, S.M.; Vettorazzi, J.F.; Nardelli, T.R.; Araujo, H.N.; Santos, G.J.; Carneiro, E.M.; Boschero, A.C.; Rezende, L.F.; et al. Acute Exercise Improves Insulin Clearance and Increases the Expression of Insulin-Degrading Enzyme in the Liver and Skeletal Muscle of Swiss Mice. *PLoS ONE* **2016**, *11*, e0160239. [CrossRef]
199. Item, F.; Konrad, D. Visceral fat and metabolic inflammation: The portal theory revisited. *Obes. Rev.* **2012**, *13* (Suppl. 2), 30–39. [CrossRef]
200. Lytrivi, M.; Castell, A.L.; Poitout, V.; Cnop, M. Recent Insights Into Mechanisms of β-Cell Lipo- and Glucolipotoxicity in Type 2 Diabetes. *J. Mol. Biol.* **2020**, *432*, 1514–1534. [CrossRef]
201. Sonne, O. Increased inhibitory potency of free fatty acid-poor albumin on the released and activity of insulin-degrading enzymes from isolated rat adipocytes and hepatocytes. *Anal. Biochem.* **1985**, *151*, 109–117. [CrossRef]
202. Juul, S.M.; Jones, R.H. Evidence for a direct effect of bacitracin on cell-mediated insulin degradation in isolated hepatocytes. *Biochem. J.* **1982**, *206*, 295–299. [CrossRef] [PubMed]
203. Svedberg, J.; Björntorp, P.; Smith, U.; Lönnroth, P. Free-fatty acid inhibition of insulin binding, degradation, and action in isolated rat hepatocytes. *Diabetes* **1990**, *39*, 570–574. [CrossRef] [PubMed]
204. Hamel, F.G.; Upward, J.L.; Bennett, R.G. In vitro inhibition of insulin-degrading enzyme by long-chain fatty acids and their coenzyme A thioesters. *Endocrinology* **2003**, *144*, 2404–2408. [CrossRef] [PubMed]
205. Du, J.; Zhang, L.; Liu, S.; Wang, Z. Palmitic acid and docosahexaenoic acid opposingly regulate the expression of insulin-degrading enzyme in neurons. *Die Pharm.* **2010**, *65*, 231–232.
206. Rosin, D.L.; Bond, J.S.; Bradley, S.G. A cysteine metalloproteinase from mouse liver cytosol. *Proc. Soc. Exp. Biol. Med.* **1984**, *177*, 112–119. [CrossRef]
207. Hsu, M.C.; Bai, J.P. Investigation into the presence of insulin-degrading enzyme in cultured type II alveolar cells and the effects of enzyme inhibitors on pulmonary bioavailability of insulin in rats. *J. Pharm. Pharmacol.* **1998**, *50*, 507–514. [CrossRef]
208. Bai, J.P.; Chang, L.L. Transepithelial transport of insulin: I. Insulin degradation by insulin-degrading enzyme in small intestinal epithelium. *Pharm. Res.* **1995**, *12*, 1171–1175. [CrossRef]
209. Abdul-Hay, S.O.; Bannister, T.D.; Wang, H.; Cameron, M.D.; Caulfield, T.R.; Masson, A.; Bertrand, J.; Howard, E.A.; McGuire, M.P.; Crisafulli, U.; et al. Selective Targeting of Extracellular Insulin-Degrading Enzyme by Quasi-Irreversible Thiol-Modifying Inhibitors. *ACS Chem. Biol.* **2015**, *10*, 2716–2724. [CrossRef]
210. Ding, L.; Becker, A.B.; Suzuki, A.; Roth, R.A. Comparison of the enzymatic and biochemical properties of human insulin-degrading enzyme and Escherichia coli protease III. *J. Biol. Chem.* **1992**, *267*, 2414–2420. [CrossRef]
211. Leissring, M.A.; Malito, E.; Hedouin, S.; Reinstatler, L.; Sahara, T.; Abdul-Hay, S.O.; Choudhry, S.; Maharvi, G.M.; Fauq, A.H.; Huzarska, M.; et al. Designed inhibitors of insulin-degrading enzyme regulate the catabolism and activity of insulin. *PLoS ONE* **2010**, *5*, e10504. [CrossRef]
212. Leroux, F.; Bosc, D.; Beghyn, T.; Hermant, P.; Warenghem, S.; Landry, V.; Pottiez, V.; Guillaume, V.; Charton, J.; Herledan, A.; et al. Identification of ebselen as a potent inhibitor of insulin degrading enzyme by a drug repurposing screening. *Eur. J. Med. Chem.* **2019**, *179*, 557–566. [CrossRef] [PubMed]
213. Meotti, F.C.; Stangherlin, E.C.; Zeni, G.; Nogueira, C.W.; Rocha, J.B. Protective role of aryl and alkyl diselenides on lipid peroxidation. *Environ. Res.* **2004**, *94*, 276–282. [CrossRef]
214. Chander, P.N.; Gealekman, O.; Brodsky, S.V.; Elitok, S.; Tojo, A.; Crabtree, M.; Gross, S.S.; Goligorsky, M.S. Nephropathy in Zucker diabetic fat rat is associated with oxidative and nitrosative stress: Prevention by chronic therapy with a peroxynitrite scavenger ebselen. *J. Am. Soc. Nephrol.* **2004**, *15*, 2391–2403. [CrossRef]
215. Nogueira, C.W.; Zeni, G.; Rocha, J.B. Organoselenium and organotellurium compounds: Toxicology and pharmacology. *Chem. Rev.* **2004**, *104*, 6255–6285. [CrossRef] [PubMed]
216. Nikawa, T.; Schuch, G.; Wagner, G.; Sies, H. Interaction of ebselen with glutathione S-transferase and papain in vitro. *Biochem. Pharmacol.* **1994**, *47*, 1007–1012. [CrossRef]
217. Chen, Z.; Jiang, Z.; Chen, N.; Shi, Q.; Tong, L.; Kong, F.; Cheng, X.; Chen, H.; Wang, C.; Tang, B. Target discovery of ebselen with a biotinylated probe. *Chem. Commun.* **2018**, *54*, 9506–9509. [CrossRef]
218. Costa, M.D.; Gai, B.M.; Acker, C.I.; Souza, A.C.; Brandao, R.; Nogueira, C.W. Ebselen reduces hyperglycemia temporarily-induced by diazinon: A compound with insulin-mimetic properties. *Chemico-Biol. Interact.* **2012**, *197*, 80–86. [CrossRef]
219. Wang, X.; Yun, J.W.; Lei, X.G. Glutathione peroxidase mimic ebselen improves glucose-stimulated insulin secretion in murine islets. *Antioxid. Redox Signal.* **2014**, *20*, 191–203. [CrossRef]

220. Park, S.; Kang, S.; Kim, D.S.; Shin, B.K.; Moon, N.R.; Daily, J.W., 3rd. Ebselen pretreatment attenuates ischemia/reperfusion injury and prevents hyperglycemia by improving hepatic insulin signaling and beta-cell survival in gerbils. *Free Radic. Res.* **2014**, *48*, 864–874. [CrossRef]
221. Suire, C.N.; Nainar, S.; Fazio, M.; Kreutzer, A.G.; Paymozd-Yazdi, T.; Topper, C.L.; Thompson, C.R.; Leissring, M.A. Peptidic inhibitors of insulin-degrading enzyme with potential for dermatological applications discovered via phage display. *PLoS ONE* **2018**, *13*, e0193101. [CrossRef]
222. Demidowich, A.P.; Levine, J.A.; Brady, S.M.; Johnson, C.D.; Soldin, S.J.; Yanovski, J.A. Bacitracin attenuates haemolysis-induced insulin degradation during insulin sensitivity testing: Repurposing an old drug for use in metabolic research. *Diabetes Obes. Metab.* **2020**, *22*, 1469–1473. [CrossRef] [PubMed]
223. Rizos, C.V.; Liberopoulos, E.N.; Mikhailidis, D.P.; Elisaf, M.S. Pleiotropic effects of thiazolidinediones. *Expert Opin. Pharmacother.* **2008**, *9*, 1087–1108. [CrossRef] [PubMed]
224. Yamamoto, Y.; Yamazaki, H.; Ikeda, T.; Watanabe, T.; Iwabuchi, H.; Nakajima, M.; Yokoi, T. Formation of a novel quinone epoxide metabolite of troglitazone with cytotoxicity to HepG2 cells. *Drug Metab. Dispos.* **2002**, *30*, 155–160. [CrossRef] [PubMed]
225. Martins, F.O.; Delgado, T.C.; Viegas, J.; Gaspar, J.M.; Scott, D.K.; O'Doherty, R.M.; Macedo, M.P.; Jones, J.G. Mechanisms by which the thiazolidinedione troglitazone protects against sucrose-induced hepatic fat accumulation and hyperinsulinaemia. *Br. J. Pharmacol.* **2016**, *173*, 267–278. [CrossRef] [PubMed]
226. Meneses, M.J.; Borges, D.O.; Dias, T.R.; Martins, F.O.; Oliveira, P.F.; Macedo, M.P.; Alves, M.G. Knockout of insulin-degrading enzyme leads to mice testicular morphological changes and impaired sperm quality. *Mol. Cell. Endocrinol.* **2019**, *486*, 11–17. [CrossRef]
227. Agbaje, I.M.; Rogers, D.A.; McVicar, C.M.; McClure, N.; Atkinson, A.B.; Mallidis, C.; Lewis, S.E. Insulin dependant diabetes mellitus: Implications for male reproductive function. *Hum. Reprod.* **2007**, *22*, 1871–1877. [CrossRef] [PubMed]
228. Gómez, O.; Ballester, B.; Romero, A.; Arnal, E.; Almansa, I.; Miranda, M.; Mesonero, J.E.; Terrado, J. Expression and regulation of insulin and the glucose transporter GLUT8 in the testes of diabetic rats. *Horm. Metab. Res.* **2009**, *41*, 343–349. [CrossRef]
229. Mirsky, I.A. The role of insulinase and insulinase-inhibitors. *Metab. Clin. Exp.* **1956**, *5*, 138–143.
230. Bennett, R.G.; Hamel, F.G.; Duckworth, W.C. An insulin-degrading enzyme inhibitor decreases amylin degradation, increases amylin-induced cytotoxicity, and increases amyloid formation in insulinoma cell cultures. *Diabetes* **2003**, *52*, 2315–2320. [CrossRef]
231. Mirsky, I.A.; Perisutti, G. Effect of insulinase-inhibitor on hypoglycemic action of insulin. *Science* **1955**, *122*, 559–560. [CrossRef]
232. Charton, J.; Gauriot, M.; Guo, Q.; Hennuyer, N.; Marechal, X.; Dumont, J.; Hamdane, M.; Pottiez, V.; Landry, V.; Sperandio, O.; et al. Imidazole-derived 2-[N-carbamoylmethyl-alkylamino]acetic acids, substrate-dependent modulators of insulin-degrading enzyme in amyloid-β hydrolysis. *Eur. J. Med. Chem.* **2014**, *79*, 184–193. [CrossRef] [PubMed]
233. Durham, T.B.; Toth, J.L.; Klimkowski, V.J.; Cao, J.X.; Siesky, A.M.; Alexander-Chacko, J.; Wu, G.Y.; Dixon, J.T.; McGee, J.E.; Wang, Y.; et al. Dual Exosite-binding Inhibitors of Insulin-degrading Enzyme Challenge Its Role as the Primary Mediator of Insulin Clearance in Vivo. *J. Biol. Chem.* **2015**, *290*, 20044–20059. [CrossRef] [PubMed]
234. Deprez-Poulain, R.; Hennuyer, N.; Bosc, D.; Liang, W.G.; Enée, E.; Marechal, X.; Charton, J.; Totobenazara, J.; Berte, G.; Jahklal, J.; et al. Catalytic site inhibition of insulin-degrading enzyme by a small molecule induces glucose intolerance in mice. *Nat. Commun.* **2015**, *6*, 8250. [CrossRef] [PubMed]
235. Drucker, D.J. The biology of incretin hormones. *Cell Metab.* **2006**, *3*, 153–165. [CrossRef] [PubMed]
236. Maianti, J.P.; Tan, G.A.; Vetere, A.; Welsh, A.J.; Wagner, B.K.; Seeliger, M.A.; Liu, D.R. Substrate-selective inhibitors that reprogram the activity of insulin-degrading enzyme. *Nat. Chem. Biol.* **2019**, *15*, 565–574. [CrossRef] [PubMed]
237. Butler, A.E.; Janson, J.; Bonner-Weir, S.; Ritzel, R.; Rizza, R.A.; Butler, P.C. Beta-cell deficit and increased beta-cell apoptosis in humans with type 2 diabetes. *Diabetes* **2003**, *52*, 102–110. [CrossRef]
238. Tomita, T. Apoptosis in pancreatic β-islet cells in Type 2 diabetes. *Bosn. J. Basic Med. Sci.* **2016**, *16*, 162–179. [CrossRef]
239. Dor, Y.; Glaser, B. beta-cell dedifferentiation and type 2 diabetes. *N. Engl. J. Med.* **2013**, *368*, 572–573. [CrossRef]
240. Cinti, F.; Bouchi, R.; Kim-Muller, J.Y.; Ohmura, Y.; Sandoval, P.R.; Masini, M.; Marselli, L.; Suleiman, M.; Ratner, L.E.; Marchetti, P.; et al. Evidence of beta-Cell Dedifferentiation in Human Type 2 Diabetes. *J. Clin. Endocrinol. Metab.* **2016**, *101*, 1044–1054. [CrossRef]
241. Thorel, F.; Népote, V.; Avril, I.; Kohno, K.; Desgraz, R.; Chera, S.; Herrera, P.L. Conversion of adult pancreatic alpha-cells to beta-cells after extreme beta-cell loss. *Nature* **2010**, *464*, 1149–1154. [CrossRef]
242. Chera, S.; Baronnier, D.; Ghila, L.; Cigliola, V.; Jensen, J.N.; Gu, G.; Furuyama, K.; Thorel, F.; Gribble, F.M.; Reimann, F.; et al. Diabetes recovery by age-dependent conversion of pancreatic delta-cells into insulin producers. *Nature* **2014**, *514*, 503–507. [CrossRef] [PubMed]
243. Brereton, M.F.; Rohm, M.; Ashcroft, F.M. beta-Cell dysfunction in diabetes: A crisis of identity? *Diabetes Obes. Metab.* **2016**, *18* (Suppl. 1), 102–109. [CrossRef] [PubMed]
244. Hunter, C.S.; Stein, R.W. Evidence for Loss in Identity, De-Differentiation, and Trans-Differentiation of Islet β-Cells in Type 2 Diabetes. *Front. Genet.* **2017**, *8*, 35. [CrossRef] [PubMed]
245. Ghosh, S.; Watanabe, R.M.; Valle, T.T.; Hauser, E.R.; Magnuson, V.L.; Langefeld, C.D.; Ally, D.S.; Mohlke, K.L.; Silander, K.; Kohtamäki, K.; et al. The Finland-United States investigation of non-insulin-dependent diabetes mellitus genetics (FUSION) study. I. An autosomal genome scan for genes that predispose to type 2 diabetes. *Am. J. Hum. Genet.* **2000**, *67*, 1174–1185. [PubMed]

246. Sladek, R.; Rocheleau, G.; Rung, J.; Dina, C.; Shen, L.; Serre, D.; Boutin, P.; Vincent, D.; Belisle, A.; Hadjadj, S.; et al. A genome-wide association study identifies novel risk loci for type 2 diabetes. *Nature* **2007**, *445*, 881–885. [CrossRef] [PubMed]
247. Pascoe, L.; Tura, A.; Patel, S.K.; Ibrahim, I.M.; Ferrannini, E.; Zeggini, E.; Weedon, M.N.; Mari, A.; Hattersley, A.T.; McCarthy, M.I.; et al. Common variants of the novel type 2 diabetes genes CDKAL1 and HHEX/IDE are associated with decreased pancreatic beta-cell function. *Diabetes* **2007**, *56*, 3101–3104. [CrossRef]
248. Snehalatha, C.; Timothy, H.; Mohan, V.; Ramachandran, A.; Viswanathan, M. Immunoreactive insulin and insulin degrading enzymes in erythrocytes. A preliminary report. *J. Assoc. Physicians India* **1990**, *38*, 558–561.
249. Standl, E.; Kolb, H.J. Insulin degrading enzyme activity and insulin binding of erythrocytes in normal subjects and Type 2 (non-insulin-dependent) diabetic patients. *Diabetologia* **1984**, *27*, 17–22. [CrossRef]
250. Fawcett, J.; Sang, H.; Permana, P.A.; Levy, J.L.; Duckworth, W.C. Insulin metabolism in human adipocytes from subcutaneous and visceral depots. *Biochem. Biophys. Res. Commun.* **2010**, *402*, 762–766. [CrossRef]
251. Pivovarova, O.; von Loeffelholz, C.; Ilkavets, I.; Sticht, C.; Zhuk, S.; Murahovschi, V.; Lukowski, S.; Döcke, S.; Kriebel, J.; de las Heras Gala, T.; et al. Modulation of insulin degrading enzyme activity and liver cell proliferation. *Cell Cycle* **2015**, *14*, 2293–2300. [CrossRef]
252. Grarup, N.; Rose, C.S.; Andersson, E.A.; Andersen, G.; Nielsen, A.L.; Albrechtsen, A.; Clausen, J.O.; Rasmussen, S.S.; Jørgensen, T.; Sandbaek, A.; et al. Studies of association of variants near the HHEX, CDKN2A/B, and IGF2BP2 genes with type 2 diabetes and impaired insulin release in 10,705 Danish subjects: Validation and extension of genome-wide association studies. *Diabetes* **2007**, *56*, 3105–3111. [CrossRef] [PubMed]
253. Zhu, X.; Hu, R.; Brissova, M.; Stein, R.W.; Powers, A.C.; Gu, G.; Kaverina, I. Microtubules Negatively Regulate Insulin Secretion in Pancreatic β Cells. *Dev. Cell* **2015**, *34*, 656–668. [CrossRef] [PubMed]
254. Schnell, A.H.; Swenne, I.; Borg, L.A. Lysosomes and pancreatic islet function. A quantitative estimation of crinophagy in the mouse pancreatic B-cell. *Cell Tissue Res.* **1988**, *252*, 9–15. [CrossRef] [PubMed]
255. Orci, L.; Ravazzola, M.; Amherdt, M.; Yanaihara, C.; Yanaihara, N.; Halban, P.; Renold, A.E.; Perrelet, A. Insulin, not C-peptide (proinsulin), is present in crinophagic bodies of the pancreatic B-cell. *J. Cell Biol.* **1984**, *98*, 222–228. [CrossRef] [PubMed]
256. Uchizono, Y.; Alarcón, C.; Wicksteed, B.L.; Marsh, B.J.; Rhodes, C.J. The balance between proinsulin biosynthesis and insulin secretion: Where can imbalance lead? *Diabetes Obes. Metab.* **2007**, *9* (Suppl. 2), 56–66. [CrossRef]
257. Lachaal, M.; Spangler, R.A.; Jung, C.Y. High Km of GLUT-2 glucose transporter does not explain its role in insulin secretion. *Am. J. Physiol.* **1993**, *265*, E914–E919. [CrossRef]
258. Thorens, B.; Wu, Y.J.; Leahy, J.L.; Weir, G.C. The loss of GLUT2 expression by glucose-unresponsive beta cells of db/db mice is reversible and is induced by the diabetic environment. *J. Clin. Investig.* **1992**, *90*, 77–85. [CrossRef]
259. Unger, R.H. Diabetic hyperglycemia: Link to impaired glucose transport in pancreatic beta cells. *Science* **1991**, *251*, 1200–1205. [CrossRef]
260. Maeda, Y.; Akazawa, S.; Akazawa, M.; Takao, Y.; Trocino, R.A.; Takino, H.; Kawasaki, E.; Yokota, A.; Okuno, S.; Nagataki, S. Glucose transporter gene expression in rat conceptus during early organogenesis and exposure to insulin-induced hypoglycemic serum. *Acta Diabetol.* **1993**, *30*, 73–78. [CrossRef]
261. Guillam, M.T.; Hümmler, E.; Schaerer, E.; Yeh, J.I.; Birnbaum, M.J.; Beermann, F.; Schmidt, A.; Dériaz, N.; Thorens, B. Early diabetes and abnormal postnatal pancreatic islet development in mice lacking Glut-2. *Nat. Genet.* **1997**, *17*, 327–330. [CrossRef]
262. Blum, B.; Hrvatin, S.; Schuetz, C.; Bonal, C.; Rezania, A.; Melton, D.A. Functional beta-cell maturation is marked by an increased glucose threshold and by expression of urocortin 3. *Nat. Biotechnol.* **2012**, *30*, 261–264. [CrossRef] [PubMed]
263. Henquin, J.C.; Nenquin, M. Immaturity of insulin secretion by pancreatic islets isolated from one human neonate. *J. Diabetes Investig.* **2018**, *9*, 270–273. [CrossRef] [PubMed]
264. Puri, S.; Roy, N.; Russ, H.A.; Leonhardt, L.; French, E.K.; Roy, R.; Bengtsson, H.; Scott, D.K.; Stewart, A.F.; Hebrok, M. Replication confers β cell immaturity. *Nat. Commun.* **2018**, *9*, 485. [CrossRef]
265. Huang, C.; Walker, E.M.; Dadi, P.K.; Hu, R.; Xu, Y.; Zhang, W.; Sanavia, T.; Mun, J.; Liu, J.; Nair, G.G.; et al. Synaptotagmin 4 Regulates Pancreatic β Cell Maturation by Modulating the Ca(2+) Sensitivity of Insulin Secretion Vesicles. *Dev. Cell* **2018**, *45*, 347–361. [CrossRef] [PubMed]
266. Talchai, C.; Xuan, S.; Lin, H.V.; Sussel, L.; Accili, D. Pancreatic β cell dedifferentiation as a mechanism of diabetic β cell failure. *Cell* **2012**, *150*, 1223–1234. [CrossRef]
267. Kropp, P.A.; Dunn, J.C.; Carboneau, B.A.; Stoffers, D.A.; Gannon, M. Cooperative function of Pdx1 and Oc1 in multipotent pancreatic progenitors impacts postnatal islet maturation and adaptability. *Am. J. Physiol. Endocrinol. Metab.* **2018**, *314*, E308–E321. [CrossRef]
268. Najjar, S.M. Regulation of insulin action by CEACAM1. *Trends Endocrinol. Metab.* **2002**, *13*, 240–245. [CrossRef]
269. Bar, R.S.; Gorden, P.; Roth, J.; Kahn, C.R.; De Meyts, P. Fluctuations in the affinity and concentration of insulin receptors on circulating monocytes of obese patients: Effects of starvation, refeeding, and dieting. *J. Clin. Investig.* **1976**, *58*, 1123–1135. [CrossRef]
270. Olefsky, J.M. The insulin receptor: Its role in insulin resistance of obesity and diabetes. *Diabetes* **1976**, *25*, 1154–1162. [CrossRef]
271. Olefsky, J.M.; Reaven, G.M. Decreased insulin binding to lymphocytes from diabetic subjects. *J. Clin. Investig.* **1974**, *54*, 1323–1328. [CrossRef]

272. Beck-Nielsen, H. The pathogenic role of an insulin-receptor defect in diabetes mellitus of the obese. *Diabetes* **1978**, *27*, 1175–1181. [CrossRef] [PubMed]
273. Ronnett, G.V.; Knutson, V.P.; Lane, M.D. Insulin-induced down-regulation of insulin receptors in 3T3-L1 adipocytes. Altered rate of receptor inactivation. *J. Biol. Chem.* **1982**, *257*, 4285–4291. [CrossRef]
274. Kono, T.; Barham, F.W. The relationship between the insulin-binding capacity of fat cells and the cellular response to insulin. Studies with intact and trypsin-treated fat cells. *J. Biol. Chem.* **1971**, *246*, 6210–6216. [CrossRef]
275. Le Marchand-Brustel, Y.; Jeanrenaud, B.; Freychet, P. Insulin binding and effects in isolated soleus muscle of lean and obese mice. *Am. J. Physiol.* **1978**, *234*, E348–E358. [CrossRef] [PubMed]
276. Hamel, F.G.; Peavy, D.E.; Ryan, M.P.; Duckworth, W.C. HPLC analysis of insulin degradation products from isolated hepatocytes. Effects of inhibitors suggest intracellular and extracellular pathways. *Diabetes* **1987**, *36*, 702–708. [CrossRef]
277. Blackard, W.G.; Ludeman, C.; Stillman, J. Role of hepatocyte plasma membrane in insulin degradation. *Am. J. Physiol.* **1985**, *248*, E194–E202. [CrossRef]
278. Hammons, G.T.; Smith, R.M.; Jarett, L. Inhibition by bacitracin of rat adipocyte plasma membrane degradation of 125I-insulin is associated with an increase in plasma membrane bound insulin and a potentiation of glucose oxidation by adipocytes. *J. Biol. Chem.* **1982**, *257*, 11563–11570. [CrossRef]
279. Caro, J.F.; Muller, G.; Glennon, J.A. Insulin processing by the liver. *J. Biol. Chem.* **1982**, *257*, 8459–8466. [CrossRef]
280. Marshall, S. Dual pathways for the intracellular processing of insulin. Relationship between retroendocytosis of intact hormone and the recycling of insulin receptors. *J. Biol. Chem.* **1985**, *260*, 13524–13531. [CrossRef]
281. Suire, C.N.; Abdul-Hay, S.O.; Sahara, T.; Kang, D.; Brizuela, M.K.; Saftig, P.; Dickson, D.W.; Rosenberry, T.L.; Leissring, M.A. Cathepsin D regulates cerebral Aβ42/40 ratios via differential degradation of Aβ42 and Aβ40. *Alzheimer's Res. Ther.* **2020**, *12*, 80. [CrossRef]
282. Gliemann, J.; Sonne, O. Binding and receptor-mediated degradation of insulin in adipocytes. *J. Biol. Chem.* **1978**, *253*, 7857–7863. [PubMed]
283. Gorden, P.J.; Carpentier, J.L.; Freychet, P.; LeCam, A.; Orci, L. Intracellular translocation of iodine-125-labeled insulin: Direct demonstration in isolated hepatocytes. *Science* **1978**, *200*, 782–785. [CrossRef] [PubMed]
284. Leissring, M.A.; Farris, W.; Chang, A.Y.; Walsh, D.M.; Wu, X.; Sun, X.; Frosch, M.P.; Selkoe, D.J. Enhanced proteolysis of beta-amyloid in APP transgenic mice prevents plaque formation, secondary pathology, and premature death. *Neuron* **2003**, *40*, 1087–1093. [CrossRef]
285. Matsumoto, M.; Pocai, A.; Rossetti, L.; Depinho, R.A.; Accili, D. Impaired regulation of hepatic glucose production in mice lacking the forkhead transcription factor Foxo1 in liver. *Cell Metab.* **2007**, *6*, 208–216. [CrossRef]
286. Leavens, K.F.; Easton, R.M.; Shulman, G.I.; Previs, S.F.; Birnbaum, M.J. Akt2 is required for hepatic lipid accumulation in models of insulin resistance. *Cell Metab.* **2009**, *10*, 405–418. [CrossRef]
287. Wan, M.; Leavens, K.F.; Saleh, D.; Easton, R.M.; Guertin, D.A.; Peterson, T.R.; Kaestner, K.H.; Sabatini, D.M.; Birnbaum, M.J. Postprandial hepatic lipid metabolism requires signaling through Akt2 independent of the transcription factors FoxA2, FoxO1, and SREBP1c. *Cell Metab.* **2011**, *14*, 516–527. [CrossRef]
288. Hummel, K.P.; Dickie, M.M.; Coleman, D.L. Diabetes, a new mutation in the mouse. *Science* **1966**, *153*, 1127–1128. [CrossRef]
289. Garris, D.R.; Garris, B.L. Cytochemical analysis of pancreatic islet hypercytolipidemia following diabetes (db/db) and obese (ob/ob) mutation expression: Influence of genomic background. *Pathobiology* **2004**, *71*, 231–240. [CrossRef]
290. Ingalls, A.M.; Dickie, M.M.; Snell, G.D. Obese, a new mutation in the house mouse. *J. Hered.* **1950**, *41*, 317–318. [CrossRef]
291. Michael, M.D.; Kulkarni, R.N.; Postic, C.; Previs, S.F.; Shulman, G.I.; Magnuson, M.A.; Kahn, C.R. Loss of insulin signaling in hepatocytes leads to severe insulin resistance and progressive hepatic dysfunction. *Mol. Cell* **2000**, *6*, 87–97. [CrossRef]
292. Cohen, S.E.; Kokkotou, E.; Biddinger, S.B.; Kondo, T.; Gebhardt, R.; Kratzsch, J.; Mantzoros, C.S.; Kahn, C.R. High circulating leptin receptors with normal leptin sensitivity in liver-specific insulin receptor knock-out (LIRKO) mice. *J. Biol. Chem.* **2007**, *282*, 23672–23678. [CrossRef] [PubMed]
293. Kubota, N.; Kubota, T.; Itoh, S.; Kumagai, H.; Kozono, H.; Takamoto, I.; Mineyama, T.; Ogata, H.; Tokuyama, K.; Ohsugi, M.; et al. Dynamic functional relay between insulin receptor substrate 1 and 2 in hepatic insulin signaling during fasting and feeding. *Cell Metab.* **2008**, *8*, 49–64. [CrossRef] [PubMed]
294. Sopasakis, V.R.; Liu, P.; Suzuki, R.; Kondo, T.; Winnay, J.; Tran, T.T.; Asano, T.; Smyth, G.; Sajan, M.P.; Farese, R.V.; et al. Specific roles of the p110alpha isoform of phosphatidylinsositol 3-kinase in hepatic insulin signaling and metabolic regulation. *Cell Metab.* **2010**, *11*, 220–230. [CrossRef] [PubMed]
295. Jia, S.; Liu, Z.; Zhang, S.; Liu, P.; Zhang, L.; Lee, S.H.; Zhang, J.; Signoretti, S.; Loda, M.; Roberts, T.M.; et al. Essential roles of PI(3)K-p110beta in cell growth, metabolism and tumorigenesis. *Nature* **2008**, *454*, 776–779. [CrossRef] [PubMed]
296. Okamoto, Y.; Ogawa, W.; Nishizawa, A.; Inoue, H.; Teshigawara, K.; Kinoshita, S.; Matsuki, Y.; Watanabe, E.; Hiramatsu, R.; Sakaue, H.; et al. Restoration of glucokinase expression in the liver normalizes postprandial glucose disposal in mice with hepatic deficiency of PDK1. *Diabetes* **2007**, *56*, 1000–1009. [CrossRef]
297. Zhang, K.; Li, L.; Qi, Y.; Zhu, X.; Gan, B.; DePinho, R.A.; Averitt, T.; Guo, S. Hepatic suppression of Foxo1 and Foxo3 causes hypoglycemia and hyperlipidemia in mice. *Endocrinology* **2012**, *153*, 631–646. [CrossRef]

298. Hagiwara, A.; Cornu, M.; Cybulski, N.; Polak, P.; Betz, C.; Trapani, F.; Terracciano, L.; Heim, M.H.; Rüegg, M.A.; Hall, M.N. Hepatic mTORC2 activates glycolysis and lipogenesis through Akt, glucokinase, and SREBP1c. *Cell Metab.* **2012**, *15*, 725–738. [CrossRef]
299. Ghadieh, H.E.; Russo, L.; Muturi, H.T.; Ghanem, S.S.; Manaserh, I.H.; Noh, H.L.; Suk, S.; Kim, J.K.; Hill, J.W.; Najjar, S.M. Hyperinsulinemia drives hepatic insulin resistance in male mice with liver-specific Ceacam1 deletion independently of lipolysis. *Metab. Clin. Exp.* **2019**, *93*, 33–43. [CrossRef]